QUANTUM NANO-PLASMONICS

With examples and clear explanation throughout, this step-by-step approach makes the quantum theory of plasmons accessible to readers without a specialized training in theory. A fully analytical formulation offers an opportunity for further development of the research and applications. The theory is focused on the random phase approximation description of plasmons in metallic nanostructures, previously defined for the bulk metal only. Particular attention is paid to the large damping of plasmons in nanostructures, including electron scattering and Lorentz friction losses; a quantum description of the plasmon photovoltaic effect is presented and there is an in-depth analysis of plasmon–polariton kinetics in metallic nano-chains. Suitable for students in the fields of plasmonics, optoelectronics, and photonics and for researchers active in the fields of photovoltaics, optoelectronics, nano-plasmonics, and nano-photonics, the book should also be useful for researchers in soft plasmonics since it includes applications to electro-signalling in neurons.

WITOLD A. JACAK is Associate Professor at Wrocław University of Science and Technology. He is a theoretical physicist active in the fields of nano-plasmonics, quantum mechanics, decoherence in quantum dots, quantum information, and communication.

QUANTUM NANO-PLASMONICS

WITOLD A. JACAK
Politechnika Wrocławska, Poland

CAMBRIDGE
UNIVERSITY PRESS

University Printing House, Cambridge CB2 8BS, United Kingdom

One Liberty Plaza, 20th Floor, New York, NY 10006, USA

477 Williamstown Road, Port Melbourne, VIC 3207, Australia

314–321, 3rd Floor, Plot 3, Splendor Forum, Jasola District Centre, New Delhi – 110025, India

79 Anson Road, #06–04/06, Singapore 079906

Cambridge University Press is part of the University of Cambridge.

It furthers the University's mission by disseminating knowledge in the pursuit of education, learning, and research at the highest international levels of excellence.

www.cambridge.org
Information on this title: www.cambridge.org/9781108478397
DOI: 10.1017/9781108777698

© Cambridge University Press 2020

This publication is in copyright. Subject to statutory exception and to the provisions of relevant collective licensing agreements, no reproduction of any part may take place without the written permission of Cambridge University Press.

First published 2020

A catalogue record for this publication is available from the British Library.

ISBN 978-1-108-47839-7 Hardback

Cambridge University Press has no responsibility for the persistence or accuracy of URLs for external or third-party internet websites referred to in this publication and does not guarantee that any content on such websites is, or will remain, accurate or appropriate.

Contents

Preface		*page* ix
1	Introduction and Description of Contents	1
2	Physics of Metals – Preliminaries	17
	2.1 Definition of a Metal	17
	2.2 Scheme of the Fermi Liquid Theory	22
	2.2.1 Phenomenological Theory of a Fermi Liquid	22
	2.2.2 Outline of the Microscopic Theory of a Normal Fermi Liquid	26
3	Quasiclassical Description of Plasmons in the Bulk Metal	35
	3.1 Random Phase Approximation Theory of Plasmons of Pines and Bohm	35
	3.2 Screening of the Coulomb Interaction in Metals	39
4	Plasmon Excitations in Nanometre-Sized Metallic Particles	41
5	Damping of Plasmons in Metallic Nanoparticles	59
	5.1 Damped Plasmonic Oscillations in Metallic Nanospheres in Dielectric Surroundings	63
	5.2 Attenuation of Dipole Surface Plasmons with Exact Inclusion of the Lorentz Friction	71
	5.3 Comparison of Surface Plasmon Oscillation Features Including Lorentz Friction with the Experimental Data and Simplified Mie Approach	77
	5.4 Numerical Modelling of Plasmon Resonances in Metallic Nanoparticles	83

6	**Plasmon Photovoltaic Effect**	97
	6.1 Semiclassical RPA Approach to Plasmons in Large Metallic Nanospheres	101
	6.2 Damping of Plasmons in Large Nanospheres	104
	6.2.1 Radiation from Dipole Surface Plasmons – Lorentz Friction for Plasmons	105
	6.3 Transfer of Sunlight Energy to a Semiconductor Mediated by Surface Plasmons through the Channel of Dipole Coupling in the Near-Field Regime	109
	6.3.1 Fermi Golden Rule for Band Electron Transitions due to Coupling with Plasmons	109
	6.3.2 Plasmon Damping Rate due to Near-Field Coupling with Semiconductor-Band Electrons	113
	6.3.3 Transfer of Light Energy via the Plasmon Channel	116
	6.4 Experimental Demonstration of the Proximity Constraints of the Plasmon Effect	120
	6.5 Calculation of the Matrix Element for the Fermi Golden Rule Expression (6.37)	124
7	**Plasmon-Induced Efficiency Enhancement of Solar Cells Modified by Metallic Nanoparticles: Material Dependence**	133
	7.1 Plasmon-Mediated Photoeffect: Probability of Electron Interband Excitation Due to Plasmons	134
	7.2 Damping Rate for Plasmons in a Metallic Nanoparticle Deposited on a Semiconductor	138
	7.3 Efficiency of the Light Absorption Channel via Plasmons for Various Materials	140
8	**Numerical Simulation of Plasmon Photoeffect**	150
	8.1 Lorentz Friction Channel for Energy Losses of Surface Plasmons in a Metallic Nanoparticle	155
	8.2 Fermi Golden Rule for Probability of Electron Interband Excitation due to Plasmons in a Metallic Nanoparticle	158
	8.3 Numerical Modelling of the Plasmon Photoeffect by COMSOL	163
	8.4 Comparison with Experiment	173
	8.5 Conclusions	176
9	**Plasmon–Polaritons in Metallic Nanoparticle Chains**	179
	9.1 Plasmon Oscillations in Metallic Nanospheres	181
	9.2 Radiative Properties of a Metallic Nanosphere in a Chain	186
	9.3 Calculation of the Radiative Damping of a Plasmon–Polariton in a Chain	190

9.4	Plasmon–Polariton Self-Modes in a Chain Propagation	195
9.5	The Self-Frequencies and Group Velocities of Plasmon–Polaritons in a Nano-Chain	198
9.6	Exact Solution of Eq. (9.20) for the Plasmon–Polariton Self-Energy	202
9.7	Collective Plasmon-Wave-Type Propagation along a Nano-Chain; Near-Field-Zone Approximation	210
	9.7.1 Near-Field-Zone Approximation of the Dipole Interaction in a Chain	211
	9.7.2 Medium- and Far-Field Corrections to the Near-Field Dipole Interaction	215
	9.7.3 Nonlinear Corrections to the Lorentz Friction	221
	9.7.4 Nonlinear Correction to the Radiation Losses of Plasmon–Polariton in a Nano-Chain	225
10	Plasmon–Polariton Kinetics in a Metallic Nano-Chain Located in Absorbing Surroundings	231
	10.1 Plasmon Oscillations in a Single Metallic Nanosphere Including Damping	233
	10.2 Radiative Properties of Plasmon–Polaritons in a Metallic Nano-Chain Embedded in a Dielectric Medium	237
	10.3 Plasmon–Polariton Self-Modes in a Chain in Dielectric Surroundings	240
	10.4 Damping of Surface Plasmons in a Single Metallic Nanoparticle Deposited on a Semiconductor Substrate	246
	10.5 Plasmon–Polariton Dynamics in a Metallic Chain Deposited on a Semiconductor	249
11	Plasmons in Finite Spherical Ionic Systems	252
	11.1 Fluctuations of the Charge Density in a Spherical Electrolyte System	253
	11.2 Model Definition	255
	11.3 Solution of RPA Plasmon Equation: Volume and Surface Plasmon Frequencies in a Spherical Finite Ion System	261
	11.3.1 Ionic Surface Ion Plasmon Frequencies for an Electrolyte Nanosphere Embedded in a Dielectric Medium with $\varepsilon_1 > 1$	262
	11.4 Damping of Plasmon Oscillations in Ionic Systems	263
	11.4.1 Exact Inclusion of the Lorentz Damping	267
	11.5 Derivation of Plasmon Frequencies for a Finite Electrolyte System	270
	11.5.1 Volume Ionic Plasmons	270
	11.5.2 Surface Ionic Plasmons	271

12	Plasmon–Polaritons in a Chain of Finite Ionic Systems; Model of Saltatory Conduction in Myelinated Axons	276
	12.1 Plasmon–Polariton Propagation in Linear Periodic Ionic Systems	279
	12.2 Plasmon–Polariton Model of Saltatory Conduction: Fitting the Kinetics to the Axon Parameters	284
	12.3 Soft Plasmonics Application: The Role of Grey and White Matter in Information Processing in the Brain	293
	12.3.1 Possible Link with the Topological Model of a Neuron Web	296
	References	303
	Index	313

Preface

Plasmonics in metallic nanostructures is a rapidly developing area with various prospective applications in photovoltaics and nano-optoelectronics. The most popular approach to plasmonics is based on the classical Fresnel problem for the propagation of electromagnetic waves with boundary conditions imposed on the Maxwell equations by the presence of metallic components. The typical solution for such a boundary problem uses the dielectric functions of both the dielectric and the metallic materials as prerequisites. The dielectric function of the metallic component is assumed to take the form of the Drude model with the parameters usually adjusted to experimental data extracted from bulk or thin-film measurements. However, this approach is not sufficient for metallic nanostructures because the dielectric functions of bulk metals and thin films differ significantly from that of a metallic nanocomponent. The strong size and shape dependences of plasmon excitations in metallic nanoparticles are caused by quantum-type effects not included in a phenomenological Drude–Lorentz approach and go beyond the classical determination of light scattering and absorption achieved by solving the Maxwell–Fresnel equations. In the present treatise, we demonstrate that the quantum properties of plasmons are of high importance for nanometre-scale metallic components and strongly influence plasmon-assisted phenomena not accounted for by the conventional electrodynamic model. We present the random phase approximation (RPA) theory of plasmons in metallic nanoparticles and detail the development of this theory for nanoplasmonic applications in plasmon-mediated photovoltaics and optoelectronics. This theory may be helpful for improving popular models of plasmonic effects based on the Mie approach and on numerical solutions of the Maxwell equations, e.g., within the COMSOL system. We describe how to upgrade conventional electrodynamic models by including quantum effects suitably due to the nanometre-scale metallic components of various plasmonic systems.

This treatise presents a completely developed, mostly analytical, quantum approach to plasmons in metallic nanostructures that can be directly applied to specific situations, such as plasmonics for photovoltaics, plasmon–polaritons in metallic arrays, and plasmons and plasmon–polaritons in ionic soft matter. To ensure completeness we present self-contained descriptions in each chapter, allowing for independent reading of the chapters. Though each chapter uses formalisms developed in the preceding chapters, a short overview of the most important information and derivations is provided in each chapter foreword, repeating and summarizing the previous derivations as necessary. Thus, the reader can begin at any arbitrary point in the book. Furthermore, the analytical formulation of the theory provides the opportunity for immediate further development and applications by readers.

1
Introduction and Description of Contents

In this introductory chapter a description of the aims of the monograph and its contents is given. The main results are briefly stated in the context of previous publications and knowledge. The applications of nano-plasmonics presented in the book are outlined.

The publication of this book is intended to meet the considerable recent interest in subdiffraction light manipulation by plasmon excitations in nanoscale metallic components [1]. The related rapid development of the new field of nano-plasmonics overlaps with nanophotonics and new-generation optoelectronics [1–6]. From the quantum physics point of view, metallic nanoparticles are quite different from semiconductor quantum dots (QDs) [7] despite their similar spatial confinement sizes. In a nutshell: QDs are manufactured nanometre-sized quantum wells that are relatively shallow and do not possess the singularities present in Coulomb confinement. Quantum dots are typically located in semiconductor surroundings and are able to trap conduction-band electrons, valence-band holes, or excitons from the surrounding material. They localize electrons in discrete quantum states analogous to those of atoms, though without the limitation caused by the instability of the atomic nucleus. Quantum dots can thus be filled with several carriers, electrons or holes; however, this is limited in practice by the depth of the QD well. The analogy between QDs and ordinary atoms can be somewhat misleading owing to the strong dephasing of states in QDs by collective excitations in the surroundings (mostly phonons). This dephasing occurs because the energy scales of the trapped carriers in QDs and of the collective phonon excitations in the surrounding crystal are similar. Hence, the decoherence of QD states (hybridization with collective band excitations) is, by several orders of magnitude, larger than that in atoms [8]. Nevertheless, the concept of a QD is physically straightforward and allows for easy and efficient numerical modelling based upon an elementary quantum mechanics scheme for a small number of particles.

For metallic nanoparticles the situation is different and is related to the much more complicated physics of metals. The typical situation for metals is a deep Fermi sea of nearly free electrons with a well-defined Fermi surface even at high temperatures; the Fermi energy in metals, of the order of 10^4 K, greatly exceeds the melting temperature. The theory of metals was developed for bulk metals in the 1960s by the application of advanced methods for multiparticle systems, mostly in terms of the Green function approach [9]. The quantum degeneracy of the dense Fermi liquid of electrons in metals (actually the degeneracy of the Landau quasiparticles, which are stable on the Fermi surface [9]) makes a finite metal sample confined at the nanoscale a quantum system very different from a QD low-populated with band carriers in semiconductors. In metals, the crystalline positive ion background that is essential for the definition of the related quantum states causes an additional complication. There have been many successful attempts to describe the excitations in metals using quantum statistical physics and many-body theory methods [9–11] or a quasiclassical phenomenological approach [12, 13].

Besides the low-energy excitations near the Fermi surface in metals (i.e., the electrons that can be expressed in terms of so-called Landau quasiparticles [9, 11]), there exist also collective high-energy excitations involving all the electrons simultaneously, not just those located close to the Fermi surface. For a bulk metal, an efficient theory for these high-energy collective excitations of electrons in metals was developed in 1952, within the random phase approximation (RPA) approach, by Pines and Bohm [14, 15]; these high-energy collective excitations were called plasmons. Owing to the large energy difference between plasmons and the Landau quasiparticles close to the Fermi surface, plasmons do not interact with the Landau quasiparticles, although all the electrons collectively create plasmons. Thus, plasmons cannot be excited just by low-energy electrons, as the plasmon energy in a bulk metal is typically of the order of 10 eV, exceeding even the Fermi energy, which is usually of the order of 7–8 eV. The plasmon energy is thus of a scale similar to that of ultraviolet (UV) photons, and so plasmons can be excited by sufficiently hard electromagnetic radiation.

However, in confined metallic nanoparticles the energies of plasmons are considerably lower [17–21], and in noble metals (Au, Ag, Cu) they fit with the visible photon energies. This observation sparked the nanotechnology revolution related to the modification and control of visible light by plasmons in metallic nanostructures [1–6] whose spatial size is much lower than the wavelength of light, referred to as subdiffraction light manipulation. It should be noted that plasmas and their excitations were the subject of interest early in the twentieth century [18] and since then have been investigated in relation to high-energy plasmas of charged particles (mostly protons) in stellar kernels, tokamaks, galaxy ion clouds, and the ionosphere. This interest has been stimulated also by the development of over-horizon radar and anti-radar technology, which has been accelerated in recent years by achievements in metamaterial construction (such as an 'invisibility cloak') allowing control over light in a manner different from ordinary reflection and refraction.

The main objective of this book is to present an effective theory of plasmons in metallic nanoparticles that is as analytical as possible, allowing universal applications that are not restricted to previously studied numerical models. The numerical methods developed in the 1980s for *ab initio* approaches to plasmons in metallic clusters were limited to approximately 300 electrons (owing to the high numerical complexity of the solution of the Kohn–Sham equations) [19, 20]. In view of these problems, insights into the plasmonics of metallic nanoparticles were confined to phenomenological approaches; these approaches used experimentally aided modelling [22] of the dielectric function in the solution of the classical Fresnel–Maxwell equations for the boundary problem of a metallic nanoparticle with spherical geometry and an incident planar wave. The analytical solution of this problem is known as the Mie approach [17, 18] and results in formulae for the scattering and extinction cross sections for light incident on a metallic sphere.

However, the Mie approach uses a phenomenological dielectric function for the metal as a prerequisite to model light absorption in the metallic nanoparticle (according to the conventional Drude–Lorentz model) [17]. The same approach is used for the numerical solution of the Maxwell equations with boundary conditions in the COMSOL system (which is not limited to spherical geometry); this utilizes the finite element method to solve the Maxwell equations. The COMSOL calculus includes the plasmon dynamics only phenomenologically, via a predefined dielectric function of the metallic compound. However, recognition of the plasmon properties of metallic nanoparticles is not available in either of the aforementioned approaches, which is a significant drawback. Therefore, progress in the description of plasmon excitations in metallic nanoparticles within a microscopic framework, preferably in an analytical form, may have considerable significance for improving upon the very popular Mie-type and COMSOL methods. Upgrading these classical methods may consist in a better declaration of the dielectric functions of the objects under study, since these functions are prerequisites for both the Mie and COMSOL calculus. The modelling of dielectric functions at the nanoscale is a source of certain discrepancies between the numerical solutions and experiments because a predefined dielectric function is usually supported by experimental data that come from bulk metals or thin films, rather than from nanoscale particles. Thus, theoretical insights into the properties of plasmons at the nanoscale that include quantum effects are necessary to improve the conventional numerical models. This can be done via the development of a microscopic quantum model for surface and volume plasmons in metallic nanoparticles using the RPA approach [16] for finite nanoscale geometries [23], in a generalization of the Pines–Bohm theory [14–16].

We will demonstrate this approach in the present text and describe an analytical description of all the multipole modes (for spherical symmetry) for surface and volume plasmons within the RPA model that agrees well with experimental observations.

Surface plasmons in a metallic nanoparticle correspond to the translational collective oscillation modes of all the electrons in the nanoparticle when the charge fluctuations that are not compensated by the static positive ion jellium occur only on the nanoparticle surface. This condition is in contrast with volume plasmons, which are compressional-type modes with charge density varying along the nanoparticle radius. Remarkably, the surface plasmons in nanoparticles have resonance energies lower than the energy of the bulk plasmons, i.e., $\hbar\omega_p = \hbar\sqrt{ne^2/(m\varepsilon_0)}$ (n is the electron density in the metal, e and m are the electron charge and mass, and ε_0 is the dielectric constant); for the simplest case, that of dipole-type surface plasmon oscillations, $\hbar\omega_1 = \hbar\omega_p/\sqrt{3}$ (the Mie energy). However, the volume plasmons in nanoparticles have resonance energies higher than those in the bulk. Neither type of plasmon excitation – the surface or the volume plasmons – occurs in the bulk metal (although for a half-space geometry there are surface plasmons at the so-called Ritchie frequency, $\omega_p/\sqrt{2}$ [5]).

The essential difference between the RPA Pines–Bohm model for bulk metals [16] and the RPA theory for metallic nanoparticles consists in the explicit definition, in the latter theory, of the finite rigid jellium (defining the shape of the nanoparticle) [23]. In the RPA Pines–Bohm theory, the infinite jellium in the bulk is renormalized out via an ideal compensation with a uniform long-wavelength coherent electron fluctuation (i.e., a uniform plasmon mode with momentum $\mathbf{q} = 0$). For a finite nanoparticle, however, such a renormalization is impossible because of the absence of translational invariance and the presence of quantum numbers that are different from momentum quantum numbers for the plasmon excitations, owing to the explicit presence of the jellium rim.

The quantum dynamics equation in the Heisenberg representation determines the self-modes for collective local charge density fluctuations [16, 23]. The gradient operator in the kinetic energy term produces Dirac delta singularities on the rim of the jellium; these arise from the derivative of the Heaviside step functions defining the jellium border. Because these singularities are located on the edge of the nanoparticle, they allow separation of the surface and volume plasmon components in the dynamics equation [23]. However, a mutual dependence of the two types of excitation is visible within the RPA approach. Moreover, the problem of the so-called spill-out of the electron liquid beyond the rim of the jellium can also be considered in the RPA theory. Nevertheless, the spill-out appears to be on the scale of the Thomas–Fermi length [19] and is thus unimportant for nanoparticles with a radius larger than about 5 nm, though for smaller nanoparticles (of size 2–3 nm) the spill-out considerably dilutes the electron density inside the cluster and redshifts the resonance plasmon frequencies, which are proportional to the square root of the charge density [19–21]. Surface effects such as spill-out and Landau damping [20] (the latter corresponds to the decay of plasmons into a pair of quasiparticles distant from the Fermi surface and thus unstable) become less important for larger nanoparticle sizes, and for nanospheres with radii larger than 5 nm these surface effects are negligible [20]. The reduction in the role of spill-out (which makes the surface fuzzy and perturbs the surface modes) for particles larger than a few nanometres in radius supports the usability of the RPA approach in a quasiclassical way: it is a sufficiently accurate quantum approach at this nanoparticle size [23] as it allows the separation of volume and surface local electron density fluctuations. In ultrasmall nanoparticles [19], mixing of the surface and volume excitations occurs for up to approximately 60 electrons, when shell effects cease to contribute [19, 21]; for larger nanoparticles the separation between the surface and volume excitations improves for greater numbers of electrons [23] and is almost ideal for nanospheres with radii larger than approximately 5 nm. The quantum dynamics equation (the Heisenberg equation for the second-order time derivative of the local electron density operator [16, 23]) has a complicated form owing to the presence of the finite jellium. However, after making the RPA simplification, including a quasiclassical averaging of the kinetic energy according to the so-called five-thirds Thomas–Fermi formula [16]), this quantum dynamics equation describes a plethora of plasmon modes in the nanoparticle [16], many more than the single volume mode of the bulk metal [16].

An important advantage of the RPA model developed in this book is the possibility of including dissipation effects such as those due to electron scattering on other electrons, on phonons, on admixtures and defects or on the boundary of the nanoparticle. The resulting shift in the plasmon resonance caused by the scattering damping of plasmons scales as $1/a$, where a is the radius of a nanoparticle [24], which agrees with experimental observations for the radius range $5 < a < 10$ nm (for Au in a vacuum), where electron scattering energy dissipation dominates other damping channels [25].

Nevertheless, plasmon oscillations are also damped by radiation, an effect that increases with the electron number in a nanoparticle and leads to pronounced crossover in the size dependence of the plasmon attenuation and the related resonance shift for nanosphere radii a of 10–12 nm (for Au in vacuum, in general the crossover depends on the metal and the dielectric surroundings). At nanoparticle sizes corresponding to this cross-over, the decrease in the plasmon damping as $1/a$ changes to a strong increase proportional to a^3 [26], the a^3 dependence is related to the fact that all the electrons participate in plasmon oscillations and their radiation properties; this is true even for surface plasmons.

Thus, the contribution of all the electrons in a nanoparticle is manifested by the volume factor a^3, which causes a similar scaling in the radiative damping of plasmons resulting from the so-called Lorentz friction [26], i.e., the energy loss of oscillating charged particles due to the radiation of electromagnetic waves. For a sufficiently large number of electrons inside the nanosphere, the Lorentz friction losses dominate other channels of plasmon damping. The large radiation energy loss expressed by the Lorentz friction damping initially grows with a^3, as mentioned above [27], but then saturates at approximately $a \sim 50$ nm (for Au in vacuum) and gradually decreases for larger nanoparticles. This behaviour is identified and described in this text.

Such behaviour reveals that the inclusion of Lorentz friction considerably changes the plasmon oscillation regime: the oscillations are not of the harmonic-oscillator type because the Lorentz friction is proportional to the third-order time derivative rather than the first-order time derivative as is the case for the ordinary friction of a harmonic oscillator. The harmonic oscillator model appears to be incorrect for plasmons in large metallic nanoparticles, and many simplified harmonic models of plasmons that are popular in the literature are misleading. In particular, the overdamped regime typical for harmonic damped oscillators, which would terminate plasmon oscillations within the harmonic model at approximately 57 nm, for Au in vacuum, does not reflect reality. The solution of the third-order dynamic differential equation is different from that for a harmonic oscillator, and the relationship between the frequency and damping of a harmonic oscillator, $\sqrt{\omega_0^2 - 1/\tau^2}$ (which defines the overdamped regime when the expression under the square root is negative), is not valid for plasmons. Beyond the harmonic model the relationship between frequency and damping also has an analytical form, which has been derived and which allows us to describe plasmon behaviour precisely.

The exact solution of the Lorentz friction problem for plasmons in metallic nanoparticles [26] gives very good agreement with experiment, with respect to the size dependence of plasmon resonances in metallic nanospheres. This result provides more precise modelling of metal dielectric functions, for specific sizes of nanostructure configurations, that includes plasmon damping and has also been applied to modify numerical studies using COMSOL and Mie-type calculation schemes [28, 22, 29], in both cases these schemes use predefined dielectric functions for the systems analysed. The results have been confirmed experimentally for Au and Ag nanoparticles.

The RPA theory of plasmons in metallic nanoparticles and their radiative properties is useful for modelling the so-called plasmon-aided photovoltaic (PV) effect. The damping of the plasmons changes radically when another electrical system is located near a metallic nanoparticle with plasmons; such a system could be a semiconductor substrate with a band electron system. In such a case, the semiconductor substrate receives an extremely strong flow of energy from the plasmons to the band electrons, which corresponds to plasmon damping greatly exceeding the Lorentz friction losses in the dielectric surroundings. This new and strong channel of energy transfer can be described by applying the Fermi golden rule to the quantum interband electron transitions induced by the near-field radiation of plasmons in a metallic nanoparticle deposited on a semiconductor substrate [30]. The mediation by plasmons in the energy harvesting of a photoactive semiconductor layer results in significant modification of the ordinary photoeffect.

This phenomenon is of high practical importance in view of the current rapid development of photovoltaics; in 2015, the total power of all PV solar cell installations worldwide was approximately 250 GW, whereas the total power produced by conventional coal or gas energy plants was approximately 29 GW in Poland. There could be an increase in the efficiency of solar cells (particularly thin-film solar cells and organic 'plastic' cells) by several per cent due to mediation by plasmons in metallic nanoparticles deposited on the cell surfaces (metallic coverings with low densities, $\sim 10^{8-10}/cm^2$ and thus low costs and easy accessibility for industry technology) [31]. This could have a large economic impact in the field of renewable energy sources [32–36].

In Au, Ag, and Cu nanoparticles with radii $a \sim 10-50$ nm the photoeffect efficiency is enhanced, in laboratory setups of photodiodes, by a factor 2–10 owing to energy transfer from the incident photons to the semiconductor substrate via plasmons in the deposited metallic components [31–36]. However, in solar cells the efficiency of the photoeffect is only one factor amongst many other factors that define the final total efficiency of a cell, and a large increase in photoeffect efficiency causes a more modest increase in the total efficiency of a solar cell.

Through the quantum calculus of the Fermi golden rule scheme, we can demonstrate [30] that the near-field coupling of the dipole mode of the surface plasmons in metallic nanoparticles deposited on top of a semiconductor is very efficient, i.e., this coupling effect causes a strong increase in the probability of interband transitions compared with the ordinary photoeffect, in which the plane-wave photons interact directly with the band electrons. The advantage of the result obtained via the Fermi golden rule approach is its analytical form (which is similar to the formula for the ordinary photoeffect in a semiconductor, though the calculus for plasmon mediation is much more complicated; however, it is analytically attainable [30]). The analytical form of this approach allows for the analysis of various competing mechanisms, and leads to an expression for the photoeffect efficiency increase due to plasmons in solar cells.

The related calculations explicitly demonstrate that the absence of translational invariance for a nanoparticle coupled in a near-field zone with a semiconductor substrate removes the constraints imposed by the momentum conservation rule.

Thus, in addition to the vertical interband transitions (conserving the electron momentum plus the negligible photon momentum) found in the ordinary photoeffect, all indirect interband transitions between arbitrary electron momenta are available, which strongly enhances the probability of interband transition. The momenta of incident photons with energies beyond the forbidden energy gap are negligible compared with the momenta of the Bloch states of the band electrons in a semiconductor, which results in a constraint to 'vertical' interband transitions in the ordinary photoeffect. This constraint is removed when the translational symmetry is violated for a small metallic nanoparticle. The coupling of the plasmon dipole in the near-field regime allows all skew interband transitions, enhancing the total probability of interband excitations of carriers in the semiconductor substrate. This effect favours smaller metallic nanoparticles but, conversely, larger nanoparticles have larger dipoles (in proportion to the electron number), which also enhances the transition probability. The resulting trade-off of these opposing tendencies defines the optimal size for a metallic nanoparticle to increase the photoeffect efficiency. Moreover, the type of nanoparticle deposition used plays an important role. Though problematic from a technological perspective, a convenient deposition method is the complete embedding of the metallic nanoparticles into the semiconductor layer. The theory has been developed to allow analysis of the aforementioned size trade-off and of the role of the nanoparticle deposition method on the photodiode surface.

Energy transfer through the coupling of the subphoton near field (i.e., the near field on a length scale 10−100 times smaller than the photon wavelength) with the surface plasmon dipoles is very efficient. This high efficiency results in high damping of the plasmons, much higher than the radiative damping due to Lorentz friction and the damping due to electron scattering [23], and elucidates the very strong plasmonic PV effect that is observed experimentally. The prospects for large-scale utilization of this effect in industry and commercial photovoltaic devices depend on the practical deposition methods for nanoparticles on cell surfaces, which usually reduce the net effect as too high a density of metallic particles results in inconvenient interference reflection effects of the metallic nanocover. Nevertheless, at low surface concentrations of the metallic nanoparticles, the theoretically determined resonance curves agree very well with the experimental results [30, 31, 33]. Another experiment elucidated by the theory concerns a two-layer structure of Si–ZnO nanopillars, consisting of a thick layer of p-Si covered with n-ZnO vertical nano-rods with diameters ranging from 200 to 300 nm and heights of approximately 1000 nm [30]. When the top of the structure was covered with silver nanoparticles (with radii of 5, 20, or 50 nm), the photoresponse doubles. This increase is caused in part by ZnO subgap transitions but also by the Si substrate; this is revealed by the characteristic size dependence. This proves that the range of the plasmonic effect is at least 1 micrometre, which makes the effect convenient for thin-layer solar cell technology [34–36]. Multi-crystalline Si solar cells sparsely covered with nanoparticles have recently shown very good agreement with the predictions of the theory; the gains in efficiency have reached 5.6% for Au nanoparticles and 4.8% for Ag nanoparticles, whereas, for CIGS (copper indium gallium diselenide) cells, gains of 1.2% (Au) and 1.4% (Ag) have been achieved [31]. Many other, even greater, experimental results have been reported. However, as mentioned, an excessive concentration of metallic covering diminishes the efficiency of solar cells (owing to screening and reflection effects).

The successful theoretical RPA description of plasmons in a single metallic nanoparticle and the recognition of their radiative properties allows for the development of the RPA model for interacting systems (arrays) of metallic nanoparticles, in particular for metallic nano-chains that could serve as low-loss plasmon–polariton wave guides for the desired miniaturization of optoelectronics. Plasmon–polaritons are collective wave-type modes of surface plasmons hybridized with an electromagnetic wave propagating, with almost lossless kinetics, along a periodic chain of metallic nanoparticles. They are analogous to the surface mode of plasmon propagation along a metal–insulator interface, also called a plasmon–polariton [1, 5].

The change in the configuration of the electromagnetic field near the metal–insulator interface around a metallic wire is utilized in high-frequency microwave techniques (e.g., in single-wire Goubau transmission lines [37]). This effect raises an interesting issue for similar propagation along discrete metallic nano-chains [38–43] – experiments confirm the lossless propagation of dipole collective wave-type oscillations in the range of several micrometres [39, 40] with group velocity at least 10 times lower than c (the velocity of light). The latter property allows for a reduction in the diffraction constraints [38] that severely limit the miniaturization of the opto-nanoelectronics, where the wavelengths of photons with energies typical for the nanoelectronics scale (meV) greatly exceed the dimensions of the nanoscale electronic elements (because of the high value of c), precluding miniaturization. The transformation of the electromagnetic signal into a plasmon–polariton, which has the same frequency but a 10 times (at least) shorter wavelength, allows the avoidance of diffraction limits. This is promising for future nanoscale plasmon optoelectronics [5, 38, 42].

An analysis has been developed of the plasmon–polariton kinetics inhmkujkj metallic nano-chains using a far-reaching analytical formulation of the RPA theory [44–46] (which has made significant progress in comparison with numerical-only studies [41, 43]). This analytical formulation allows the precise identification of various factors that were previously resistant to insight within complex numerical approaches [42, 43]. These factors are related to detailed identification of how the damping and the group velocity of plasmon–polaritons depend on the material and geometry parameters.

In particular, it has been demonstrated that the accurate inclusion of all terms in the dipole interaction in the near-, medium-, and far-field zones, together with retardation effects (frequently neglected in the literature), leads to ideal compensation of the Lorentz friction in each nanosphere in a chain by the radiation energy income from the remaining nanoparticles in the chain [46]: thus the propagation of plasmon–polaritons in the chain is radiatively lossless. Taking into account that the Lorentz friction in a large metallic nanosphere, with radius > 12 nm, for Au in vacuum, greatly exceeds the electron scattering energy losses (i.e., the irreversible dissipation of plasmon energy into Joule heating due to scattering from electrons, phonons, admixtures, defects, and boundaries), a metallic nano-chain behaves as an almost lossless ideal wave guide for plasmon–polaritons [46]; this agrees with experimental observations [39, 38] and indicates possible applications in sub-diffraction plasmon-optoelectronics due to the low group velocity (and wavelength) of plasmon–polaritons.

The RPA model developed for plasmon–polaritons [46] shows that this ideal compensation of the Lorentz friction occurs only inside the light cone (on which the plasmon–polariton phase velocity equals the velocity of light); outside the light cone the damping of plasmon–polaritons is strengthened to values above the Lorentz friction scale (this damping outside the light cone increases steadily with increasing plasmon–polariton wave number for longitudinal plasmon–polariton polarization, when the dipoles oscillate along the chain direction, and increases stepwise for transverse polarization [47, 45]).

On the light cone (in one dimension there are only two points in the Brillouin zone for a periodic chain – the light cone here is a triangle as a function of the chain separation), a logarithmic singularity of the dynamics equation occurs [45]. This singularity is caused by constructive interference of the radiation from the nanospheres in the chain in the far-field zone and occurs only for transverse polarizations of the plasmon–polaritons; it causes a similar divergence at all orders in the perturbation series for the dispersion [45]. The logarithmic perturbative divergence in the dispersion of the plasmon–polaritons in turn causes a hyperbolic singularity in the group velocity for transversely polarized plasmon–polaritons, which is observed in many numerical simulations presented in the literature (in a numerical solution of the dynamics, a certain kind of perturbative solution is linked with an unavoidable truncation of the exact infinite Green function series, which produces a numerical artefact that has been erroneously interpreted as superluminal propagation).

The aforementioned error has been explained in detail and is not present in an exact solution of the dynamic equation for plasmon–polaritons obtained using a special nonperturbative method (i.e., by separate solution of the nonlinear problem at approximately 20,000 points of the Brillouin zone by a Newton-type procedure) which allows for determination of the dispersion and the damping of the plasmon–polaritons beyond any perturbative step [45]. This solution has revealed the precise removal of the 'perturbative' singularity in the group velocity (which is hyperbolic for transverse polarizations) via its truncation at the velocity of light in the dielectric surroundings, which agrees with Lorentz invariance.

Other singularities in the group velocity of the plasmon–polariton were also identified for both polarizations; these singularities occur also on the light cone, are of logarithmic character, and are caused by constructive interference in the medium-field zone in the dynamics equation for plasmon–polaritons [45] (again manifesting themselves at any perturbation order, these singularities correspond to the derivative with respect to the wave number of the medium-field-zone contribution to the self-frequency, which is logarithmically divergent at the light cone). Via the exact nonperturbative solution, it was demonstrated that these group velocity singularities (without any dispersion singularities) are excluded and truncated exactly at the velocity of light. This analysis elucidates the misleading interpretations of numerical approximate simulations as being due to artefacts caused by the perturbative approach. An interesting analogy might be drawn: in quantum field theory, singularities occur at any order of the perturbation series but these singularities can be removed by renormalization procedures. Hence, the problem analysed above of plasmon–polariton dynamics is actually an example of explicitly performed renormalization via the exact solution of the highly nonlinear singular problem, regularized by Lorentz invariance.

In a similar manner, the problem of the numerically observed long-range plasmon–polariton modes close to the light cone is explained. In contrast with the explanation in the literature, of an extraordinary mode with lowered damping, the longer-range modes can be identified with a local increase in the group velocity for a thin wave packet close to the truncation singularity; this gives a range of propagation several times longer at the same level of damping. By analysis of the convergence of an appropriate series related to the radiation effects in the chain, the small impact of the finiteness of the chain on the plasmon–polariton kinetics was demonstrated and thus the usability of the infinite-chain results for finite-length periodic structures (even 10 periodic elements in the chain yield almost the same plasmon–polariton behaviour as an infinite chain, owing to the very quick convergence of the radiation series, except for those series resulting in singularities but these types of series do not occur in finite chains).

The theory developed for plasmon–polaritons in a metallic nano-chain demonstrates the propagation of a collective surface plasmon mode with the whole electromagnetic field compressed along the chain (which agrees with COMSOL simulations and near-field scanning optical microscopy (SNOM)) [39, 40]. Owing to the incommensurability of photon wavelengths and the wavelengths of plasmon–polaritons with the same energies, any mutual perturbation of both these excitations is precluded; thus the detection, excitation, or perturbation of plasmon–polaritons by free photons is not possible, which makes plasmon–polaritons immune to electromagnetic perturbations. Moreover, the lossless subdiffraction plasmon–polariton mode controlled by the chain parameters might be able to sense nanodeformations due to strain in various mechanical constructions. This could be done by the use of elastic substrates with metallic nano-chains simply pasted onto the strained elements, stepwise changing the regime of the plasmon–polariton kinetics, which are very sensitive to small changes in nanoparticle separation due to length deformation of the substrate.

This effective microscopic RPA-type theory of plasmons and plasmon–polaritons in metallic nanostructures creates an opportunity for the development of similar approaches for other (ionic) charged multiparticle systems. The plasmon and plasmon–polariton RPA theory of metallic nanostructures has been applied to ionic plasmons and plasmon–polaritons in finite electrolyte systems (confined by the dielectric membranes that are frequently found in bio-cell organization) [48]. This generalization involves the following steps. (a) The accommodation of the RPA model defined for a finite quantum degenerate Fermi liquid of electrons in a metal to a Boltzmann-type nondegenerate liquid of ions (fermions or bosons) requires substitution of the so-called 5/3 Thomas–Fermi formula for a degenerated electron gas by a classical estimation of the average kinetic energy according to the Maxwell–Boltzmann distribution; simultaneously, the Fermi velocity must be substituted by the Boltzmann average velocity of ions. These changes introduce a specific temperature dependence that is absent in the metallic RPA model. (b) The core element of the ionic theory of plasmons is the introduction of an auxiliary fictitious two-component 'jellium' for a binary electrolyte. In any electrolyte, ions of both signs are dynamical components of the system, unlike in metals, where the positive jellium in the form of the crystalline lattice defines a stiff shape for the nanoparticle. Nevertheless, by modelling the Hamiltonian for a binary electrolyte, a fictitious jellium with two components whose charges mutually cancel can be introduced without any change in the total energy. For sufficiently small local charge excitations in the electrolyte, the two-component fictitious jellium model allows for a description of plasmon excitations that is in analogy with that for metals, in the form of two mutually coupled ionic excitations.

This model admits the classification of volume and surface ionic plasmons (compressional and translational modes, respectively) with various polarizations for spherical symmetry and the estimation of their energy and damping in analogy to that for metallic nanoparticles. Because ions have larger masses than electrons and typically have lower concentrations in electrolytes compared with the free electron concentrations in metals, the characteristic plasmonic size scale (defined as the size of a finite system with maximal plasmon radiation) is on the scale of micrometres for ions rather than the nanometre scale of electrons in metals, and the plasmon frequency is strongly reduced by many orders of magnitude depending on the ion concentration and the ion mass and charge. All the elements of the RPA model generalized to finite ion systems have counterparts in metals, including the characteristic and typical behaviours of surface and volume plasmons (however, the latter exhibit a temperature dependence for ions that is absent for electrons in metals). There are similar radiation effects, including a Lorentz friction loss that is larger than ordinary scattering losses (for ions it depends on the temperature in a quite different way than that for the electron archetype owing to the substitution of the practically temperature-independent Fermi velocity of metals by the temperature-dependent averaged velocity of the ions in an electrolyte).

In addition to the plasmon excitations in a separated single finite electrolyte system, a theory of ionic plasmons on finite periodic ionic systems with ionic plasmon–polaritons was developed [49]. In analogy to plasmon–polaritons in metallic nano-chains, similar properties were described for ionic plasmon–polaritons in periodic ionic microchains. The wave guide kinetics of plasmon–polaritons in ionic chains was analysed with respect to a wide range of ion parameters and confinement scales including overlap with additional specific vibrational and rotational excitations of the electrolyte's solvent molecules (of water).

The plasmon–polariton model of a chain of finite electrolyte systems was next applied to the original explanation for so-called saltatory conduction in myelinated axons. The intention was to explain using ionic plasmon–polariton kinetics the unknown and mysterious mechanism of the quick and efficient nervous signal transduction observed in periodically myelinated axons in neurons of the peripheral nervous system and in the white matter of the brain and spinal cord [50, 51]. The standard cable model [52] (originated by William Thomson in the nineteenth century) describes well the diffusion-type kinetics of the ionic electrical signal along the dendrites and nonmyelinated axons (in the grey matter in the central nervous system). According to the cable theory, the velocity of a signal depends on the conductivity of the inner-neuron cytoplasm and on the capacity across the cell membrane between the inner and outer cytoplasm; it is stiffly confined, and reaches at most 1–3 m/s. This signal velocity is too low for the transduction of nerve signals over longer distances; moreover, in myelinated axons, which are periodically wrapped with a white lipid called myelin, the action potential somehow jumps between the consecutive so-called Ranvier nodes separating the segments wrapped with the myelin sheath. In this way, the signal accelerates and the velocity reaches the 100–200 m/s required for proper signalling and functioning of the body. The mechanism of these jumps is unknown, having been researched for a long time without progress, and is phenomenologically designated 'saltatory conduction'. It is agreed [50, 51] that the myelin sheath is crucial for saltatory conduction: the myelin is much thicker than would be needed for isolation only; moreover, any deficit in the thickness of the sheaths results in severe demyelination diseases such as multiple sclerosis, manifesting itself in slowing saltatory conduction and motor dysfunction. Myelin is produced for neurons in the peripheral nervous system by Schwann cells, whereas in the central nervous system oligodendrocytes produce the necessary myelin [51, 50]. Remarkably, the length of myelinated sectors is typically 100 μm and, as mentioned above, the myelinated sectors are separated by very short nonmyelinated fragments called Ranvier nodes [51, 53]. In the Ranvier nodes spikes of the action potential form, according to a well-known mechanism, on a time scale of a few milliseconds [50], but the mechanism by which the ignition signal jumps to the neighbouring Ranvier nodes cannot be explained by any accepted model based on the cable theory. We proposed an original assumption, that saltatory conduction in neurons periodically wrapped with the thick myelin sheath has an ionic plasmon–polariton character. We then obtained a satisfactory prediction for the signal velocity at realistic parameters for the electrolyte of the neuron cytoplasm. The model achieved good consistency with the various features of saltatory conduction observed in myelinated axons [49], including the following. (a) A high group velocity for the plasmon–polariton wave packet in agreement with the velocity observed in myelinated axons and real electrolyte parameters inside the axon. (b) There is an absence of radiative damping of the plasmon–polaritons and the reduction of attenuation to the level of only ohmic losses (lower than those for ordinary conduction); with only low energy supplementation (which occurs at the Ranvier nodes) a forced undamped plasmon–polariton mode can propagate without deformation over arbitrary long distances. (c) The energy supplementation takes place residually, at the generation of action potential spikes on consecutive Ranvier nodes, according to a known [50] mechanism. This mechanism employs ion channels across the unmyelinated axon cell membrane at the Ranvier node. It is the final phase of a cycle during which steady conditions are restored, and so an external energy supply is needed to actively transfer ions (Na^+ and K^+) across the cell membrane against the concentration gradient. The energy is supplied by the ADP/ATP mechanism in the cell and residually also covers the ohmic losses (ultimately transferred to Joule heat) of plasmon–polaritons. (d) The wave nature of saltatory conduction is confirmed by the observation that axon ignition is maintained even if the axon is broken into two separate pieces located near each other (this property is impossible to explain within the cable theory or on the basis of other electrical-current-type mechanisms but fits reasonably well with properties of plasmon–polaritons, which can propagate either along a continuous fibre or along a discontinuous chain) [50]. That plasmon–polaritons have a wave nature also agrees well with the observed one-way direction of propagation when an axon is initiated from the synapse or axon hillock and two-way propagation when a passive neuron is ignited from its centre. The plasmon–polariton mechanism is also consistent with the observation that axon firing continues even if several Ranvier nodes are damaged and inactive [50]. (e) The model gives a temperature dependence that is typical for ionic plasmon–polaritons, and it also gives the dependence of the plasmon–polariton group velocity on the cross section of the thin inner cord of the axon. The role of myelin, different other than giving insulation, is explained by the plasmon–polariton model (a sufficiently thick myelin sheath is required to build the dielectric tunnel around the inner cord needed for the formation of a plasmon–polariton; decreasing the thickness of this myelin tunnel results in the deceleration of plasmon–polariton propagation, which is actually observed in multiple sclerosis) [50]. (f) The model is also consistent with the exclusive property that plasmon–polaritons cannot be ignited, perturbed, or detected by electromagnetic waves (since photons with low energies identical to those of ionic plasmon–polaritons do not interact with plasmon–polaritons), and therefore, plasmon–polariton signalling is immune to electromagnetic perturbations (note that ordinary current conduction in the grey matter can be detected by electromagnetic means (encephalography) and can even be perturbed electromagnetically, but signalling in the white matter cannot). Application of plasmon and plasmon–polariton effects in ionic systems opens a new field, that of the soft plasmonics of confined electrolytic systems in relation to bio-matter organization and its functionality. The example presented above illustrates the usefulness of the developed RPA model of plasmons and plasmon–polaritons.

Author contributions

In this treatise the original results of the author with coauthors are utilized with permission of the coauthors and the publishers. To be specific: in Chapter 4, reference [23]; in Chapter 5, references [26–28]; in Chapters 6 and 7, references [30, 31]; in Chapter 8, references [145, 159]; in Chapter 9, references [44–46]; in Chapter 10, reference [47], in Chapters 11 and 12, references [48, 49]. The references are listed below.

[23] J. Jacak, J. Krasnyj, W. Jacak, R. Gonczarek, A. Chepok, and L. Jacak, 'Surface and volume plasmons in metallic nanospheres in semiclassical RPA-type approach; near-field coupling of surface plasmons with semiconductor substrate', *Phys. Rev. B* **82**, 035418, 2010 (with permission from APS).

[26] W. A. Jacak, 'Lorentz friction for surface plasmons in metallic nanospheres', *J. Phys. Chem. C* **119**(12), 6749–6759, 2015 (author reuse in the book with permission from the American Chemical Society).

[27] W. A. Jacak, 'Size-dependence of the Lorentz friction for surface plasmons in metallic nanospheres', *Opt. Express* **23**, 4472–4481, 2015 (author reuse in the book with permission from the Optical Society).

[28] K. Kluczyk and W. Jacak, 'Damping-induced size effect in surface plasmon resonance in metallic nano-particles: comparison of RPA microscopic model with numerical finite element simulation (COMSOL) and Mie approach', *J. Quant. Spectrosc. Radiat. Transfer.* **78**, 168, 2016 (with permission from Elsevier for the author reuse in the book).

[30] W. Jacak, E. Popko, A. Henrykowski, E. Zielony, G. Luka, R. Pietruszka *et al.*, 'On the size dependence and the spatial range for the plasmon effect in photovoltaic efficiency enhancement', *Sol. Energy Mater. Sol. Cells* **147**, 1, 2016 (with permission from Elsevier for author reuse in the book).

[31] M. Jeng, Z. Chen, Y. Xiao, L. Chang, J. Ao, Y. Sun, *et al.*, 'Improving efficiency of multicrystalline silicon and CIGS solar cells by incorporating metal nanoparticles', *Materials* **8**(10), 6761–6771, 2015 (with permission from MDPI for unrestricted use and reproduction as an open access publication).

[44] W. Jacak, 'On plasmon polariton propagation along metallic nano-chain', *Plasmonics* **8**, 1317, 2013. DOI: 10.1007/s11468-013-9528-8 (with permission for author reuse from Springer Nature).

[45] W. Jacak, 'Exact solution for velocity of plasmon–polariton in metallic nano-chain', *Opt. Express* **22**, 18958, 2014 (author reuse in the book with permission from the Optical Society).

[46] W. Jacak, J. Krasnyj, and A. Chepok, 'Plasmon–polariton properties in metallic nanosphere chains', *Materials* **8**, 3910, 2015 (with permission from MDPI for unrestricted use and reproduction as open access publication).

[47] W. Jacak, 'On plasmon polariton propagation along metallic nano-chain', *Plasmonics* **8**, 1317, 2013 (with permission for author reuse from Springer Nature).

[48] W. Jacak, 'Plasmons in finite spherical electrolyte systems: RPA effective jellium model for ionic plasma excitations', *Plasmonics*, 2015. DOI: 10.1007/s11468-015-0064-6 (with permission for author reuse from Springer Nature).

[49] W. Jacak, 'Propagation of collective surface plasmons in linear periodic ionic structures: plasmon polariton mechanism of saltatory conduction in axons', *J. Phys. Chem. C* **119**(18), 10015, 2015 author (reuse in the book with permission from the American Chemical Society).

[145] K. Kluczyk, C. David, J. Jacak, and W. Jacak, 'On modeling of plasmon-induced enhancement of the efficiency of solar cells modified by metallic nano-particles', *Nanomaterials* **9**, 3, 2019. DOI: 10.3390/nano9010003 (with permission from MDPI for unrestricted use and reproduction as open access publication).

[159] K. Kluczyk, C. David, and W. Jacak, 'On quantum approach to modeling of plasmon photovoltaic effect', *J. Optical Soc. America B* **34**, 2115, 2017 (author reuse in the book with permission from the Optical Society).

2
Physics of Metals – Preliminaries

A concise description of the theory of metals in the normal bulk state is given. The concept of Landau quasiparticles close to the Fermi surface is briefly described. A microscopic formulation of the normal Fermi liquid theory in terms of the kinetic equation and Green function approach is sketched.

2.1 Definition of a Metal

Using single-electron-band Bloch theory the Schrödinger equation in a periodic potential can be solved. The periodicity corresponds to a crystal lattice. A periodic potential in a crystal lattice can be expressed as follows:

$$V(\mathbf{r}) = V(\mathbf{r} + \mathbf{a}_n). \tag{2.1}$$

with $\mathbf{a}_n = n_1\mathbf{a}_1 + n_2\mathbf{a}_2 + n_3\mathbf{a}_3$, where the Bravais vectors $\mathbf{a}_i$, $i = 1,2,3$, span the elementary crystal lattice cell and $\mathbf{a}_n$ is an arbitrary lattice vector, i.e., a linear combination of the three basis vectors $\mathbf{a}_i$ defining the unit cell, with integer coefficients n_i. The choice of the vectors $\mathbf{a}_i$ is sometimes ambiguous for a particular periodic crystal structure, which can result in different so-called Bravais webs; these are, however, equivalent [54]. The Schrödinger equation for an electron in a periodic crystalline potential thus attains the form

$$-\frac{\hbar^2 \nabla^2}{2m}\Psi(\mathbf{r}) + V(\mathbf{r})\Psi(\mathbf{r}) = E\Psi(\mathbf{r}). \tag{2.2}$$

This equation is invariant with respect to translations in the crystal lattice, and so the wave functions $\Psi(\mathbf{r})$ and $\Psi(\mathbf{r} + \mathbf{a}_n)$ satisfy the same equation as each other.[1] Hence, for nondegenerate states,

$$\Psi(\mathbf{r} + \mathbf{a}_n) = e^{i\alpha(\mathbf{a}_n)}\Psi(\mathbf{r}). \tag{2.3}$$

---

[1] Because $\nabla_\mathbf{r} = \nabla_{\mathbf{r}+\mathbf{a}_n}$.

In the case of degeneracy of the states for a particular energy E, a wave function with a shifted argument can be expressed as a linear combination of unshifted wave functions which span the degenerate subspace of the Hilbert space. Owing to the unitarity of the related representation one can diagonalize the coefficient matrix, obtaining phase factors as the diagonal elements, which resolves the problem to one that is analogical to the case of no degeneracy. It is easy to see that

$$\alpha(\mathbf{a}_1 + \mathbf{a}_2) = \alpha(\mathbf{a}_1) + \alpha(\mathbf{a}_2), \qquad (2.4)$$

since the sum of lattice vectors is also a lattice vector. Next, taking into account the uniqueness of the solution of Eq. (2.4) in the form of a linear function, we get

$$\alpha(\mathbf{a}_n) = \mathbf{k} \cdot \mathbf{a}_n, \qquad (2.5)$$

where $\mathbf{k}$ is associated with the energy E of the stationary state and with the corresponding wave function; hence both $E_\mathbf{k}$ and $\Psi_\mathbf{k}(\mathbf{r})$ are assigned by $\mathbf{k}$ (this mapping $E \to \mathbf{k}$ is single-valued, but its inverse is in general multi-valued). The phase factor $e^{i\mathbf{k}\cdot\mathbf{a}_n}$ is invariant with respect to the translation of the vector $\mathbf{k}$ by an arbitrary vector of the form,

$$\mathbf{k}_m = m_1 \mathbf{K}_1 + m_2 \mathbf{K}_2 + m_3 \mathbf{K}_3, \qquad (2.6)$$

where m_i are integers, and

$$\begin{aligned}
\mathbf{K}_1 &= 2\pi \frac{\mathbf{a}_2 \times \mathbf{a}_3}{\mathbf{a}_1 \cdot (\mathbf{a}_2 \times \mathbf{a}_3)}, \\
\mathbf{K}_2 &= 2\pi \frac{\mathbf{a}_3 \times \mathbf{a}_2}{\mathbf{a}_1 \cdot (\mathbf{a}_2 \times \mathbf{a}_3)}, \\
\mathbf{K}_3 &= 2\pi \frac{\mathbf{a}_1 \times \mathbf{a}_2}{\mathbf{a}_1 \cdot (\mathbf{a}_2 \times \mathbf{a}_3)}.
\end{aligned} \qquad (2.7)$$

Indeed, we see that $e^{i\mathbf{k}\cdot\mathbf{a}_n} = e^{i(\mathbf{k}+\mathbf{k}_m)\cdot\mathbf{a}_n}$. The vectors $\mathbf{K}_i$, $i = 1, 2, 3$, define the so-called reciprocal lattice for a crystal and its elementary cell is the domain for the vectors $\mathbf{k}$. The vectors $\mathbf{k}$ forming the elementary cell of the reciprocal lattice thus enumerate the phase factors caused by translations of the wave functions in the Bravais lattice. These $\mathbf{k}$-vectors enumerate the stationary states of an electron in a periodic potential. There exist as many different stationary states as the $\mathbf{k}$-vectors that are available. The number of available $\mathbf{k}$-vectors in the elementary cell of the reciprocal lattice is finite, because of the discrete structure of $\mathbf{k}$ caused by the normalization condition for the wave functions $\Psi_\mathbf{k}(\mathbf{r})$ (which is similar to that for the stationary states of a free particle with momentum $\mathbf{p} = \hbar\mathbf{k}$, with periodic Born–Karman conditions imposed [55]).

2.1 Definition of a Metal

The translation properties of a stationary wave function for an electron in a crystal are thus defined by the vectors **k** taken from an elementary cell in the reciprocal lattice:

$$\Psi_{\mathbf{k}}(\mathbf{r} + \mathbf{a}_n) = \Psi_{\mathbf{k}}(\mathbf{r}) e^{i\mathbf{k}\cdot\mathbf{a}_n}. \tag{2.8}$$

One can rewrite the wave function as follows: $\Psi_{\mathbf{k}}(\mathbf{r}) = e^{i\mathbf{k}\cdot\mathbf{r}} u(\mathbf{r})$ (taking advantage of the fact that $e^{i\mathbf{k}\cdot\mathbf{r}} \neq 0$). By virtue of the formula (2.8) we see that

$$u(\mathbf{r} + \mathbf{a}_n) = u(\mathbf{r}). \tag{2.9}$$

The periodic function $u(\mathbf{r})$ is called a Bloch function and its periodicity property is referred to as the Bloch theorem.

One can notice that the factor $e^{i\mathbf{k}\cdot\mathbf{r}}$ is the unnormalized eigenfunction of the momentum operator, with eigenvalue $\hbar\mathbf{k}$. Hence, one can say that the Bloch function periodically modulates the momentum states of an electron. The latter momentum states are, however, not arbitrary as for a free particle: in a crystal $\hbar\mathbf{k}$ can attain only values corresponding to the **k**-vectors from the single elementary cell of the reciprocal lattice. This difference is expressed in the term 'pseudomomentum' that is used for $\hbar\mathbf{k}$ in a crystal rather than 'momentum', the term used for a free particle.

The Bohr–Karman periodic conditions allowing for normalization of the wave function $\Psi_{\mathbf{k}}(\mathbf{r}) = e^{i\mathbf{k}\cdot\mathbf{r}} u(\mathbf{r})$ confined to a box with volume $V = \mathbf{L}_1 \cdot (\mathbf{L}_2 \times \mathbf{L}_3)$ (this box can be oriented along the Bravais lattice vectors $\mathbf{a}_i$) are the same for the pseudomomentum as for the ordinary momentum: $k_i = 2\pi l_i / L_i$, where l_i are integers. Hence, one can assess the total number of **k**-vectors in an elementary cell of a reciprocal lattice, by dividing the volume of this cell, $v_o = \mathbf{K}_1 \cdot (\mathbf{K}_2 \times \mathbf{K}_3) = (2\pi)^3/v$ (here $v = \mathbf{a}_1 \cdot (\mathbf{a}_2 \times \mathbf{a}_3)$ is the volume of an elementary cell in the corresponding Bravais lattice) by the **k**-vector density; this equals the volume of the parallelepiped falling at a single **k** in the Bohr–Karman discretization,

$$\frac{(2\pi)^3}{L_1 L_2 L_3} = \frac{(2\pi)^3}{V}.$$

The result of this division gives the number of possible **k**-states, $N = 2V/v$ (the factor 2 accounts for additional possibility of two distinct orientations of the spin of an electron). One can conclude that the number of stationary states of an electron in a crystal is equal to double the number of elementary Bravais cells in the volume of the crystal sample, $2V/v$.

In order to better reflect a particular crystal symmetry, instead of the elementary cell in a reciprocal lattice one can select the Wigner–Seitz cell, which has the same volume as the elementary cell but is symmetrical according to the point group of the crystal. The Wigner–Seitz cell is defined by the planes perpendicular to all reciprocal lattice vectors that can be drawn from a selected node in this lattice, when these planes divide the lattice vectors at their centres [54]. The Wigner–Seitz cell contains a set of vectors **k** equivalent to those of the elementary reciprocal cell and is called the first Brillouin zone.

The mapping $E_\mathbf{k} \to \mathbf{k}$ is usually not one-to-one because the mapping $\mathbf{k} \to E_\mathbf{k}^i$ is multi-valued; this is denoted by an additional index i enumerating 'energy bands'. In each band there is the same number of different **k**-states – the number of states in the elementary cell of a reciprocal lattice or, equivalently, in the first Brillouin zone. This number (including spin) equals twice the number of Bravais elementary cells in the crystal sample.

In each Bravais elementary cell there is the same complex of atoms with the same number of collectivized electrons. If only one electron in each Bravais cell is collectivized then we have a half-filled single band; if two electrons are collectivized in each Bravais cell then the single band is completely filled. When three electrons are collectivized per cell then one and a half bands are full in the crystal (because each state in a band may be occupied by only a single electron, since electrons are fermions). The following distinct situations are possible. The last-filled band is separated from the next band by an energy gap, either large or small; alternatively, the succeeding bands overlap in some region of the first Brillouin zone. If the last band is completely filled and the gap to the next empty band is large then the crystal is insulating. If the gap to the next empty band is small (compared with $k_B T$, where $k_B = 1.38 \times 10^{23}$ J/K is the Boltzmann constant and T is the temperature in kelvins) then we have a semiconductor crystal. If the last band is half-filled, which happens when there is an odd number of collectivized electrons in each Bravais cell, then the crystal is a metal. In the case of an overlap between the last completely filled band and the next empty band, the crystal is a so-called semi-metal [54].

Metals are thus characterized by a half-filled last band, and electrons can hop to empty **k**-states in this band when the crystal is subject to an external electric field. In this field the half-filled band will be reorganized in such a manner that an electric current will flow in the crystal. The related nonzero conductivity is thus a result of the specific filling of bands with electrons. In the case of a completely filled band any reorganization of the electrons, enabling a current, is blocked by the energy gap to the next empty band. A large gap blocks the current even at a strong electric field – this is an insulator. In semiconductors, where the gap is small, the temperature 'chaos' allows for a small proportion of the electrons in the completely filled band (called the valence band) to hop to the originally empty conduction band. This small amount of electrons at the bottom of the conduction band gives a small current if there is an applied electric field.

Turning back to the case of a metal, we note that a half-filled band still contains a huge number of electrons, the number of elementary Bravais cells in the crystal sample. The latter number is of the order of Avogadro's number (6.02×10^{23}) for a crystal sample with mass equal to the atomic mass number (in grams) of the metal element. The Fermi sphere (the ground state in momentum space) of so huge a number of electrons has a large radius p_F, the Fermi radius.[2] The Fermi energy (the energy of the uppermost electrons) $\varepsilon_F = p_F^2/(2m)$ is typically of the order of $10^4 - 10^5$ K (in units where the Boltzmann constant $k_B = 1$) and is thus much larger than room temperature or melting temperature (the melting temperature for a crystal corresponds to the thermal decomposition of the ion web; the energy of the hot electrons on the Fermi surface cannot, however, be used to melt the crystal because these electrons are not able to lose their energy since the lower-energy states are completely filled by other electrons). The Fermi momentum and energy may thus be found from the relation

$$2\frac{4\pi p_F^3}{3} \bigg/ \frac{(2\pi)^3}{V} = N. \tag{2.10}$$

In a metal the Fermi surface accommodates itself to the crystal iso-energy surfaces in the Brillouin zone, and only in an idealized case does it have the shape of a sphere. When the Fermi surface crosses the borders of the first Brillouin zone, it must be perpendicular to these borders (as is required by time-reversal symmetry) [12]. In such a case the whole 'Fermi ball' spills out beyond the borders of the first Brillouin zone and partly fills the second (and then the third) Brillouin zone. The shapes of the Fermi surfaces for various crystalline symmetries are known in detail for all metals, both experimentally and theoretically [12]. Despite the fact that the true Fermi surface in a real metal is strongly deformed and complicated, the idealized isotropic spherical model is frequently helpful.

The electrons in a crystal interact strongly with the background ions and with each other; thus the single-electron gas model is a very crude approximation to a real electron-interacting Fermi liquid. Interactions also modify to some extent the picture of the Fermi surface: the sharp ideal Fermi surface at $T = 0$ K is blurred due to interactions, but the expressions given above for the Fermi energy and the Fermi momentum still hold (according to the Luttinger theorem [9]).

The existence of a free Fermi surface beyond which are empty states in the band can be thus identified with the definition of a metal in its normal phase. Because of the large depth of the Fermi sea, greatly exceeding the temperature and energy of an applied electric or magnetic field, the majority of experimentally available

---

[2] In the idealized spherically symmetrical case fermions fill the discrete **k**-states one by one starting from zero (the origin of the first Brillouin zone); the energy of an electron in the gas model is $\hbar^2(k_x^2 + k_y^2 + k_z^2)/(2m)$ and the ground state corresponds to a filled sphere in **k**-space with radius given by $p_F/\hbar = (\sqrt{k_x^2 + k_y^2 + k_z^2})_F$.

thermal and galvano-magnetic phenomena in metals involve only electrons located very close to the Fermi surface. The deeply located electrons are usually passive in thermal and transport effects because their excitation requires too high an energy. An exception is provided by plasmon effects, to which all electrons contribute, including those located deep below the Fermi surface.

The description of the interacting Fermi liquid of electrons in metals, including the electron–electron Coulomb interaction and the interaction of the electrons with the ion background, was the subject of intensive studies in the second half of the twentieth century. Very precise methods have been developed within the Green function approach to metals [9] (earlier investigations were in terms of the kinetic equation method [56, 57], which has been proved to be equivalent to the Green function picture [9, 10, 58]). The exceptional precision of the microscopic theory of a Fermi liquid is linked with the existence, for metals, of a small parameter expressed as the ratio of the temperature and the energy of the experimentally accessible electric and magnetic field values, with respect to the Fermi energy, large in metals. This small parameter ensures the quick convergence of all perturbation series with respect to the interaction, which results in the exceptionally high accuracy of metal models. In the case of semiconductors the Fermi energy is small and the small parameter identified above disappears; this means that the semiconductor description is less accurate and less predictable than that for metals (in semiconductors the widths of the bands and of the interband gaps are important instead) [59].

2.2 Scheme of the Fermi Liquid Theory
2.2.1 Phenomenological Theory of a Fermi Liquid

The phenomenological theory of a normal Fermi liquid,[3] formulated by Landau [56, 60, 61, 13] is based on the assumption that the ground state and low lying excited states in a multiparticle Fermi liquid in its so-called normal phase may be mapped one-to-one onto the well-defined Fermi surface of the gas of electrons and the excited free electrons beyond this surface, without interaction. In this way one can introduce the powerful concept of quasiparticles – excited states low lying in energy – for a strongly interacting multi-electron system. The concentration of these quasiparticles can be expressed in terms of the concentration of their noninteracting partners; thus the change in concentration is $\delta n(\mathbf{p}) = n(\mathbf{p}) - n^0(\mathbf{p})$. At $T = 0$, $n^0(\mathbf{p})$ corresponds here to the sharp Fermi surface for a noninteracting electron gas (at higher temperatures the Fermi surface is blurred on the scale of $k_B T$, as shown in Fig. 2.1).

---

[3] Not superconducting and not magnetic.

2.2 Scheme of the Fermi Liquid Theory

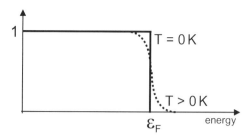

Figure 2.1 One-particle distribution function for a fermion gas, $f(\varepsilon) = \frac{1}{e^{(\varepsilon-\mu)/k_BT}+1}$ (the Fermi–Dirac function). For $T = 0$ K the chemical potential $\mu = \varepsilon_F$; at temperatures $T > 0$ K but small in comparison to ε_F/k_B (k_B is the Boltzmann constant) the Fermi surface blurs and also μ shifts slightly [13].

Next, Landau introduced an effective Hamiltonian for quasiparticles in order to derive a kinetic equation of Boltzmann type. This effective Hamiltonian has the form

$$H = \text{Tr}\sum_{\mathbf{p}} \varepsilon_{\mathbf{p}} \delta n(\mathbf{p}) + \text{Tr}\,\text{Tr}' \sum_{\mathbf{p},\mathbf{p}'} f_{\mathbf{pp}'} \delta n(\mathbf{p}) \delta n(\mathbf{p}'), \quad (2.11)$$

where $f_{\mathbf{pp}'} = f^s_{\mathbf{pp}'}\sigma^0\sigma^{0'} + f^a_{\mathbf{pp}'}\vec{\sigma}\cdot\vec{\sigma}'$ expresses the effective interaction of quasiparticles, additionally including the spin structure; σ^0 and $\vec{\sigma}$ are the unit 2×2 matrix and the vector of Pauli matrices, respectively, which together span the spin structure $\delta n(\mathbf{p}) = \delta n^s(\mathbf{p})\sigma^0 + \delta\mathbf{n}^a(\mathbf{p})\cdot\vec{\sigma}$, and $\varepsilon_{\mathbf{p}} = v_F(p - p_F)$, $v_F = p_F/m^*$, where m^* is the effective mass of a quasiparticle. The single-quasiparticle energy can be thus written as $\tilde{\varepsilon}_{\mathbf{p}} = \varepsilon_{\mathbf{p}} + Tr'\sum_{\mathbf{p}'} f_{\mathbf{pp}'}\delta n(\mathbf{p}')$ and, assuming that the density (concentration) of quasiparticles varies in time and space, one can derive a Boltzmann-type kinetic equation, $\{\tilde{\varepsilon},\delta n\} + \partial\delta n/\partial t = I[\delta n]$ (here $\{,\}$ denotes a Poisson bracket).[4]

Keeping only the linear terms in the above equation, we obtain,

$$\frac{\partial \delta n}{\partial t} + \frac{\partial \varepsilon_{\mathbf{p}}}{\partial \mathbf{p}}\frac{\partial \delta n}{\partial \mathbf{r}} - \frac{\partial\,\text{Tr}'\sum_{\mathbf{p}'} f_{\mathbf{pp}'}\delta n(\mathbf{p}')}{\partial \mathbf{r}}\frac{\partial n_0}{\partial \mathbf{p}} = I[\delta n]. \quad (2.12)$$

Taking into account that $\partial n_0/\partial\mathbf{p} = -\hat{\mathbf{p}}\delta(p - p_F) = -v_F\hat{\mathbf{p}}\delta(\varepsilon - \varepsilon_F)$ (for $n_0(\varepsilon_{\mathbf{p}}) = 1 - \Theta(\varepsilon_{\mathbf{p}} - \varepsilon_F)$, where Θ is the Heaviside step function) and assuming that $\delta n = \delta(\varepsilon - \varepsilon_F)\nu(\hat{\mathbf{p}})$, for Fourier space–time components $e^{-i\omega+i\mathbf{q}\cdot\mathbf{r}}$, we get

$$(\omega - v_F\hat{\mathbf{p}}\cdot\mathbf{q})\nu(\hat{\mathbf{p}}) = \hat{\mathbf{p}}\cdot\mathbf{q}\frac{p_F^2}{(2\pi\hbar)^3}\text{Tr}'\int d\Omega' f_{\hat{\mathbf{p}}\hat{\mathbf{p}}'}\nu(\hat{\mathbf{p}}'), \quad (2.13)$$

---

[4] Let us note that the effective Hamiltonian for quasiparticles depends in an implicit manner on the temperature via the temperature-dependent quasiparticle density – thus this Hamiltonian is of a phenomenological, not a microscopic, character.

where $d\Omega = \sin\Theta\, d\Theta\, d\phi$ is the differential in spherical coordinates;[5] the collision integral, on the right-hand side of Eq. (2.12), does not contribute in the case $\omega\tau \gg 1$, where $\tau \sim T^{-2}$ is the mean intercollision time for quasiparticles, since the damping of quasiparticles is proportional to square of the separation from the Fermi surface and is zero on the Fermi surface itself [13].

The kinetic equation for quasiparticles in metals (2.13), called the Landau kinetic equation, allows the identification of the energy $\omega(q)$ of the so-called zeroth sound, of spin waves, and of Landau damping as well as the expression of parameters for these effects in terms of Landau amplitudes, i.e., the coefficients in the decomposition of $f_{\hat{\mathbf{p}}\hat{\mathbf{p}}'}^{s(a)}$ (see the text below Eq. (2.11)). In the isotropic case this function depends only on the angle between the versors $\hat{\mathbf{p}}$ and $\hat{\mathbf{p}}'$, which admits the following decomposition in the Legendre polynomial basis: $f_{\hat{\mathbf{p}}\hat{\mathbf{p}}'}^{s(a)} = \sum_l (2l+1) f_l^{s(a)} P_l(\hat{\mathbf{p}} \cdot \hat{\mathbf{p}}')$. The coefficients in this decomposition of the quasiparticle interaction are the Landau amplitudes.

Utilizing the definition of the group velocity, $\mathbf{v} = \partial \tilde{\varepsilon}/\partial \mathbf{p}$, one can express the total flow of the quasiparticle mass (taking into account that the number of quasiparticles agrees with the number of bare noninteracting electrons) as follows:

$$\int \frac{d^3\mathbf{p}}{(2\pi\hbar)^3} n(\mathbf{p}) \mathbf{p} = m \int \frac{d^3\mathbf{p}}{(2\pi\hbar)^3} n(\mathbf{p}) \frac{\partial \tilde{\varepsilon}}{\partial \mathbf{p}}. \qquad (2.14)$$

Using the expression for the quasiparticle energy (see the paragraph after (2.11)) one can rewrite the change in flow induced by a variation δn of the quasiparticle density (taking the variation of both sides of Eq. (2.14)):

$$\int \frac{d^3\mathbf{p}}{(2\pi\hbar)^3} \delta n(\mathbf{p}) \frac{\mathbf{p}}{m} = \int \frac{d^3\mathbf{p}}{(2\pi\hbar)^3} \delta n \frac{\partial \varepsilon(\mathbf{p})}{\partial \mathbf{p}}$$
$$+ \operatorname{Tr}\operatorname{Tr}' \int \frac{d^3\mathbf{p}'}{(2\pi\hbar)^3} \int \frac{d^3\mathbf{p}}{(2\pi\hbar)^3} \frac{\partial}{\partial \mathbf{p}} f_{\mathbf{p}\mathbf{p}'} n(\mathbf{p}) \delta n(\mathbf{p}'). \qquad (2.15)$$

Now using integration by parts in the second term on the right-hand side, and exchanging in this term the notations $\mathbf{p}$ and $\mathbf{p}'$, we finally obtain

$$\int \frac{d^3\mathbf{p}}{(2\pi\hbar)^3} \delta n \frac{\mathbf{p}}{m} = \int \frac{d^3\mathbf{p}}{(2\pi\hbar)^3} \delta n \frac{\partial \varepsilon(\mathbf{p})}{\partial \mathbf{p}}$$
$$- \operatorname{Tr}\operatorname{Tr}' \int \frac{d^3\mathbf{p}'}{(2\pi\hbar)^3} \int \frac{d^3\mathbf{p}}{(2\pi\hbar)^3} f_{\mathbf{p}\mathbf{p}'} \frac{\partial}{\partial \mathbf{p}'} n(\mathbf{p}') \delta n(\mathbf{p}). \qquad (2.16)$$

Owing to the arbitrariness of $\delta n(\mathbf{p})$ we get

$$\frac{\mathbf{p}}{m} = \frac{\partial \varepsilon(\mathbf{p})}{\partial \mathbf{p}} - \operatorname{Tr}\operatorname{Tr}' \int \frac{d^3\mathbf{p}'}{(2\pi\hbar)^3} f_{\mathbf{p}\mathbf{p}'} \frac{\partial n(\mathbf{p}')}{\partial \mathbf{p}'}. \qquad (2.17)$$

---

[5] The reader should note that here Θ refers to a spherical coordinate.

2.2 Scheme of the Fermi Liquid Theory

For $n(\mathbf{p}) = 1 - \Theta(p - p_F)$, we have $\partial n/\partial \mathbf{p} = -\hat{\mathbf{p}}\delta(p - p_F)$; thus

$$\frac{1}{m} = \frac{1}{m^*} + \frac{p_F}{(2\pi\hbar)^3} \int d\Omega f^s(\cos\vartheta) \cos\vartheta, \tag{2.18}$$

giving for the effective mass

$$m^* = m(1 + F_1). \tag{2.19}$$

Here we employed decomposition in the Legendre polynomial basis, $f^s(\cos\vartheta) = \sum_l (2l+1) P_l(\cos\vartheta) f_l^s$ and

$$\int \frac{d\Omega}{4\pi} P_l(\cos\vartheta) \cos\vartheta = \frac{1}{2l+1}\delta_{l,1}, \qquad F_l = \frac{p_F m^*}{2\pi^2 \hbar^3} f_l^s.$$

The Landau kinetic equation (2.13) allows analysis of the collective excitations in a normal Fermi liquid. For spinless excitations (when δn does not depend on the spin) the kinetic equation (2.13) can be rewritten in the form

$$(\omega - \mathbf{k}\cdot\mathbf{v})\nu(\hat{\mathbf{p}}) = \mathbf{k}\cdot\mathbf{v}\int F(\vartheta)\nu(\hat{\mathbf{p}}')\frac{d\Omega'}{4\pi}, \tag{2.20}$$

where $\text{Tr}' f = (2\pi^2\hbar^3/(p_F m^*))F(\hat{\mathbf{p}}\cdot\hat{\mathbf{p}}')$. Using the notation $s = (\omega/(v_F k))$ we obtain

$$(s - \cos\Theta)\nu(\Theta,\varphi) = \cos\Theta\int \frac{d\Omega'}{4\pi} F(\cos\vartheta)\nu(\Theta',\varphi'). \tag{2.21}$$

The above equation can be solved for specific forms of $F(\cos\vartheta)$, where $\cos\vartheta = \hat{\mathbf{p}}\cdot\hat{\mathbf{p}}'$. For $F = F_0 = \text{const.}$, owing to the absence of any dependence on the angle ϑ under the integral, we get

$$\nu = \text{const.} \times \frac{\cos\Theta}{s - \cos\Theta}, \tag{2.22}$$

and substitution into (2.21) gives

$$\frac{F_0}{2}\int_0^\pi \frac{\cos\Theta}{s - \cos\Theta} \sin\Theta\, d\Theta = 1. \tag{2.23}$$

After integrating we obtain

$$\frac{s}{2}\ln\frac{s+1}{s-1} - 1 = \frac{1}{F_0}. \tag{2.24}$$

This equation has a real solution for s, i.e., for the zero sound velocity, when $F_0 > 0$. For $-1 < F_0 < 0$ the solution is complex and corresponds to damping – the so-called Landau damping. For $F_0 \leq -1$ the complex solution is unstable; the normal Fermi liquid loses its stability as the ground state of interacting fermions (this instability was described by Pomeranchuck [62]). The sound identified above as a spinless collective mode arising from the interaction of Landau quasiparticles is called the 'zero sound'. The zero sound corresponds to oscillations or deformations of the quasiparticle density on the Fermi surface without any change in the volume of the Fermi sphere – thus without any density fluctuation in position space. This is the difference from ordinary acoustic sound, which corresponds to fluctuations of the density in position space and is linked with oscillations of the Fermi sphere size (the Fermi sphere radius depends on the particle density). The zero sound is a collective excitation which can propagate in a Fermi liquid even at $T = 0$ K, whereas ordinary sound does not propagate at zero temperature (at zero temperature, phonons, the carriers of ordinary sound, disappear). In analogy to the zero sound in the Fermi liquid, spin waves corresponding to fluctuations of the spin part of the quasiparticle density on the Fermi surface may also occur; the velocity of spin waves is governed by the spin part of the Landau interaction function [10, 9, 13].

The Landau theory of the normal Fermi liquid, though based on phenomenological assumptions, agrees very well with experimental observations of electrons in metals (some generalizations towards inclusion of the Coulomb interaction and its screening by positive ion background in a crystalline lattice were developed by Silin [57]). The most important aspect of the Landau theory is the introduction of quasiparticles and of their effective interaction, expressed by the Landau function and different from the Coulomb interaction of electron charges, the latter interaction having been largely included in the definition of quasiparticles. The notion of quasiparticles and their effective interaction found complete confirmation in the microscopic theory of the normal Fermi liquid, which was developed next. Worth emphasizing is the fact that the Landau theory of quasiparticles is not of a perturbation character and thus is applicable for strongly interacting electron systems, such as electrons in metals. The confirmation of this theory using Green function formalism requires the exact summation of an infinite perturbation series and thus is independent of the perturbation scheme and demonstrates in detail the structure of quasiparticles and their effective interaction [58, 9, 63, 64]. It is surprising that electrons in metals are quasiparticles on the Fermi surface and are actually different from ordinary free electrons (they also have a different effective mass). These quasiparticles interact residually in a way that is quite distinct from that of strongly interacting ordinary electrons. The quasiparticle picture in metals agrees perfectly with experiment.

2.2.2 Outline of the Microscopic Theory of a Normal Fermi Liquid

The fundamental assumption of the Fermi liquid theory is the existence of a well-defined Fermi surface for noninteracting fermions at $T = 0$ K. The interaction can

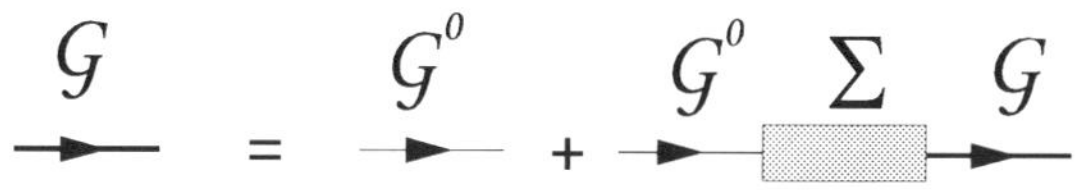

2.2 Scheme of the Fermi Liquid Theory

$$\mathcal{G} = \mathcal{G}^0 + \mathcal{G}^0 \Sigma \mathcal{G}$$

Figure 2.2 Graphical presentation of the Dyson equation. The mass operator Σ is built from all multi-line graphs which cannot be cut across a single line; the thin lines represent the zeroth (bare – without interaction) Matsubara Green functions; the thick lines represent the full Matsubara Green functions with the interaction included.

be switched on via a mass operator and vertex functions conveniently expressed in terms of Matsubara Green functions (of imaginary time) [9, 11]. Next, assuming that the mass operator is a continuous and smooth, i.e., differentiable, function of its own parameters (momentum and energy or the Matsubara frequency after an analytical continuation) near the Fermi surface, a so-called 'normal' metal can be defined. With the above conditions one can develop the mass operator up to linear terms (the imaginary part of the mass operator is of second order in the momentum with respect to the Fermi surface). Such a development of the mass operator allows for the definition of Landau quasiparticles as poles of a retarded single-particle Green function. Poles of a two-particle retarded Green function define collective excitations: the zero sound and spin waves. An important achievement of the microscopic Fermi liquid theory is the expression of the Landau amplitudes (the Legendre components of the quasiparticle interaction, in the isotropic case) as an appropriate limit of the vertex function for Matsubara Green functions. The advantage of the Matsubara functions is the inclusion of zero and nonzero temperatures simultaneously in the same formalism.

The mass operator is defined by the Dyson equation for the Matsubara Green function (see Fig. 2.2),

$$\mathcal{G}(\mathbf{k}, i\omega_\nu) = \mathcal{G}^0(\mathbf{k}, i\omega_\nu) + \mathcal{G}^0(\mathbf{k}, i\omega_\nu) \Sigma(\mathbf{k}, i\omega_\nu) \mathcal{G}(\mathbf{k}, i\omega_\nu). \quad (2.25)$$

Then (taking into account that $\mathcal{G}^0 = 1/(i\omega_\nu - \varepsilon_\mathbf{k})$, $\varepsilon_\mathbf{k} = p^2/2m - \mu$, where μ is the chemical potential and $\mathbf{p} = \hbar\mathbf{k}$; $i\omega_\nu$ is the Masubara frequency),

$$\mathcal{G}(\mathbf{k}, i\omega_\nu) = \frac{1}{i\omega_\nu - \varepsilon_\mathbf{k} - \Sigma(\mathbf{k}, i\omega_\nu)}, \quad (2.26)$$

or after analytical continuation of the discrete imaginary Matsubara frequency onto the whole complex plane ($i\omega_\nu \to E$),

$$G(k, E) = \frac{1}{E - \varepsilon_\mathbf{k} - \Sigma(\mathbf{k}, E)}, \quad E = \text{Re}\, E + i\,\text{Im}\, E. \quad (2.27)$$

The above function is the Fourier transform of the retarded or advanced Green function (correspondingly to sign of the pole) [10] and its pole defines the quasiparticles in the system with interactions. The real part of the pole defines the energy of the quasiparticle whereas the imaginary part defines the quasiparticle lifetime (the lifetime of the quasiparticle arises owing to the scattering of interacting bare particles). The imaginary part of the pole equals the imaginary part of the mass operator. Hence, when the mass operator is real according to some approximation, the quasiparticles are stable under this approximation. This is the case for a normal Fermi liquid, where the mass operator is real in the vicinity of the Fermi surface in an approximation that is linear with respect to the distance from the Fermi surface (the Fermi surface is defined here for a system without interactions at $T = 0$ K) [9]. Upon assuming the continuity of the derivative of the mass operator near the Fermi surface, we have

$$\Sigma(\mathbf{k}, i\omega_\nu) = \Sigma(p_F, 0) + \frac{\partial \Sigma}{\partial p}|_{(p_F, 0)}(p - p_F) + \frac{\partial \Sigma}{\partial i\omega_\nu}|_{(p_F, 0)} i\omega_\nu,$$

$$\mu - \Sigma(p_F, 0) = \frac{p_F^2}{2m}$$

(from the Luttinger theorem, the relation between the Fermi momentum p_F and the particle density is not perturbed by interactions [9]). Then

$$\mathcal{G}(p, i\omega_\nu) = \frac{a}{i\omega_\nu - v_F(p - p_F)}, \qquad (2.28)$$

where the residuum,

$$a = \left(1 - \frac{\partial \Sigma}{\partial i\omega_\nu}|_{(p_F, 0)}\right)^{-1}$$

and

$$v_F = \frac{p_F}{m^*} = a\left(\frac{p_F}{m} + \frac{\partial \Sigma}{\partial p}|_{(p_F, 0)}\right).$$

The analytical properties of retarded and advanced Green functions (related to their analytical continuation (2.28)) justify the definition of the quasiparticle energy in the form $\varepsilon_p = v_F|p - p_F|$ both for $p > p_F$ and for $p < p_F$.

Derivatives of the mass operator can be expressed, utilizing the Ward identities [9], via the vertex function corresponding to the sum of the ladder graphs in the particle–hole channel (with oppositely directed Green functions). The integral kernel related to this ladder sum has the form $(\mathbf{k} \cdot \mathbf{q}/(\omega - \mathbf{k} \cdot \mathbf{q}))$, where ω and $\mathbf{k}$ are the energy and momentum transfer at scattering. One notices that the long-wave limit at $\omega = 0$ is equal to -1, whereas it is 0 at $\omega \neq 0$, which gives rise to the introduction of limits of the vertex function: Γ^k when $k \to 0, \omega = 0$, and Γ^ω when $k \to 0, \omega \neq 0$, respectively. The interaction of the Landau quasiparticles is expressed via $a^2 \Gamma^\omega$ [9]. Hence, one can find the effective mass, using the quasiparticle interaction function.

2.2 Scheme of the Fermi Liquid Theory

Figure 2.3 The two-particle Green function can be expressed on an effective vertex.

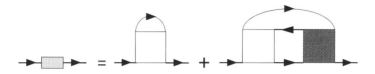

Figure 2.4 For the two-particle interaction (the open square, i.e., the bare interaction vertex) the mass operator satisfies the equation in the figure, which follows from the fact that the two ends of the mass operator belong either to one bare vertex or to two distinct bare vertices. In the latter case one bare vertex must be substituted by a full vertex (all lines in the figure are full Green functions).

Simultaneously the constant a (the numerator in Eq. (2.28)), which gives the jump in the single-particle distribution function for particles, rather than quasiparticles [10], on the Fermi surface at $T = 0$ K, disappears from the formalism (it has entered the definition of the Landau interaction). This jump a is caused by the presence of interactions (without interactions $a = 1$; in the presence of interactions $a < 1$). The vertex function is linked with the two-particle Green function, as shown in Fig. 2.3, where the lines represent Matsubara Green functions that are diagonal in the indices $k_i = (i\omega_i, \mathbf{p}_i)$, i.e., they are diagonal in the Matsubara frequency and momentum as in a space–time translationally invariant system.

The mass operator satisfies the equation in Fig. 2.4, where the solid lines represent the Matsubara Green functions and the open square is the bare interaction whereas the shaded square is the full vertex (the vertex function). The equation for the vertex function cannot be written in terms of the bare interaction (the bare vertex) in the form of a ladder sum because the repeating element of a pair of Green functions may be of two types in general. The two Green functions in a repeating pair may be oriented in parallel (the particle–particle scattering channel) or in opposite directions (the particle–hole channel). In the latter case the Green functions may differ slightly in four-momentum (energy and momentum) and either Green function may depend on the four-momentum of the vertex function. Hence, as a result we are dealing with three different types of repeating ladders; see Fig. 2.5.

Each of these three channels leads to a different Bethe–Salpeter equation for the appropriate part of the vertex function with the irreducible vertex function accommodated to each scattering channel. For antiparallel Green functions with a small four-momentum transfer, the Bethe–Salpeter equation has the form shown in

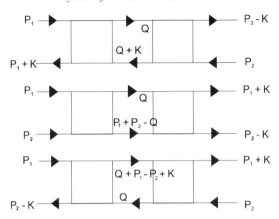

Figure 2.5 Various channels for scattering in the ladder structure of the full vertex; if the Green functions linking vertices are oriented in the same way, the channel is of the particle–particle type, but when these Green functions are in opposite directions, the channel is of the particle–hole type.

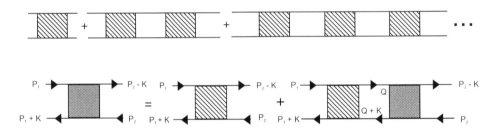

Figure 2.6 (Upper) The scheme of the ladder structure which leads to an equation of the Bethe–Salpeter type. Each scatttering type (as shown in Fig. 2.5) leads to a distinct Bethe–Salpeter equation for the full vertex (filled square) with different irreducible vertices (hatched square) associated with the distinct scattering channel. (Lower) The Bethe–Salpeter equation for the particle–hole channel with the small four-momentum transfer allows for a description of the Landau Fermi liquid in the normal phase.

Fig. 2.6. This channel and the corresponding equation is of particular significance because the kernel of the integral equation for this channel is nonanalytical, which manifests itself in collective excitations in the normal Fermi liquid: the zero sound, spin waves, and Landau damping – the same effects as are described by the phenomenological Landau kinetic equation, as presented above. The particle–particle channel, however, leads to a different phenomenon – an instability of the normal Fermi liquid to the coherent scattering of particle pairs. This instability is assigned by a pole whose imaginary part agrees in sign with the bare interaction. For bare-particle attraction the normal Fermi liquid is thus unstable against the formation of Cooper pairs, resulting in another ground state – the superfluid or superconducting

2.2 Scheme of the Fermi Liquid Theory

Fermi liquid (for uncharged or charged carriers, respectively, e.g., He^3 atoms or electrons in metal). The third channel (again of the electron–hole type) is not assigned by any singularity.

The particle–hole channel with a small transfer of four-momentum can be taken in two different limits (as the kernel is nonanalytical), $\mathbf{k} \to 0$ at $\omega - 0$ or $\omega \to 0$ at $\mathbf{k} = 0$. The former is denoted the k limit and the latter as the ω limit. Correspondingly, appropriate limits can be defined for the vertex function. Taking into account the form of the integral kernel in this channel for small transfers of the four-momentum $k = (\mathbf{k}, \omega)$, we obtain

$$\frac{2\pi i a^2}{v_F} \frac{v_F \hat{\mathbf{q}} \cdot \mathbf{k}}{\omega - v_F \hat{\mathbf{q}} \cdot \mathbf{k}} \delta(\varepsilon) \delta(|\mathbf{q}| - q_F) + \varphi_{reg}, \quad (2.29)$$

where $q = (\varepsilon, \mathbf{q})$ and φ_{reg} is the regular part of the product of two antiparallel Green functions in this channel,[6] one can write the k and ω limits of the Bethe–Salpeter equation (Fig. 2.6):

$$\Gamma^\omega = \Pi|_{k=0} - i \int dq \, \Pi|_{k=0} \, \varphi_{reg} \Gamma^\omega, \quad (2.30)$$

and

$$\Gamma^k = \Pi|_{k=0} - i \int dq \, \Pi|_{k=0} \left[-\frac{2\pi i a^2}{v_F} \delta(\varepsilon) \delta(|\mathbf{q}| - p_F) + \varphi_{reg} \right] \Gamma^k. \quad (2.31)$$

Equation (2.30) together with the initial Bethe–Salpeter equation can be rewritten in the following form (eliminating the irreducible vertex $\Pi|_{k=0} \simeq \Pi$):

$$\Gamma = \Gamma^\omega + \frac{a^2 p_F^2}{(2\pi)^3 v_F} \int d\Omega_q \Gamma^\omega \frac{v_F \hat{\mathbf{q}} \cdot \mathbf{k}}{\omega - v_F \hat{\mathbf{q}} \cdot \mathbf{k}} \Gamma \quad (2.32)$$

or, in a more developed way, as

$$\Gamma_{\alpha\beta,\gamma\delta}(p_1, p_2, k) = \Gamma^\omega_{\alpha\beta,\gamma\delta}(p_1, p_2)$$
$$+ \frac{a^2 p_F^2}{(2\pi)^3 v_F} \int d\Omega_q \Gamma^\omega_{\alpha\nu,\gamma\mu}(p_1, q) \frac{v_F \hat{\mathbf{q}} \cdot \mathbf{k}}{\omega - v_F \hat{\mathbf{q}} \cdot \mathbf{k}} \Gamma_{\nu\beta,\mu\delta}(q, p_2); \quad (2.33)$$

here p_2 and β, δ play the role of parameters only and the essential dependence is $\Gamma(\hat{\mathbf{p}}_1, k)$, or $\Gamma(\hat{\mathbf{q}}, k)$ in the integral over direction, i.e.,

$$\Gamma_{\alpha\beta,\gamma\delta}(p_1, p_2, k) = \chi_{\alpha\gamma}(p_1, k) \chi_{\beta\delta}(p_2, k)$$

---

[6] For nonzero temperatures, Matsubara Green function integration over the four-vector is substituted by summation over the Matsubara frequency and integration over the momentum [9, 10].

and the second factor appears on both sides of (2.32). Next, one can rewrite Eq. (2.32) in the following form, suitable for determination of the poles of the vertex function:

$$(\omega - v_F \mathbf{p} \cdot \hat{\mathbf{k}}_1) v_{\alpha\gamma}(\hat{\mathbf{p}}_1, k) = \frac{a^2 p_F^2}{(2\pi)^3} \hat{\mathbf{p}} \cdot \mathbf{k}_1 \int d\Omega_q \Gamma^\omega_{\alpha\xi;\gamma\eta}(\hat{\mathbf{p}}_1, \hat{\mathbf{q}}) v_{\eta\xi}(\hat{\mathbf{q}}, k). \quad (2.34)$$

The following notation has been introduced above:

$$v_{\alpha\gamma}(\hat{\mathbf{p}}_1, k) = \frac{\hat{\mathbf{p}}_1 \cdot \mathbf{k}}{\omega - v_F \hat{\mathbf{p}}_1 \cdot \mathbf{k}} \chi_{\alpha\gamma}(p_1, k).$$

Moreover, Γ^ω has been omitted – this function does not depend on k and is small in comparison with Γ, which is large close to the singularity at a pole. The poles of the vertex function are simultaneously poles of the two-particle Green function [9, 11], which, owing to the linear response theory of Kubo, defines a collective excitation via its singularities.

Equation (2.34), as derived in the manner presented above, is identical (on setting $\hbar = 1$) with the Landau kinetic equation, the Landau quasiparticle interaction being given by $a^2 \Gamma^\omega$.

This is a central result of the microscopic Fermi liquid theory – the proof that the phenomenological Landau interaction of quasiparticles is the vertex function $\Gamma^\omega a^2$ and the Landau kinetic equation is in fact the Bethe–Salpeter equation in the particle–hole channel for ladder-type coherent scattering. Factoring out the residuum a through identities for the derivatives of the mass operator, which in fact result from conservation rules in the system (and called Ward identities), we arrive at a complete derivation, in a microscopic manner, of the previously formulated phenomenological theory of a Fermi liquid, with the heuristic concept of quasiparticles near the Fermi surface instead of the original electrons. The microscopic Green function theory is accurate and confirms the validity of the quasiparticle concept independently of the interaction strength, though it is limited to the so-called normal Fermi liquid[7] and has the proviso that the interaction is of finite range (which is true for the Coulomb interaction of electrons screened by a positive ion background, as is the case for metals).

Electrons in a normal Fermi liquid are identified with quasiparticles that are well defined near the Fermi surface and stable on it. The strong interaction of electrons screened by the ion background in metals is included in the definition of the quasiparticles; they are dressed with this interaction although some residual interaction between quasiparticles still exists. Via this residual interaction the effective mass of the quasiparticles and the characteristics of collective excitations are expressed (the zero sound and spin wave velocities). The effective mass is renormalized by the first

---

[7] The theory of a normal Fermi liquid can be generalized to the cases of a superfluid or magnetic Fermi liquid [9, 11, 10].

Landau amplitude of this interaction (in the isotropic case): $m^* = m(1 + F_1)$. It can be found experimentally from the heat capacity of the metal (its electron part can be distinguished from the ionic part by their specific temperature dependences [12]).

In a normal Fermi liquid the notion of the Fermi surface is utilized, although it is accurately defined for a fermion gas only. In an interacting system the Fermi surface for the original fermions is fuzzy. Nevertheless, a Fermi surface that is still well defined holds for quasiparticles. This Fermi surface is a map of the Fermi surface in a gas and the quasiparticles near this ground state are numbered by the low-energy excitations in the gas [10]. Such a picture is general and can be applied to electrons in metals or to other fermionic systems. Quasiparticles are thus the original particles, dressed with the interaction. Their residual interaction is weak, which explains the common approximate picture of electrons in metals as free particles, despite their strong interaction as charged particles. The Fermi surface in gas is correctly defined at $T = 0$ K. With increasing temperature the Fermi surface blurs. For $T \ll \varepsilon_F$ this blurring is relatively small and the Fermi surface picture may be maintained (in metals ε_F is large, $\sim 10^4 - 10^5$ K, hence the Fermi surface picture is quite accurate even up to the melting temperature). Quasiparticles are stable states only on the Fermi surface, but when their distance from the Fermi surface increases, they become damped in proportion to $(p - p_F)^2$ [65, 66].

In the case of transport phenomena in metals, the related energies of excitations beyond the Fermi surface are small in comparison with the Fermi energy, and in practice the whole of the galvano-magnetic physics in metals takes place on the Fermi surface [12]. Thus, in related effects, only a relatively small fraction of electrons participate – those on the Fermi surface (the rest are hidden below the Fermi surface and are inactive). That there is a relatively small amount of active electrons in metals explains the experimentally observed values of the conductivity and other observed kinetic characteristics of metals, which could not be properly understood if all the electrons participated [12].

For other than a normal Fermi liquid, different collective states of fermions are also observed. They are superfluid or superconducting states and magnetic states. For these cases also, accurate microscopic descriptions are available. With regard to a superfluid or superconducting Fermi liquid (the latter for charged fermions such as electrons and the former for neutral fermions such as He^3 atoms), we are dealing with a specific reorganization of the Fermi surface due to the instability of coherent scattering in the particle–particle channel. This instability occurs as an imaginary pole of the vertex function in the particle–particle channel at temperatures lower than the critical temperature and for attracting fermions, indicating another ground state. This new ground state of interacting fermions corresponds to the pairing of particles (as for the particle–particle channel) and must be defined beyond the normal Fermi liquid paradigm. The new state is of an emergent character, as is always the case for spontaneously broken symmetries. Here the gauge symmetry of the Hamiltonian of interacting (attracting) fermions breaks down. Despite the gauge symmetry of the Hamiltonian and the related particle number conservation, a ground state of the superfluid or superconducting type is not gauge invariant – the nonzero density of Cooper pairs (pairs of fermions on the Fermi surface with opposite momenta) cannot be present for a gauge invariant Hamiltonian and occurs via a spontaneous symmetry breaking and simultaneous locking of the wave function's phase. For a wave function with fixed phase the particle number is not determined, owing to the appropriate uncertainty principle. The fixed phase corresponds, however, to a fixed inertial reference system in which the Cooper pairs flow without any resistance and thus are superfluid or superconducting. The pairs of fermions constitute bosons, which can all condense into the same single-particle ground state at $T = 0$ K. If this state is of zero energy and zero momentum then it is impossible to determine the number of particles in the condensate. Hence, it is by virtue of the particle-number–phase uncertainty principle that the phase is fixed, and this means that the reference frame is selected by quantum rules. In this system we are dealing with a superflow or supercurrent, respectively, without viscosity or resistance, unless we consider that the particles are excited to higher levels at higher temperatures (beyond the critical temperature). The pairs of fermions have an associated spin structure. Two 1/2 spins can be added to achieve a total spin of 0 (the singlet state) or 1 (the triplet state). Both situations are observed in experiments – in superconductors in metals the Cooper pairs are usually in a singlet spin state, whereas for superfluid He^3 triplet pairs are present.

A precise microscopic description of a Fermi liquid in the superfluid or superconducting phase has been developed [9, 10], though a simultaneous ladder-type sum in the particle–hole and particle–particle scattering channels is technically nontrivial. The stability of the superfluid or superconducting state depends, however, on the quasiparticle interaction [65]. Gauge invariance breaking in the superfluid phase requires the inclusion, in terms of Green functions, of an anomalous type of Green function representing the Cooper pair condensate. Such a formalism has been developed both for a singlet Bardeen–Cooper–Schrieffer (BCS) superfluid [67] and for a triplet superfluid [68, 69].

The phenomena sketched above occur close to the Fermi surface in metals and do not exhaust the physics of these systems. In the energy scale exceeding the Fermi energy the picture of quasiparticles near the Fermi surface is irrelevant, but electrons may take part in the response to such energetic perturbations. This is the case for plasmons, which can be excited in metals by high-energy photons (typically of UV radiation type for bulk metals). The description of plasmons needs an approach different from the standard Fermi liquid model; this will be presented in what follows.

3
Quasiclassical Description of Plasmons in the Bulk Metal

The random phase approximation (RPA) theory of plasmons in bulk metals of Pines and Bohm is described. Details of the derivation are included. The role of the electron–electron interaction for coherent plasmon excitations of electrons in metal is emphasized; this interaction is essential for plasmon formation, indicating that plasmons do not exist in gaseous systems (in which there is no interaction). The screening of unbalanced charges by the electrons in a normal metal is briefly presented alongside the Thomas–Fermi scheme.

3.1 Random Phase Approximation Theory of Plasmons of Pines and Bohm

Plasmons are fluctuations in the local electron density on the background of an inertial positive ion distribution. In order to describe the electron liquid oscillations corresponding to charge density fluctuations one considers a Hamiltonian for N electrons interacting with each other and with the positive ions, the oscillations of which can be neglected owing to their much greater mass in comparison with that of electrons. Thus the system of ions can be accounted for as a static positive 'jellium', i.e., a uniform positively charged background balancing the total charge of the electrons. The electronic part of this Hamiltonian can be written as

$$H = \sum_i \frac{p_i^2}{2m} + \frac{1}{2}\sum_{i \neq j} \frac{e^2}{|\mathbf{r}_i - \mathbf{r}_j|}, \tag{3.1}$$

where m and e are the electron mass and charge, respectively. The fluctuations in the electron density can be represented as follows:

$$\rho(\mathbf{r}) = \sum_i \delta(\mathbf{r} - \mathbf{r}_i) = \sum_\mathbf{k} \rho_\mathbf{k} e^{i\mathbf{k}\cdot\mathbf{r}}, \tag{3.2}$$

where $\rho(\mathbf{r})$ is the electron density at the point $\mathbf{r}$, whereas the Fourier components $\rho_\mathbf{k}$, given in (3.3) below, represent particular fluctuations of the electron density with respect to the uniform distribution $\rho_0 = N$. It should be emphasized that the definition (3.2) carries a quasiclassical image of the electron system with defined positions $\mathbf{r}_i$ of particles which in a fully quantum sense can only be referred to as variables (coordinates) of the multiparticle wave function $\Psi(\mathbf{r}_1, \ldots, \mathbf{r}_N)$. It is easy to see that

$$\rho_\mathbf{k} = \sum_i e^{-i\mathbf{k}\cdot\mathbf{r}_i}, \qquad (3.3)$$

which follows from the equality $\sum_\mathbf{k} e^{i\mathbf{k}\cdot(\mathbf{r}-\mathbf{r}_i)} = \delta(\mathbf{r}-\mathbf{r}_i)$ (we assume here that the system volume $V = 1$). From the above it can be seen that, indeed $\rho_0 = \sum_i 1 = N$.

The Hamiltonian (3.1) may be rewritten in the following form:[1]

$$H = \sum_i \frac{p_i^2}{2m} + \sum_\mathbf{k}{}' \frac{V_\mathbf{k}}{2} \left(\rho_\mathbf{k}^+ \rho_\mathbf{k} - N\right), \qquad (3.4)$$

where $V_\mathbf{k} = 4\pi e^2/k^2$ is the Fourier transform of the Coulomb potential and the superscript 'plus' means the hermitian conjugate. The prime on the sum indicates that the component with $\mathbf{k} = 0$ is omitted here, which corresponds to the cancellation of the uniform jellium and the uniform electron distribution with $\mathbf{k} = 0$, i.e., ρ_0. It means that the uniform jellium has no impact on the electron fluctuations for $\mathbf{k} \neq 0$.

For the electron density fluctuations one can write out dynamic equation in the Heisenberg representation:

$$\dot{\rho}_\mathbf{k} = \frac{1}{i\hbar}[\rho_\mathbf{k}, H], \qquad (3.5)$$

where the brackets indicate the commutator of $\rho_\mathbf{k}$ and H.

---

[1] Here we employ the following calculation:

$$\frac{1}{2}\sum_{i\neq j}\frac{e^2}{|\mathbf{r}_i - \mathbf{r}_j|} = \frac{1}{2}\sum_{i\neq j}\int d^3r_1 \int d^3r_2 \delta(\mathbf{r}_1 - \mathbf{r}_i)\delta(\mathbf{r}_2 - \mathbf{r}_j)\frac{e^2}{|\mathbf{r}_1 - \mathbf{r}_2|}$$

$$= \frac{1}{2}\sum_{i\neq j}\int d^3r_1 \int d^3r_2 \sum_{\mathbf{k}_1} e^{i\mathbf{k}_1\cdot(\mathbf{r}_1-\mathbf{r}_i)} \sum_{\mathbf{k}_2} e^{i\mathbf{k}_2\cdot(\mathbf{r}_2-\mathbf{r}_j)}\frac{e^2}{|\mathbf{r}_1 - \mathbf{r}_2|}$$

(for Dirac function representation, $V = 1$ is assumed). One can change variables to $\mathbf{r}_1 = \mathbf{R} + \mathbf{r}/2$, $\mathbf{r}_2 = \mathbf{R} - \mathbf{r}/2$ (the Jacobian for this variables change is equal to 1). The integral with respect to d^3R gives $\delta_{\mathbf{k}_1, -\mathbf{k}_2}$. The sum over $\mathbf{k}_2$ can be thus performed and one can change indices so that $\mathbf{k}_1 \to \mathbf{k}$. Taking into account that $k^2 \phi_\mathbf{k} = 4\pi q$ (the Fourier version of the Poisson equation, i.e., the first Maxwell equation for a pointlike charge q, $\mathrm{div}(-\mathrm{grad}\,\phi) = 4\pi q\delta(\mathbf{r})$, in Gauss units), we arrive finally at an expression with only one sum over $\mathbf{k}$. The sums over $i \neq j$ of $e^{\pm i\mathbf{k}\cdot\mathbf{r}_i(j)}$ give $\rho_\mathbf{k}^+ \rho_\mathbf{k} - N$ (N must be subtracted because the sum is over $i \neq j$) multiplied by the Fourier transform of the Coulomb potential, $e^2 4\pi/k^2$, after performing the integration with respect to d^3r.

3.1 Random Phase Approximation Theory of Plasmons of Pines and Bohm

For the Hamiltonian (3.4) we obtain[2]

$$\dot{\rho}_{\mathbf{k}} = i \sum_i e^{-i\mathbf{k}\cdot\mathbf{r}_i} \left(-\frac{\mathbf{k}\cdot\mathbf{p}_i}{m} + \frac{\hbar k^2}{2m} \right). \qquad (3.6)$$

When looking for the second-order derivative of the local electron density in the Heisenberg representation, a subsequent commutator, $[\dot{\rho}_{\mathbf{k}}, H]$, has to be found. The result is[3]

$$\ddot{\rho}_{\mathbf{k}} = -\sum_i e^{-i\mathbf{k}\cdot\mathbf{r}_i} \left(-\frac{\mathbf{k}\cdot\mathbf{p}_i}{m} + \frac{\hbar k^2}{2m} \right)^2 - \sum_{\mathbf{q}}' \frac{4\pi e^2}{mq^2} \mathbf{k}\cdot\mathbf{q}\, \rho_{\mathbf{k}-\mathbf{q}} \rho_{\mathbf{q}}. \qquad (3.7)$$

---

[2] In the calculation of the commutator in (3.5) we take into account that

$$[\rho_{\mathbf{k}}, H] = \left[\rho_{\mathbf{k}}, \sum_j \frac{-\hbar^2 \nabla_j^2}{2m} \right]$$

$$= \hbar^2 \sum_j \nabla_j \cdot \left((-i\mathbf{k}) e^{-i\mathbf{k}\cdot\mathbf{r}_j}/2m + \sum_i e^{-i\mathbf{k}\cdot\mathbf{r}_i} \nabla_j/2m \right)$$

$$- \hbar^2 \sum_{i,j} e^{-i\mathbf{k}\cdot\mathbf{r}_i} \nabla_j^2/2m,$$

which gives

$$[\rho_{\mathbf{k}}, H] = -\sum_j e^{-i\mathbf{k}\cdot\mathbf{r}_j} \left(\frac{\hbar^2 k^2}{2m} - \frac{\hbar \mathbf{k}\cdot(-i\hbar\nabla_j)}{m} \right).$$

[3] In the calculation of this subsequent commutator one must take into account that the operator ∇_j acts on the position of the jth particle, $\mathbf{r}_j$; in the case of the commutator $[\dot{\rho}_{\mathbf{k}}, H]$ this operator is present both in $\dot{\rho}_{\mathbf{k}}$ (see Eq. (3.6)) and in H (in the former commutator (3.5) the operator ∇_j is not present in $\rho_{\mathbf{k}}$). Hence, the first term on the right-hand side of Eq. (3.7) is obtained as the result of the action of ∇_j on the kinetic energy term in the Hamiltonian and gives the square of the expression, $i(-\mathbf{k}\cdot(-i\hbar\nabla_i)/m + \hbar k^2/2m)$, the same as in Eq. (3.6), but here divided by $i\hbar$. The second term on the right-hand side of Eq. (3.7) results from the commutator of $\dot{\rho}_{\mathbf{k}}$ with the interaction term in the Hamiltonian, i.e.,

$$\left[-i \sum_i e^{-i\mathbf{k}\cdot\mathbf{r}_i} \frac{\mathbf{k}\cdot(-i\hbar\nabla_i)}{m}, \sum_{\mathbf{q}}' \frac{e^2 2\pi}{q^2} \rho_{\mathbf{q}}^+ \rho_{\mathbf{q}} \right]$$

$$= \left[-i \sum_i e^{-i\mathbf{k}\cdot\mathbf{r}_i} \frac{\mathbf{k}\cdot(-i\hbar\nabla_i)}{m}, \sum_{\mathbf{q}}' \frac{e^2 2\pi}{q^2} \sum_j e^{i\mathbf{q}\cdot\mathbf{r}_j} \sum_l e^{-i\mathbf{q}\cdot\mathbf{r}_l} \right].$$

This can be rewritten in the form

$$-\hbar \sum_i \sum_{\mathbf{q}}' \frac{e^2 2\pi}{q^2 m} e^{-i\mathbf{k}\cdot\mathbf{r}_i} \left(\mathbf{k}\cdot(i\mathbf{q}) e^{i\mathbf{q}\cdot\mathbf{r}_i} \sum_l e^{-i\mathbf{q}\cdot\mathbf{r}_l} + \sum_j e^{i\mathbf{q}\cdot\mathbf{r}_j} \mathbf{k}\cdot(-i\mathbf{q}) e^{-i\mathbf{q}\cdot\mathbf{r}_i} \right).$$

Changing the sign of $\mathbf{q}$ in the sum in the second term we get

$$-i\hbar \sum_{\mathbf{q}}' \frac{e^2 4\pi}{mq^2} \mathbf{k}\cdot\mathbf{q}\, \rho_{\mathbf{k}-\mathbf{q}} \rho_{\mathbf{q}}.$$

The first term on the right-hand side of the equation represents the contribution of the kinetic energy of the particles, whereas the second term is linked with the particle interaction. Let us rewrite the equation and shift the last term, taken for $\mathbf{k} = \mathbf{q}$, to the left-hand side of the equation (for $\mathbf{k} = \mathbf{q}$, the large factor $\rho_0 = N$ occurs in this term), which gives the formula [14–16],

$$\ddot{\rho}_\mathbf{k} + \omega_p^2 \rho_\mathbf{k} = -\sum_i e^{-i\mathbf{k}\cdot\mathbf{r}_i}\left(-\frac{\mathbf{k}\cdot\mathbf{p}_i}{m} + \frac{\hbar k^2}{2m}\right)^2 - \sum_{\mathbf{q}\neq\mathbf{k}}' \frac{4\pi e^2}{mq^2}\mathbf{q}\cdot\mathbf{k}\,\rho_{\mathbf{k}-\mathbf{q}}\rho_\mathbf{q}. \quad (3.8)$$

where $\omega_p = \left(4\pi n e^2/m\right)^{1/2}$ is the plasmon frequency of the electron gas (in Gauss units).

If one sets the right-hand side of Eq. (3.8) to zero, it can be seen that the equation then describes, for each Fourier component $\rho_\mathbf{k}$, harmonic oscillations at the frequency ω_p. One can assess when it is possible to neglect the right-hand side of Eq. (3.8) as follows. The first term on the right-hand side is of order $k^2 v_0^2 \rho_\mathbf{k}$, where v_0 is the Fermi velocity ($v_0 = p_F/m^*$). The second term contains the product of two fluctuations in the local electron density that have *nonzero* momenta (i.e., wave vectors). Because the positions $\mathbf{r}_i$ of the electrons are random (and arbitrary), it follows that $\rho_{\mathbf{q}\neq 0} = \sum_i e^{-i\mathbf{q}\cdot\mathbf{r}_i}$ is small (for a translationally invariant system, the expected value $\langle \rho_{\mathbf{q}\neq 0}\rangle$ should be zero). Therefore this term, with the product of two such small quantities, may be omitted in the linear approximation – this is the random phase approximation (RPA) [16]. The random contributions of the phase factors $e^{-i\mathbf{q}\cdot\mathbf{r}_i}$ cancel out in the sum for $\mathbf{q}\neq 0$, in contrast with the coherent sum for $q = 0$ (this property in fact reflects the behaviour of the Dirac delta function when represented as the integral of an exponential factor).

Neglecting the first term on the right-hand side of Eq. (3.8) is formally equivalent to taking the long-wave oscillation limit $k \to 0$. In this limit one can imagine a uniform oscillating shift of the whole electron liquid with respect to the static jellium – in this case the oscillations have the plasmon frequency ω_p. For nonzero $\mathbf{k}$, the kinetic energy contribution should be taken into account; this leads to a dispersion of the plasmon oscillations that is quadratic in k.

One can assess that the kinetic-term contribution will be small when $k^2 v_0^2/\omega_p^2 \ll 1$, i.e., for $k \ll \omega_p/v_0$. For small k the kinetic term does not play an important role, and in the system in question the coherent plasmon oscillating excitation dominates. In the opposite case, for large k, the kinetic energy dominates and this leads to individual excitations of the electrons.

For free electron concentrations $n \sim 10^{23}/\text{cm}^3$ that are typical for metals, one obtains $\omega_p \sim 10^{16}/\text{s}$, and so the energy of plasmon oscillations attains very high values, $\hbar\omega_p \sim 12$ eV (even higher than the Fermi energy, which indicates that in plasmon oscillations all electrons take part, not just those close to the Fermi surface). If it were possible to excite a plasmon $\hbar\omega_p$ by a single electron (i.e., a quasiparticle) with momentum $\hbar\mathbf{k}$ then, by virtue of energy and momentum conservation,

$$\frac{p^2}{2m} - \frac{(\mathbf{p} - \hbar\mathbf{k})^2}{2m} = \hbar\omega_p. \tag{3.9}$$

Because $p \leq p_0$ (where p_0 is the Fermi momentum), for small $\mathbf{k}$ this process is impossible as $\hbar\omega_p > \varepsilon_F$ (it is impossible to excite a plasmon by a single electron close to the Fermi surface; the excitations close to the Fermi surface, i.e., the Landau quasiparticles, do not interact directly with plasmons owing to the large energy incommensurability). Thermal fluctuations in the metal, which are close to the Fermi surface in energy, also do not interact with plasmons.

One might suppose that plasmon excitations have a similar character to sound waves. Nevertheless, while for sound waves it is crucial that there are many scatterings on the scale of the sound period, $\omega\tau \ll 1$ where τ is the scattering time rate, for plasmons, which are coherent fluctuations of all the electrons, collisions are harmful (they lead to the damping of plasmons); hence for plasmons the inverse inequality, $\omega\tau \geq 1$, is required.

3.2 Screening of the Coulomb Interaction in Metals

The electric interaction of charges of the electron and ion charges in metals reflects two important aspects of these systems; the high density of electrons, of the order of the Avogadro number per cm^3, with the same density of ion charges; and the high mobility of the light electron liquid in comparison to the massive and thus inertial positive ions. To consider the problem of screening by the electron fluid of some uncompensated positive charge q located at point $\mathbf{r}_0$, one must solve the Poisson equation (i.e., the first Maxwell equation),

$$\mathrm{div}(-\nabla\psi(\mathbf{r})) = -\Delta\psi(\mathbf{r}) = 4\pi q\delta(\mathbf{r} - \mathbf{r}_0) - 4\pi e\delta\rho(\mathbf{r}), \tag{3.10}$$

taking into account the presence of the charge q and of a screening electron cloud $\delta\rho(\mathbf{r})$ (for the sake of simplicity we can choose $\mathbf{r}_0 = 0$). Upon the assumption that local changes in the electron density are small on the scale of the interatomic distance one can reasonably expect that such small changes do not cause local changes in the Fermi energy. The local Fermi energy (or Fermi momentum) will equal the constant Fermi energy shifted by the potential $\psi(\mathbf{r})$, i.e.,

$$\frac{\hbar^2 k_F^2}{2m} = \varepsilon_F + e\psi(\mathbf{r}), \tag{3.11}$$

where ε_F is the Fermi energy in the absence of the charge q. Taking into account the relation between the local density of electrons $\rho(\mathbf{r})$ and the local Fermi momentum, one can write

$$\rho(\mathbf{r}) = n + \delta\rho(\mathbf{r}) = \frac{(k_F(\mathbf{r}))^3}{3\pi^2} = \frac{1}{3\pi^2}\left(\frac{2m}{\hbar^2}(\varepsilon_F + e\psi(\mathbf{r}))\right)^{3/2}, \quad (3.12)$$

where n is the mean electron density. Provided that $e\psi(\mathbf{r})/\varepsilon_F \ll 1$, one can linearize Eq. (3.12) with respect to this small parameter, which gives the formula

$$\delta\rho(\mathbf{r}) = \frac{3}{2}n\frac{e\psi(\mathbf{r})}{\varepsilon_F}. \quad (3.13)$$

If one now substitutes Eq. (3.13) into Eq. (3.10), one obtains the Poisson equation in the form

$$\left(\Delta - \frac{6\pi ne^2}{\varepsilon_F}\right)\psi(\mathbf{r}) = -4\pi q\delta(\mathbf{r}). \quad (3.14)$$

The Fourier transform of the above equation is

$$\left(-k^2 - \frac{6\pi ne^2}{\varepsilon_F}\right)\psi_{\mathbf{k}} = -4\pi q. \quad (3.15)$$

Hence,

$$\psi_{\mathbf{k}} = \frac{4\pi q}{k^2 + k_{TF}^2}, \quad (3.16)$$

where $k_{TF} = \sqrt{6\pi ne^2/\varepsilon_F}$ is the Thomas–Fermi wave vector [70, 16].

The reverse Fourier transform of the expression (3.16) determines the potential originating from the charge q and screened by the electron fluid. This potential has the form

$$\psi(\mathbf{r}) = \frac{q}{r}e^{-k_{TF}r}. \quad (3.17)$$

Thus, the screening of the charge q takes place on a spatial scale $1/k_{TF}$, which for metals (where n is of the order of $10^{23}/\text{cm}^3$ and $\varepsilon_F \sim 10\,\text{eV}$) is comparable with the interatomic separation. Therefore the screening in metals is very effective, which, on the other hand, supports the accuracy of the jellium model.

4
Plasmon Excitations in Nanometre-Sized Metallic Particles

This chapter presents the random phase approximation for plasmons in the case of finite nanometre-sized metallic particle. The calculation details are presented for a spherically symmetric particle. The stiff jellium defining the shape of a nanoparticle is accounted for in an explicit manner. The dynamic RPA equations for surface and volume plasmons in a metallic nanoparticle are derived and solved, resulting in the determination of the frequencies of the corresponding plasmon modes.

In the case of a metallic particle the problem of plasmon excitations is more complicated than in the bulk metal because of the presence of a particle surface. The lower the size of the particle, the more important is the role of its surface (the surface scales as a^2, where a is the radius of the particle, whereas the volume scales as a^3, and so for small values of a surface effects start to dominate).[1]

In the case of a finite-sized metallic particle, the positive jellium can be assumed to be a uniform distribution of positive charge whose shape and volume coincide with those of the particle. For a spherical particle with radius a this jellium can be defined by the distribution function

$$n(\mathbf{r}) = n\Theta(a - r), \quad (4.1)$$

where n is the equilibrium positive charge density, $n = \rho Z$ (ρ is the number of ions per unit volume and Z is the ionization level of the ions in the metal). Hence $n(\mathbf{r})|e|$ is the positive charge density of the uniform jellium in a ball with radius a.

---

[1] For example, owing to the domination of surface effects for small particles their melting temperature decreases with decreasing particle radius because melting is a surface phenomenon involving the destruction of the internal crystalline structure; for few-nanometre-sized nanoparticles of, e.g., Au their melting temperature is reduced to room temperature or lower, and so small pieces of metal can be melted even by the focusing of light.

It should be emphasized that in the case of a finite-sized metallic particle, the electron liquid always spills out beyond the jellium rim to some extent. This creates an equilibrium state, in which mobile electrons try to escape beyond the jellium but then are attracted back by the locally uncompensated positive charge of the jellium. The equilibrium state thus corresponds to a electron cloud larger than jellium and with a diluted density due to its larger volume than that of the jellium. This spill-out effect is important in ultrasmall metallic nanospheres (i.e., 'clusters') and has been studied for nanospheres with radii 1–2 nm, containing around 200–300 electrons [19, 21]. When the radius of a nanoparticle reaches larger values, e.g., 10 nm or greater, spill-out loses its significance and can be neglected. The spill-out effect may be assessed by the ratio of the volume of the spilled electrons and the volume of the particle defined by its jellium. The thickness of a spilled layer is of the order of the interatomic distance (the Thomas–Fermi radius), hence the spill-out scales as $1/a$ and this surface-type effect is negligible for a larger than about 5 nm.

The Hamiltonian for a metallic nanoparticle can be written as follows:

$$\hat{H} = \sum_{i=1}^{N_e} \frac{\hat{p}_i^2}{2m} + \sum_{j=1}^{N} \frac{\hat{P}_j^2}{2M} + \sum_{j,j',j\neq j'} u(\mathbf{R}_j - \mathbf{R}_{j'})$$
$$+ \frac{1}{2} \sum_{i,i',i\neq i'} \frac{e^2}{|\mathbf{r}_i - \mathbf{r}_{i'}|} + \sum_{i,j} w(\mathbf{r}_i - \mathbf{R}_j), \qquad (4.2)$$

where $\hat{p}_i, \hat{P}_j$ are the momentum operators for the electrons and ions, respectively; m and M denote the masses of the electrons and ions, N is the number of ions, $N_e = ZN$ is the number of electrons, Z is the dissociation level of ions, u denotes the effective potential (pseudopotential) of mutual interaction of the ions (the ions are usually of an irregular shape and thus can be treated only approximately as spherically symmetrical pointlike charges with interaction potential $\frac{1}{2}\sum_{j,j',j\neq j'} (Ze)^2/|\mathbf{R}_j - \mathbf{R}_{j'}|$), and w is the interaction potential of the electrons and the ions (using the spherical-ion approximation it equals $\sum_{i,j} -Ze^2/|\mathbf{r}_i - \mathbf{R}_j|$).

Now we can explicitly include the jellium model by adding and subtracting the terms

$$\pm e^2 Z \sum_j \int \frac{n(\mathbf{R})d^3 R}{|\mathbf{R}_j - \mathbf{R}|} \qquad (4.3)$$

and

$$\pm \sum_i e^2 \int \frac{n(\mathbf{r})d^3 r}{|\mathbf{r}_i - \mathbf{r}|}, \qquad (4.4)$$

which describe the interaction of the jellium with the ions and electrons, respectively.

In this way one can rewrite the Hamiltonian as the following sum, $\hat{H} = \hat{H}_e + \hat{H}_i + \hat{H}_{ie}$, where the terms are respectively the Hamiltonians of the electrons, the ions and the electron–ion interaction:

$$\hat{H}_e = \sum_{i}^{N_e} \left(\frac{\hat{p}_i^2}{2m} - e^2 \int \frac{n(\mathbf{r})d^3r}{|\mathbf{r}_i - \mathbf{r}|} \right) + \frac{1}{2} \sum_{i,i'.i\neq i'} \frac{e^2}{|\mathbf{r}_i - \mathbf{r}_{i'}|}, \quad (4.5)$$

$$\hat{H}_i = \sum_{j=1}^{N} \left(\frac{\hat{P}_j^2}{2M} - e^2 Z \int \frac{n(\mathbf{R})d^3R}{|\mathbf{R}_j - \mathbf{R}|} \right) + \sum_{j,j'.j\neq j'} u(\mathbf{R}_j - \mathbf{R}_{j'}), \quad (4.6)$$

and

$$\hat{H}_{ie} = \sum_{i,j}^{N_e,N} w(\mathbf{R}_j - \mathbf{r}_i) + \sum_{j} e^2 Z \int \frac{n(\mathbf{R})d^3R}{|\mathbf{R}_j - \mathbf{R}|} + e^2 \sum_{i} \int \frac{n(\mathbf{r})d^3r}{|\mathbf{r}_i - \mathbf{r}|}. \quad (4.7)$$

For so-called normal metals [16] (including the noble metals, alkali and transition metals) the pseudopotential in Eq. (4.6) of the ion interaction, u, shifted by interaction with the jellium, is negligibly small. The dynamic effect of the ions in $\hat{H}_i$ is also small because of the large mass M of the ions and additionally because they are immobilized in a crystal lattice; thus $\hat{H}_i$ can be neglected. When the interaction of electrons with ions is expressed by the jellium as in Eq. (4.7), $\hat{H}_{ie}$ may also be omitted (this has been verified in numerous models, including numerical *ab initio* calculations for metallic clusters, by the application of the Kohn–Sham approach [19]).

The remaining Hamiltonian for the electrons $\hat{H}_e$ may be represented by the use of the quasiclassical local electron density in an analogous manner to the Pines–Bohm approach for the bulk metal (Eq. (3.4)). By a direct calculation one can show that

$$\hat{H}_e = \sum_i \frac{\hat{p}_i^2}{2m} - \frac{1}{(2\pi)^3} \int d^3k \frac{e^2 2\pi}{k^2} n(\mathbf{k}) \left(\rho^+(\mathbf{k}) + \rho(\mathbf{k}) \right)$$
$$+ \frac{1}{(2\pi)^3} \int d^3k \frac{e^2 2\pi}{k^2} \left(\rho^+(\mathbf{k})\rho(\mathbf{k}) - N_e \right), \quad (4.8)$$

where $n(\mathbf{k}) = \int d^3r e^{-i\mathbf{k}\cdot\mathbf{r}} n(\mathbf{r}) = n \int d^3r e^{-i\mathbf{k}\cdot\mathbf{r}} \Theta(a - r)$ is the Fourier transform of the jellium distribution in the nanosphere, and $4\pi/k^2 = \int d^3r e^{-i\mathbf{k}\cdot\mathbf{r}} 1/r$ is the Fourier transform of the Coulomb potential.[2]

---

[2] If we change to a discrete spectrum then

$$\frac{1}{(2\pi)^3} \int d^3k \cdots \to \frac{1}{V} \sum_{\mathbf{k}} \cdots$$

(note that previously $V = 1$ was assumed in formula (3.4)).

In distinction to the *bulk* case (3.4), for a nanosphere the jellium is introduced explicitly (it can be generalized for other, nonspherical, nanoparticle shapes; then the isotropic Θ-function would be substituted by an anisotropic version adjusted to the particular nanoparticle shape). Moreover, the summation over $\mathbf{k}$ is substituted here by integration. In the expression (4.8) the local electron density (its Fourier component) is expressed by $\rho(\mathbf{k}) = \sum_j e^{-i\mathbf{k}\cdot\mathbf{r}_j}$, where $\mathbf{r}_j$ is the position of the jth electron (in the quasiclassical sense), i.e., $\rho(\mathbf{r}) = \sum_j \delta(\mathbf{r} - \mathbf{r}_j)$. Remembering that $(1/(2\pi)^3)\int d^3k e^{i\mathbf{k}\cdot\mathbf{r}} = \delta(\mathbf{r})$ we have

$$\frac{1}{(2\pi)^3}\int d^3k e^{i\mathbf{k}\cdot\mathbf{r}}\sum_j e^{-i\mathbf{k}\cdot\mathbf{r}_j} = \sum_j \delta(\mathbf{r} - \mathbf{r}_j). \tag{4.9}$$

Moreover, $\rho^+(\mathbf{k}) = \rho(-\mathbf{k})$.[3]

An electron Hamiltonian in the form (4.8) allows us to find the first- and second-order derivatives of the operator for local fluctuations of the electron density (in the Heisenberg picture), according to the formulae

$$\frac{d\rho(\mathbf{k})}{dt} = \frac{1}{i\hbar}[\rho(\mathbf{k}), \hat{H}_e], \qquad \frac{d^2\rho(\mathbf{k})}{dt^2} = \frac{1}{i\hbar}[\dot{\rho}(\mathbf{k}), \hat{H}_e]. \tag{4.10}$$

The first commutator leads to an expression analogous to Eq. (3.6),

$$\frac{d\rho(\mathbf{k})}{dt} = i\sum_j e^{-i\mathbf{k}\cdot\mathbf{r}_j}\left(-\frac{\mathbf{k}\cdot\hat{\mathbf{p}}_j}{m} + \frac{\hbar k^2}{2m}\right). \tag{4.11}$$

---

[3] In the derivation of the second term in the expression (4.8) we have initially $-e^2\sum_i \int d^3r n(\mathbf{r})/|\mathbf{r}_i - \mathbf{r}|$, which can be rewritten as $-\int d^3r_1 e^2 \sum_i \delta(\mathbf{r}_1 - \mathbf{r}_i) \int d^3r n(\mathbf{r})/|\mathbf{r}_1 - \mathbf{r}|$ and then

$$-\int d^3r_1 d^3r e^2 \sum_i \frac{1}{(2\pi)^3}\int d^3k e^{i\mathbf{k}\cdot(\mathbf{r}_1-\mathbf{r}_i)}\frac{1}{(2\pi)^3}\int d^3k_1 e^{i\mathbf{k}_1\cdot\mathbf{r}}n(\mathbf{k}_1)\frac{1}{|\mathbf{r}_1-\mathbf{r}|};$$

after the variable, exchange $\mathbf{r}_1 - \mathbf{r} = \mathbf{q}$, $\mathbf{r}_1 + \mathbf{r} = 2\mathbf{p}$, we obtain

$$-e^2\frac{1}{(2\pi)^6}\int d^3p d^3q \int d^3k d^3k_1 e^{i\mathbf{p}\cdot(\mathbf{k}+\mathbf{k}_1)}n(\mathbf{k}_1)\sum_i e^{i\mathbf{k}\cdot(\mathbf{q}/2-\mathbf{r}_i)}e^{i\mathbf{k}_1\cdot(-\mathbf{q}/2)}\frac{1}{|\mathbf{q}|};$$

the integral with respect to $d^3p/(2\pi)^3$ gives $\delta(\mathbf{k}+\mathbf{k}_1)$ and the integral with respect to d^3k_1 can be performed. Next, with respect to d^3q, we arrive at

$$-\frac{1}{(2\pi)^3}\int d^3k \frac{e^2 4\pi}{k^2} n(-\mathbf{k})\rho(\mathbf{k}) = -\frac{1}{(2\pi)^3}\int d^3k \frac{e^2 2\pi}{k^2} n(\mathbf{k})(\rho^+(\mathbf{k}) + \rho(\mathbf{k}))$$

(in the last step the exchange of variables $\mathbf{k} \to -\mathbf{k}$ can be done in the first term in parentheses, provided that $n(-\mathbf{k}) = n(\mathbf{k})$ which is satisfied for a nanosphere). The last term in Eq. (4.8) can be obtained via an analogous calculation (the term $-N_e$ in the last parentheses arises from the condition $i \neq i'$ in the last term in Eq. (4.5)).

The second commutator, in a similar way to the bulk case, leads to the expression

$$\frac{d^2\rho(\mathbf{k})}{dt^2} = -\sum_j e^{-i\mathbf{k}\cdot\mathbf{r}_j}\left(-\frac{\mathbf{k}\cdot\hat{\mathbf{p}}_j}{m} + \frac{\hbar k^2}{2m}\right)^2$$

$$+ \frac{e^2}{m(2\pi)^3}\int d^3q\rho(\mathbf{k}-\mathbf{q})\frac{4\pi\mathbf{k}\cdot\mathbf{q}}{q^2}n(\mathbf{q}) \quad (4.12)$$

$$- \frac{e^2}{m(2\pi)^3}\int d^3q\frac{4\pi\mathbf{k}\cdot\mathbf{q}}{q^2}\rho(\mathbf{k}-\mathbf{q})\rho(\mathbf{q}).$$

The first term on the right-hand side of Eq. (4.12) is related to the kinetic energy of the electrons, whereas the two last terms arise due to the commutation of the gradient operator in $\dot{\rho}$ with the interaction terms in the Hamiltonian (4.8) (in analogy to Eq. (3.7)). We now represent the operator for local electron density fluctuations by the amount of this fluctuation above the uniform density distribution (charged oppositely to the jellium), i.e.,

$$\rho(\mathbf{r}) = n(\mathbf{r}) + \delta\rho(\mathbf{r}) \quad (4.13)$$

and, similarly, for the corresponding Fourier components (owing to the linearity of the Fourier picture),

$$\rho(\mathbf{k}) = n(\mathbf{k}) + \delta\rho(\mathbf{k}). \quad (4.14)$$

Then we can easily rewrite Eq. (4.12) in the following form (note that $n(\mathbf{k})$ does not depend on time; in the Heisenberg representation $[n(\mathbf{k}), \hat{H}_e] = 0$):

$$\frac{d^2\delta\rho(\mathbf{k})}{dt^2} = -\sum_j e^{-i\mathbf{k}\cdot\mathbf{r}_j}\left\{-\frac{\hbar^2}{m^2}(\mathbf{k}\cdot\nabla_j)^2 + \frac{\hbar^2 k^2}{m^2}i\mathbf{k}\cdot\nabla_j + \frac{\hbar^2 k^4}{4m^2}\right\}$$

$$- \frac{4\pi e^2}{m(2\pi)^3}\int d^3q n(\mathbf{k}-\mathbf{q})\frac{\mathbf{k}\cdot\mathbf{q}}{q^2}\delta\rho(\mathbf{q}) \quad (4.15)$$

$$- \frac{4\pi e^2}{m(2\pi)^3}\int d^3q\delta\rho(\mathbf{k}-\mathbf{q})\frac{\mathbf{k}\cdot\mathbf{q}}{q^2}\delta\rho(\mathbf{q}).$$

Note that for $\mathbf{k} = 0$ the density fluctuation beyond the uniform distribution, $\delta\rho(\mathbf{k})$, in contrast to $\rho(\mathbf{k})$ is not proportional to the number of electrons but is equal to zero. This means that the last term in Eq. (4.15) does not contain a coherent component with limiting long-wave plasmon oscillations, as was the case for the bulk system. This coherent component can be identified as the penultimate term, however, where, for $\mathbf{q} = \mathbf{k}$, the term $n(\mathbf{k}-\mathbf{q})$ attains a macroscopically large value equal to the number of electrons in the system.[4]

---

[4] Indeed, this follows because $n(\mathbf{k}) = \int d^3r e^{-i\mathbf{k}\cdot\mathbf{r}}n\Theta(a-r)$ and so $n(0) = nV = N$.

Another aspect concerns the singular point in the Fourier transform of the Coulomb interaction $4\pi e^2/q^2$. In the bulk case this singular point in the sum over $\mathbf{q}$ is removed via implicit inclusion of the screening jellium, which for a uniform distribution of electrons at $\mathbf{q} = 0$ completely compensates the jellium. Thus implicit inclusion of the jellium in the bulk case is equivalent to the avoidance of the sum component with $\mathbf{q} = 0$ (this fact is indicated by the prime on the sum over $\mathbf{q}$ in the bulk case). If the above integral representation is applied to describe a nanoparticle, this problem does not occur, however. The singularity at zero of the Fourier transform of the Coulomb interaction disappears upon integration, and the jellium is introduced explicitly and independently of this singularity.

The variable $\delta\rho(\mathbf{k})$ is an operator (in particular, its derivatives with respect to time include the momentum operator) whose time dependence is expressed in the Heisenberg representation as derived above. In fact we are interested in a measurable quantity, the average of the density fluctuation operator with respect to the eigenfunction of the Hamiltonian. Taking such an average of both sides of Eq. (4.15), we obtain

$$\frac{\partial^2 \delta\rho(\mathbf{k},t)}{\partial t^2} = -\left\langle \Psi \left| \sum_j e^{-i\mathbf{k}\cdot\mathbf{r}_j} \left\{ -\frac{\hbar^2}{m^2}(\mathbf{k}\cdot\nabla_j)^2 + \frac{\hbar^2 k^2}{m^2} i\mathbf{k}\cdot\nabla_j + \frac{\hbar^2 k^4}{4m^2} \right\} \right| \Psi \right\rangle$$

$$- \frac{4\pi e^2}{m(2\pi)^3} \int d^3 q\, n(\mathbf{k}-\mathbf{q}) \frac{\mathbf{k}\cdot\mathbf{q}}{q^2} \delta\rho(\mathbf{q},t)$$

$$- \left\langle \Psi \left| \frac{4\pi e^2}{m(2\pi)^3} \int d^3 q\, \delta\rho(\mathbf{k}-\mathbf{q}) \frac{\mathbf{k}\cdot\mathbf{q}}{q^2} \delta\rho(\mathbf{q}) \right| \Psi \right\rangle,$$

(4.16)

where $\delta\rho(\mathbf{k},t) = \langle \Psi | \delta\rho(\mathbf{k}) | \Psi \rangle$.

The last term in Eq. (4.16) does not contain a coherent component (according to the above discussion) and wholly of second order with respect to the small departure $\delta\rho(\mathbf{k})$ of the density from the uniform distribution, which is a small quantity (note that the coherent contribution is not present in this term). Thus in the RPA method we can omit this term with the same justification concerning the summation of random phase factors as in the bulk case.

The next approximation to be used in Eq. (4.16) has a quasiclassical character: we assume that the electron density changes only weakly on an interatomic distance scale. With such an assumption for the first term of Eq. (4.16), only the first component remains (for small k this term dominates). Hence, the contribution of the kinetic energy attains the form

$$-\left\langle \Psi \left| \sum_j e^{-i\mathbf{k}\cdot\mathbf{r}_j} \left\{ -\frac{\hbar^2}{m^2}(\mathbf{k}\cdot\nabla_j)^2 + \frac{\hbar^2 k^2}{m^2} i\mathbf{k}\cdot\nabla_j + \frac{\hbar^2 k^4}{4m^2} \right\} \right| \Psi \right\rangle$$

$$\simeq \left\langle \Psi \left| \sum_j e^{-i\mathbf{k}\cdot\mathbf{r}_j} \left\{ \frac{\hbar^2}{m^2}(\mathbf{k}\cdot\nabla_j)^2 \right\} \right| \Psi \right\rangle \simeq \frac{2k^2}{3} \left\langle \Psi \left| \sum_j e^{-i\mathbf{k}\cdot\mathbf{r}_j} \frac{\hbar^2 \nabla_j^2}{2m} \right| \Psi \right\rangle, \quad (4.17)$$

where the last approximation is justified for a nanoparticle with spherical symmetry.[5]

In order to assess the contributions of the particular terms in Eq. (4.16) it is convenient to invert the Fourier transform, i.e., to apply to both sides of this equation the operation $(1/(2\pi)^3) \int d^3k\, e^{i\mathbf{k}\cdot\mathbf{r}} \cdots$. The expression (4.17) then attains the form of the mean value of the kinetic energy,[6]

$$\frac{1}{(2\pi)^3} \int d^3k\, e^{i\mathbf{k}\cdot\mathbf{r}} \left(\frac{2k^2}{3m} \left\langle \Psi \left| \sum_j e^{-i\mathbf{k}\cdot\mathbf{r}_j} \frac{\hbar^2 \nabla_j^2}{2m} \right| \Psi \right\rangle \right)$$

$$= -\frac{1}{(2\pi)^3} \frac{2\nabla_r^2}{3m} \int d^3k\, e^{i\mathbf{k}\cdot\mathbf{r}} \left(\left\langle \Psi \left| \sum_j e^{-i\mathbf{k}\cdot\mathbf{r}_j} \frac{\hbar^2 \nabla_j^2}{2m} \right| \Psi \right\rangle \right)$$

$$= -\frac{2\nabla_r^2}{3m} \left\langle \Psi \left| \sum_j \delta(\mathbf{r}-\mathbf{r}_j) \frac{\hbar^2 \nabla_j^2}{2m} \right| \Psi \right\rangle, \quad (4.18)$$

which can be estimated using the conventional Thomas–Fermi formula, the so-called $\rho^{5/3}$ formula [16],

$$-\left\langle \Psi \left| \sum_j \delta(\mathbf{r}-\mathbf{r}_j) \frac{\hbar^2 \nabla_j^2}{2m} \right| \Psi \right\rangle \simeq \frac{3}{5}(3\pi^2)^{2/3} \frac{\hbar^2}{2m} (\rho(\mathbf{r},t))^{5/3}. \quad (4.19)$$

Because $\rho(\mathbf{r},t) = n(\mathbf{r}) + \delta\rho(\mathbf{r},t)$, one can develop the above expression in a Taylor series and confine it to the term linear with respect to $\delta\rho(\mathbf{r},t)$. This development has the form

$$\frac{3}{5}(3\pi^2)^{2/3} \frac{\hbar^2}{2m} [\rho(\mathbf{r},t)]^{5/3} \simeq \frac{3}{5}(3\pi^2)^{2/3} \frac{\hbar^2}{2m} n^{5/3} \Theta(a-r) \left(1 + \frac{5}{3}\frac{\delta\rho(\mathbf{r},t)}{n} + \cdots \right)$$

$$= \left(\frac{3}{5}\varepsilon_F n + \varepsilon_F \delta\rho(\mathbf{r},t) \right) \Theta(a-r), \quad (4.20)$$

---

[5] It is linked with averaging over the spherical angles in the product of two versors, $\int (d\Omega/(4\pi)) \hat{k}_i \hat{k}_j = (1/3)\delta_{ij}$, where $d\Omega = \sin\Theta\, d\Theta d\varphi$.
[6] Because $(1/(2\pi)^3) \int d^3k\, e^{i\mathbf{k}\cdot(\mathbf{r}-\mathbf{r}_j)} = \delta(\mathbf{r}-\mathbf{r}_j)$.

where $n(\mathbf{r}) = n\Theta(a - r)$ and $(\Theta(a - r))^{5/3} = \Theta(a - r)$, the Fermi energy $\varepsilon_F = p_F^2/(2m) = 3^{2/3}\pi^{4/3}\hbar^2 n^{2/3}/(2m)$, and the Fermi momentum $p_F = (3\pi^2\hbar^3 n)^{1/3}$.

The above approximation has a quasiclassical character and was formerly discussed widely in comparison with *ab initio* numerical studies of metallic ultrasmall nanoparticles (clusters) of sizes of 1–2 nm with a limitingly small number of electrons [19, 71]. The reason for this limitation was the efficiency constraints of the numerical local density approximation (LDA) and time-dependent LDA (TDLDA), both based on the Kohn–Sham approach.[7] Nevertheless, it was proved [19] that going beyond the quasiclassical approximation does not bring any important corrections, especially in the case of nanoparticles. For nanoparticles the dominating term of highest order with respect to the gradient (which is actually included in the quasiclassical approximation) is even more significant in finite systems than in the bulk material because of the greater variation in the density induced by a particle border.

Within the RPA scheme we neglect the second-order term in interaction of electron fluctuations but the second-order term in the interaction of the electrons with the jellium is maintained:

$$-\frac{4\pi e^2}{m(2\pi)^3}\int d^3q n(\mathbf{k} - \mathbf{q})\frac{\mathbf{k}\cdot\mathbf{q}}{q^2}\delta\rho(\mathbf{q},t). \quad (4.21)$$

The inverse Fourier transform of this term can be written in the form

$$-\frac{1}{(2\pi)^3}\int d^3k e^{i\mathbf{k}\cdot\mathbf{r}}\frac{4\pi e^2}{m(2\pi)^3}\int d^3q n(\mathbf{k} - \mathbf{q})\frac{\mathbf{k}\cdot\mathbf{q}}{q^2}\delta\rho(\mathbf{q},t)$$

$$= -\frac{1}{(2\pi)^3}\int d^3k e^{i\mathbf{k}\cdot\mathbf{r}}\frac{4\pi e^2}{m(2\pi)^3}\int d^3q \int d^3r_1$$

$$\times e^{-i(\mathbf{k}-\mathbf{q})\cdot\mathbf{r}_1} n(\mathbf{r}_1)\frac{\mathbf{k}\cdot\mathbf{q}}{q^2}\int d^3r_2 e^{-i\mathbf{q}\cdot\mathbf{r}_2}\delta\rho(\mathbf{r}_2,t) \quad (4.22)$$

$$= -\frac{1}{(2\pi)^3}\frac{1}{i}\nabla_\mathbf{r}\cdot\int d^3k e^{i\mathbf{k}\cdot\mathbf{r}}\frac{4\pi e^2}{m(2\pi)^3}$$

$$\times\int d^3q\int d^3r_1 e^{-i(\mathbf{k}-\mathbf{q})\cdot\mathbf{r}_1} n\Theta(a - r_1)\frac{\mathbf{q}}{q^2}\int d^3r_2 e^{-i\mathbf{q}\cdot\mathbf{r}_2}\delta\rho(\mathbf{r}_2,t).$$

However, $(1/(2\pi)^3)\int d^3k e^{i\mathbf{k}\cdot(\mathbf{r}-\mathbf{r}_1)} = \delta(\mathbf{r} - \mathbf{r}_1)$, and it is possible to perform the integration with respect to d^3r_1, which leads to the following expression (the dot after the gradient operator denotes the scalar product $\nabla\cdot\mathbf{q}$),

---

[7] Recently there have appeared TDLDA simulations for larger number of electrons – even up to 10^5, but with some additional simplifications.

Plasmon Excitations in Nanometre-Sized Metallic Particles

$$-\frac{1}{i}\nabla_{\mathbf{r}} \cdot \frac{4\pi e^2}{m(2\pi)^3} \int d^3q\, e^{i\mathbf{q}\cdot\mathbf{r}} n\Theta(a-r)\frac{\mathbf{q}}{q^2} \int d^3r_2\, e^{-i\mathbf{q}\cdot\mathbf{r}_2} \delta\rho(\mathbf{r}_2, t)$$

$$= -\frac{1}{i}\nabla_{\mathbf{r}} \cdot \frac{1}{m} n\Theta(a-r) \frac{1}{i}\nabla_{\mathbf{r}} \int d^3r_2 \left(\frac{1}{(2\pi)^3} \int d^3q\, e^{i\mathbf{q}\cdot(\mathbf{r}-\mathbf{r}_2)} \frac{4\pi e^2}{q^2} \right) \delta\rho(\mathbf{r}_2, t)$$

$$= \frac{\omega_p^2}{4\pi} \nabla_{\mathbf{r}} \cdot \left[\Theta(a-r) \nabla_{\mathbf{r}} \int d^3r_1 \frac{1}{|\mathbf{r}-\mathbf{r}_1|} \delta\rho(\mathbf{r}_1, t) \right]. \tag{4.23}$$

In the penultimate equality the variable of integration $\mathbf{r}_2$ has been renamed as $\mathbf{r}_1$, and the inverse Fourier transform of the Coulomb interaction has been taken; moreover, in the last line the quantity $\omega_p = \sqrt{4\pi n e^2/m}$ has been introduced, the frequency of volume bulk plasmons.

Summarizing, after the steps described above,

- as in the RPA, we neglect terms of second order in the fluctuations,
- we take the inverse Fourier transform of the electron–jellium interaction (4.23),
- we use a quasiclassical approximation for the kinetic energy contribution (4.18), (4.19), and (4.20).

We then arrive at the following equation:

$$\frac{\partial^2 \delta\rho(\mathbf{r},t)}{\partial t^2} = -\frac{2}{3m}\nabla_{\mathbf{r}}^2 \left\langle \Psi \left| \sum_j \delta(\mathbf{r}-\mathbf{r}_j) \frac{\hbar^2 \nabla_j^2}{2m} \right| \Psi \right\rangle$$

$$+ \frac{\omega_p^2}{4\pi} \nabla_{\mathbf{r}} \cdot \left[\Theta(a-r)\nabla_{\mathbf{r}} \int d^3r_1 \frac{1}{|\mathbf{r}-\mathbf{r}_1|} \delta\rho(\mathbf{r}_1,t) \right]$$

$$= \frac{2}{3m}\nabla_{\mathbf{r}}^2 \left[\frac{3}{5}\varepsilon_F n + \varepsilon_F \delta\rho(\mathbf{r},t) \right] \Theta(a-r)$$

$$+ \frac{\omega_p^2}{4\pi} \nabla_{\mathbf{r}} \cdot \left[\Theta(a-r)\nabla_{\mathbf{r}} \int d^3r_1 \frac{1}{|\mathbf{r}-\mathbf{r}_1|} \delta\rho(\mathbf{r}_1,t) \right]. \tag{4.24}$$

This equation can be transformed by utilizing the formula

$$\nabla_{\mathbf{r}} \Theta(a-r) = -\hat{\mathbf{r}} \delta(a-r), \tag{4.25}$$

where $\hat{\mathbf{r}} = \mathbf{r}/r$ is the versor of the vector $\mathbf{r}$. Thus the first term on the right-hand side of Eq. (4.24) becomes

$$\frac{2}{3m}\nabla_{\mathbf{r}}^2 \left[\frac{3}{5}\varepsilon_F n + \varepsilon_F \delta\rho(\mathbf{r},t) \right] \Theta(a-r)$$

$$= -\frac{2}{3m}\nabla_{\mathbf{r}} \left[\frac{3}{5}\varepsilon_F n + \varepsilon_F \delta\rho(\mathbf{r},t) \right] \frac{\mathbf{r}}{r} \delta(a-r)$$

$$- \frac{2}{3m}\varepsilon_F \delta(a-r) \frac{\mathbf{r}}{r} \nabla_{\mathbf{r}} \delta\rho(\mathbf{r},t) + \frac{2}{3m}\varepsilon_F \Theta(a-r) \nabla_{\mathbf{r}}^2 \delta\rho(\mathbf{r},t). \tag{4.26}$$

The second term on the right-hand side of Eq. (4.24) becomes

$$\frac{\omega_p^2}{4\pi} \nabla_\mathbf{r} \cdot \left[\Theta(a-r) \nabla_\mathbf{r} \int d^3 r_1 \frac{1}{|\mathbf{r}-\mathbf{r}_1|} \delta\rho(\mathbf{r}_1, t) \right]$$

$$= -\frac{\omega_p^2}{4\pi} \frac{\mathbf{r}}{r} \delta(a-r) \nabla_\mathbf{r} \int d^3 r_1 \frac{1}{|\mathbf{r}-\mathbf{r}_1|} \delta\rho(\mathbf{r}_1, t) - \omega_p^2 \Theta(a-r) \delta\rho(\mathbf{r}, t).$$
(4.27)

The last term in the above equality follows from the relation

$$\omega_p^2 \Theta(a-r) \frac{\nabla_\mathbf{r}^2}{4\pi} \int d^3 r_1 \frac{1}{|\mathbf{r}-\mathbf{r}_1|} \delta\rho(\mathbf{r}_1, t) = -\omega_p^2 \Theta(a-r) \delta\rho(\mathbf{r}, t),$$
(4.28)

because from Poisson's equation we have $-\nabla_\mathbf{r}^2 (1/|\mathbf{r}-\mathbf{r}_1|) = 4\pi \delta(\mathbf{r}-\mathbf{r}_1)$.

Taking into account Eqs. (4.26) and (4.27), one can rewrite the dynamic equation for the electron liquid in the nanoparticle metal in the following form:

$$\frac{\partial^2 \delta\rho(\mathbf{r}, t)}{\partial t^2} = \left[\frac{2}{3m} \varepsilon_F \nabla_\mathbf{r}^2 \delta\rho(\mathbf{r}, t) - \omega_p^2 \delta\rho(\mathbf{r}, t) \right] \Theta(a-r)$$

$$- \frac{2}{3m} \nabla_\mathbf{r} \left[\left\{ \frac{3}{5} \varepsilon_F n + \varepsilon_F \delta\rho(\mathbf{r}, t) \right\} \frac{\mathbf{r}}{r} \delta(a-r) \right]$$

$$- \left[\frac{2}{3m} \varepsilon_F \frac{\mathbf{r}}{r} \nabla_\mathbf{r} \delta\rho(\mathbf{r}, t) + \frac{\omega_p^2}{4\pi} \nabla_\mathbf{r} \int d^3 r_1 \frac{1}{|\mathbf{r}-\mathbf{r}_1|} \delta\rho(\mathbf{r}_1, t) \right] \delta(a-r).$$
(4.29)

The structure of this equation forces the shape of the solution, including a part distinguished by the Dirac delta function, $\sim \delta(r-a)$, at the nanoparticle surface and a finite part inside the nanoparticle. The general form of the solution can be thus written as follows:

$$\delta\rho(\mathbf{r}, t) = \begin{cases} \delta\rho_1(r, t) & \text{for } r < a, \\ \delta\rho_2(r, t) & \text{for } r = a \ (r \geq a, r \to a+); \end{cases}$$
(4.30)

the indices 1, 2 indicate disjoint parts of the domain of Eq. (4.29).

After substitution of Eq. (4.30) into Eq. (4.29) one obtains the equations for both parts of the domain:

$$\frac{\partial^2 \delta\rho_1(\mathbf{r}, t)}{\partial t^2} = \frac{2}{3m} \varepsilon_F \nabla_\mathbf{r}^2 \delta\rho_1(\mathbf{r}, t) - \omega_p^2 \delta\rho_1(\mathbf{r}, t)$$
(4.31)

for $r < a$ and

$$\frac{\partial^2 \delta\rho_2(\mathbf{r}, t)}{\partial t^2} = -\frac{2}{3m} \nabla_\mathbf{r} \left[\left\{ \frac{3}{5} \varepsilon_F n + \varepsilon_F \delta\rho_2(\mathbf{r}, t) \right\} \frac{\mathbf{r}}{r} \delta(a + \epsilon - r) \right]$$

$$- \left[\frac{2}{3m} \varepsilon_F \frac{\mathbf{r}}{r} \nabla_\mathbf{r} \delta\rho_2(\mathbf{r}, t) + \frac{\omega_p^2}{4\pi} \nabla_\mathbf{r} \int d^3 r_1 \frac{1}{|\mathbf{r}-\mathbf{r}_1|} \right.$$

$$\left. \times \{\delta\rho_1(\mathbf{r}_1, t) \Theta(a-r_1) + \delta\rho_2(\mathbf{r}_1, t) \Theta(r_1-a)\} \right] \delta(a + \epsilon - r)$$
(4.32)

for $r = a$, ($r \geq a$, $r \to a+$). Here $\epsilon = 0+$ is an infinitely small shift of the argument applied to satisfy the formal definition of the Dirac delta (as the generalized distribution function is a linear functional in Hilbert space), which requires the singular point to be located inside an open set, not on its border. This formal definition via the limit $\epsilon = 0+$ does not introduce any asymmetry with respect to the nanosphere surface, and is only a rigorous mathematical clarification ensuring the uniqueness of the Dirac delta definition.[8] Thus, on the basis of the quasiclassical approximation we get two types of plasmon excitation in a metallic nanosphere:

- surface plasmons with delta-shape localization on the nanosphere surface,
- volume plasmons defined inside the nanosphere.

In the case of ultrasmall clusters, when spill-out is relatively important, the surface is not sharply defined and is blurred by the spill-out. For such small nanoparticles, with fuzzy surfaces, the surface and volume plasmon modes are coupled to some extent and their separation may be impossible. The emergence of a distinctly separate surface plasmon mode has been demonstrated numerically by the TDLDA method for around 60 electrons in the cluster, though this mode is still coupled to the volume modes. For a larger number of electrons, $\sim 10^{5-7}$, as in nanoparticles with radii around $5-10$, nm the spill-out effects are negligible and the quasiclassical RPA approach with a sharply defined nanoparticle surface is fully justified.

However, it should be noted that some special situations must be considered with caution. For example, when a metallic nanosphere is embedded in a dielectric medium, these surroundings diminish the Coulomb interaction of electrons on the surface but certainly also in some nonzero-thickness layer;[9] this could be hard to include in the developed quasiclassical RPA model for a sharp surface (this situation will be addressed below).

The solution of Eqs. (4.31) and (4.32) is performed taking into account the boundary conditions and the continuity conditions as well as the charge conservation constraint. It is convenient to represent the volume type of plasmon excitation in the form

$$\delta\rho_1(\mathbf{r},t) = n(f_1(r) + F(\mathbf{r},t)) \qquad \text{for } r < a, \qquad (4.33)$$

which inserted into Eq. (4.31) results in the conditions

$$\nabla^2 f_1(r) - k_T f_1(r) = 0 \qquad (4.34)$$

and

$$\frac{\partial^2 F(\mathbf{r},t)}{\partial t^2} = \frac{v_F^2}{3} \nabla_\mathbf{r}^2 F(\mathbf{r},t) - \omega_p^2 F(\mathbf{r},t), \qquad (4.35)$$

---

[8] It is well known that, in a nonrigorous sense, $\int_0^\infty \delta(x)dx = \alpha$ where $\alpha = \Theta(0)$ can be chosen arbitrarily.
[9] The interaction on the surface is influenced by the adjacent regions on both sides owing to the three-dimensional electrodynamics.

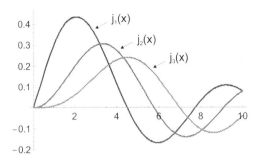

Figure 4.1 The positions of the Bessel function nodes depend on the function order, as shown for $j_l(x)$, $l = 1, 2, 3$.

where $k_T = \sqrt{6\pi ne^2/v_F} = \sqrt{3\omega_p^2/v_F^2}$ is the reciprocel of the Thomas–Fermi radius and $\omega_p = \sqrt{4\pi ne^2/m}$ is the volume-plasmon frequency in the bulk material. The solution for f_1 (regular at zero) is as follows:

$$f_1(r) = \alpha \frac{e^{-k_T a}}{k_T r} \left(e^{-k_T r} - e^{k_T r} \right), \qquad \alpha = \text{const}. \tag{4.36}$$

For the function $F(\mathbf{r},t)$ we assume the initial condition $F(\mathbf{r},0) = 0$, which leads to $F(\mathbf{r},t) = F_\omega(\mathbf{r})\sin(\omega t)$. Thus the function F_ω satisfies a Helmholtz-type equation,

$$\nabla^2 F_\omega(\mathbf{r}) + k^2 F_\omega(\mathbf{r}) = 0, \tag{4.37}$$

where $k^2 = 3(\omega^2 - \omega_p^2)/v_F^2$ and the solution regular at the origin has the form

$$F_\omega(\mathbf{r}) \sim j_l(kr) Y_{lm}(\Omega), \tag{4.38}$$

where $j_l(x) = \sqrt{(\pi/2x)} I_{l+1/2}(x)$ is a spherical Bessel function ($I_n(x)$ is a type-1 Bessel function), $Y_{lm}(\Omega)$ is a spherical harmonic, and Ω is a spherical angle in spherical coordinates. The boundary condition $\lim_{r \to a} F(\mathbf{r},t) = 0$ (in agreement with the quasiclassical approximation) leads to the constraint $j_l(ka) = 0$, and thus to discrete values of the parameter $k = k_{nl} = x_{nl}/a$, where x_{nl} are consecutive nodes ($n = 1, 2, 3, \ldots$) of the Bessel function j_l (Fig. 4.1). In this way we obtain the frequency spectrum of the volume-plasmon self-oscillations in a metallic nanosphere,

$$\omega_{nl}^2 = \omega_p^2 \left(1 + \frac{x_{nl}^2}{k_T^2 a^2} \right). \tag{4.39}$$

Thus the general solution for $F(\mathbf{r},t)$ has the shape

$$F(\mathbf{r},t) = \sum_{l=0}^{\infty} \sum_{m=-l}^{l} \sum_{n=1}^{\infty} A_{lmn} j_l(k_{nl} r) Y_{lm}(\Omega) \sin(\omega_{nl} t). \tag{4.40}$$

It is noticeable that all the mode frequencies ω_{nl} are larger than ω_p, as follows from Eq. (4.39), i.e., in a metallic nanoparticle these volume plasmons are more energetic than in the bulk material, and this difference increases for lower radii a. This resembles the behaviour of energy levels in a narrow quantum well.

For the surface plasmon modes we assume that

$$\delta\rho_2(\mathbf{r},t) = nf_2(r) + \sigma(\Omega,t)\delta(a+\epsilon-r), \tag{4.41}$$

for $r \geq a$, $\epsilon = 0+$. The neutrality condition $\int d^3r\rho(\mathbf{r},t) = nV = N_e$ leads to the relations

$$-\int_0^a dr r^2 f_1(r) = \int_a^\infty dr r^2 f_2(r),$$

$$\int d^3r\Theta(a-r)F(\mathbf{r},t) = 0, \tag{4.42}$$

$$\int d\Omega\sigma(\Omega,t) = 0,$$

from which one can determine that $f_2(r) = \beta e^{-k_T(r-a)}/(k_T r)$ and, taking into account the continuity condition $1 + f_1(r)|_{r\to a} = f_2(r)|_{r\to a}$, one can determine the constant $\beta = k_T a - (k_T a + 1/2)(1 - e^{-2k_T a})$ and the constant $\alpha = -(k_T a + 1)/2$ (the latter is needed for Eq. (8.9)). The functions describing the static spill-out, with the assumed approximation, have the form

$$f_1(r) = -\frac{k_T a + 1}{2}e^{-k_T(a-r)}\frac{1 - e^{-2k_T r}}{k_T r}, \qquad \text{for } r < a,$$

$$f_2(r) = \left[k_T a - \frac{k_T a + 1}{2}(1 - e^{-2k_T a})\right]\frac{e^{-k_t(r-a)}}{k_T r} \qquad \text{for } r \geq a. \tag{4.43}$$

The scale of this static spill-out of charge is thus governed by the Thomas–Fermi radius, i.e., it is of the order of the interparticle distance, and for large nanoparticles ($a \sim 5$ nm) the spill-out plays a marginal role. For nanoparticles of this size or greater, the spill-out may be safely neglected without affecting the analysis of the plasmon excitations. In the framework of the semiclassical RPA model, the electrical balance is therefore kept separately for the volume modes and for the surface modes. In ultrasmall nanoparticles, especially for $N_e < 60$, this condition is not fulfilled because of the fuzzy surface, and the surface- and volume-plasmon modes are not separable; for $N_e > 60$ this volume–surface coupling effect gradually disappears and for $N_e \sim 200$ does not play a significant role despite the still pronounced spill-out; for $N_e > 10^5$ (for nanoparticles with $a > 5$ nm) both the spill-out and the surface–volume coupling are completely negligible.

One can see that from the neutrality condition, $\int d^3r\Theta(a-r)F(\mathbf{r},t) = 0$, it follows that $A_{00n} = 0$, because $\int d\Omega Y_{lm}\Omega = 4\pi\delta_{l0}\delta_{m0}$.

Let us now consider an explicit solution for the surface plasmons. In order to determine the self-frequencies for the surface plasmons, we need to consider Eq. (4.32) and the solution to it given by Eq. (4.41).

The first term on the right-hand side of Eq. (4.32) can be rewritten as

$$\frac{2\varepsilon_F}{3m}\nabla\left(\frac{3}{5}n+\delta\rho_2\right)\nabla\Theta(a-r)+\frac{2\varepsilon_F}{3m}\left(\frac{3}{5}n+\delta\rho_2\right)\Delta\Theta$$

$$=-\frac{2\varepsilon_F}{3m}\delta(a-r)\frac{\partial}{\partial r}\left(\frac{3}{5}n+\delta\rho_2\right)-\frac{2\varepsilon_F}{3m}\left(\frac{3}{5}n+\delta\rho_2\right)\frac{1}{r^2}\frac{\partial}{\partial r}(r^2\delta(a-r))$$

$$=-\frac{2\varepsilon_F}{3m}\frac{1}{r^2}\frac{\partial}{\partial r}\left[\left(\frac{3}{5}n+\delta\rho_2\right)r^2\delta(a-r)\right], \tag{4.44}$$

where we have used the formulae

$$\nabla\Theta(a-r)=-\frac{\mathbf{r}}{r}\delta(a-r), \qquad \frac{\mathbf{r}}{r}\nabla=\frac{\partial}{\partial r}$$

and, in spherical coordinates,

$$\Delta=\frac{1}{r^2}\frac{\partial}{\partial r}r^2\frac{\partial}{\partial r}+\frac{1}{\sin\theta}\frac{\partial}{\partial\theta}\sin\theta\frac{\partial}{\partial\theta}+\frac{1}{\sin^2\theta}\frac{\partial^2}{\partial\phi^2}$$

(note that for the function $\Theta(r)$ only the first term contributes to $\Delta\Theta$).

The next term on the right-hand side of Eq. (4.32) can be transformed into

$$-\frac{2\varepsilon_F}{3m}\delta(a-r)\frac{\mathbf{r}}{r}\nabla\delta\rho_2-\frac{\omega_p^2}{4\pi}\delta(a-r)\frac{\mathbf{r}}{r}\nabla\int\frac{d^3r_1\delta\rho(\mathbf{r}_1)}{|\mathbf{r}-\mathbf{r}_1|}$$

$$=-\frac{3\varepsilon_F}{3m}\delta(a-r)\frac{\partial}{\partial r}\delta\rho_2-\frac{\omega_p^2}{4\pi}\delta(a-r)\frac{\partial}{\partial r}\int\frac{d^3r_1\delta\rho(\mathbf{r}_1)}{|\mathbf{r}-\mathbf{r}_1|}. \tag{4.45}$$

Equation (4.32) thus attains the form

$$\frac{\partial^2\rho_2}{\partial t^2}=-\frac{2\varepsilon_F}{3m}\frac{1}{r^2}\frac{\partial}{\partial r}\left[\left(\frac{3}{5}n+\delta\rho_2\right)r^2\delta(a-r)\right]$$

$$-\frac{2\varepsilon_F}{3m}\delta(a-r)\frac{\partial}{\partial r}\delta\rho_2-\frac{\omega_p^2}{4\pi}\delta(a-r)\frac{\partial}{\partial r}\int\frac{d^3r_1\delta\rho(\mathbf{r}_1)}{|\mathbf{r}-\mathbf{r}_1|}. \tag{4.46}$$

Now we suppose that the solution of the above equation has the form $\delta\rho_2=\sigma(\Omega,t)\delta(a+0^+-r)$ and multiply both sides of the equation by r^2. Then we find the integral with respect to r beween arbitrary limits, i.e., $\int_l^L r^2 dr\cdots$, such that $a\in(l,L)$ (this integration removes the Dirac deltas), which leads to the equation

$$a^2 \frac{\partial^2 \sigma(\Omega, t)}{\partial t^2} = -\frac{2\varepsilon_F}{3m} \int_l^L dr \frac{\partial}{\partial r} \left[\left(\frac{3}{5} n + \delta\rho_2 \right) r^2 \delta(a-r) \right]$$

$$- \frac{2\varepsilon_F}{3m} \sigma(\Omega, t) \int_l^L r^2 dr \delta(a-r) \frac{\partial}{\partial r} \delta(a-r)$$

$$- \frac{\omega_p^2}{4\pi} \int_l^L r^2 dr \delta(a-r) \frac{\partial}{\partial r} \int_a^\infty r_1^2 dr_1 \int d\Omega \frac{\delta\rho_2(\mathbf{r}_2)}{|\mathbf{r}-\mathbf{r}_1|}$$

$$- \frac{\omega_p^2}{4\pi} \int_l^L r^2 dr \delta(a-r) \frac{\partial}{\partial r} \int_0^a r_1^2 dr_1 \int d\Omega \frac{\delta\rho_1(\mathbf{r}_1)}{|\mathbf{r}-\mathbf{r}_1|}. \quad (4.47)$$

The first two terms on the right-hand side of the above equation vanish, because

$$-\frac{2\varepsilon_F}{3m} \int_l^L dr \frac{\partial}{\partial r} \left[\left(\frac{3}{5} n + \delta\rho_2 \right) r^2 \delta(a-r) \right]$$

$$= -\frac{2\varepsilon_F}{3m} \left[\left(\frac{3}{5} n + \delta\rho_2 \right) r^2 \delta(r-a) \right]\bigg|_l^L = 0 \quad (4.48)$$

and

$$\int_l^L r^2 dr \delta(a-r) \frac{\partial}{\partial r} \delta(a-r) = a^2 \int_l^L dr \frac{1}{2} \frac{\partial}{\partial r} \delta^2(a-r) = \frac{a^2}{2} \delta^2(a-r) \bigg|_l^L$$

$$= -\frac{kT}{m} \frac{a^2}{2} \lim_{\mu \to 0} \frac{1}{\pi} \frac{\mu}{\mu^2 + (a-r)^2} \delta(a-r) \bigg|_l^L = 0. \quad (4.49)$$

The two last terms on the right-hand side of Eq. (4.47) can be transformed using the formula [72],

$$\frac{1}{\sqrt{1+z^2-2z\cos\gamma}} = \sum_{l=0}^\infty P_l(\cos\gamma) z^l \quad \text{for } z < 1,$$

where

$$P_l(\cos\gamma) = \frac{4\pi}{2l+1} \sum_{m=-l}^l Y_{lm}(\Omega) Y_{lm}^*(\Omega)$$

are Legendre polynomials. This formula leads to the following:

$$\frac{\partial}{\partial a} \frac{1}{|\mathbf{a} - \mathbf{r}_1|} = \begin{cases} \sum_{l=0}^{\infty} \frac{l a^{l-1}}{r_1^{l+1}} P_l(\cos\gamma) & \text{for } a < r_1, \\ -\sum_{l=0}^{\infty} \frac{(l+1) r_1^l}{a^{l+2}} P_l(\cos\gamma) & \text{for } a > r_1, \end{cases} \quad (4.50)$$

where $\mathbf{a} = a\mathbf{r}/r$, $\cos\gamma = \mathbf{a} \cdot \mathbf{r}_1/(ar_1)$. Employing Eq. (4.50), the last two terms in Eq. (4.47) can be transformed as follows:

$$-\frac{\omega_p^2}{4\pi} \int_l^L r^2 dr \delta(a-r) \frac{\partial}{\partial r} \int_a^\infty r_1^2 dr_1 \int d\Omega_1 \frac{\delta\rho_2(\mathbf{r}_1)}{|\mathbf{r} - \mathbf{r}_1|}$$

$$= -\frac{\omega_p^2}{4\pi} a^2 \int d\Omega_1 \int_a^\infty r_1^2 dr_1 \delta\rho_2(\mathbf{r}_1) \frac{\partial}{\partial a} \frac{1}{\sqrt{a^2 + r_1^2 - 2ar_1\cos\gamma}}$$

$$= -\frac{\omega_p^2}{4\pi} a^2 \int d\Omega_1 \int_a^\infty r_1^2 dr_1 \sigma(\Omega_1) \delta(a+0+-r_1) \sum_{l=0}^{\infty} \frac{l a^{l-1}}{r_1^{l+1}} P_l(\cos\gamma)$$

$$= -\frac{\omega_p^2}{4\pi} a^2 \int d\Omega_1 \sigma(\Omega_1) \frac{1}{a^2} \sum_{l=0}^{\infty} \frac{4\pi l}{2l+1} \sum_{m=-l}^{l} Y_{lm}(\Omega) Y_{lm}^*(\Omega_1)$$

$$= -\omega_p^2 a^2 \sum_{l=0}^{\infty} \sum_{m=-l}^{l} \frac{l}{2l+1} Y_{lm}(\Omega) \int d\Omega_1 \sigma(\Omega_1) Y_{lm}^*(\Omega_1) \quad (4.51)$$

and

$$-\frac{\omega_p^2}{4\pi} \int_l^L r^2 dr \delta(a-r) \frac{\partial}{\partial r} \int_0^a r_1^2 dr_1 \int d\Omega_1 \frac{\delta\rho_1(\mathbf{r}_1)}{|\mathbf{r} - \mathbf{r}_1|}$$

$$= -\frac{\omega_p^2}{4\pi} a^2 \int d\Omega_1 \int_0^a r_1^2 dr_1 n F(\mathbf{r}_1, t) \frac{\partial}{\partial a} \frac{1}{\sqrt{a^2 + r_1^2 - 2ar_1\cos\gamma}}$$

$$= \frac{\omega_p^2}{4\pi} a^2 \int d\Omega_1 n F(\mathbf{r}_1, t) \sum_{l=0}^{\infty} \frac{(l+1) r_1^l}{a^{l+2}} P_l(\cos\gamma)$$

$$= \omega_p^2 n \sum_{l=0}^{\infty} \frac{l+1}{2l+1} Y_{lm}(\Omega) \int_0^a r_1^2 dr_1 \frac{r_1^l}{a^l}$$

$$\times \sum_{l_1=1}^{\infty} \sum_{m_1=-l_1}^{l_1} \sum_i A_{lmi} j_{l_1}(k_{l_1 i} r_1) \sin(\omega_{l_1 i} t) \int d\Omega_1 Y_{lm}^*(\Omega_1) Y_{l_1 m_1}(\Omega_1)$$

$$= \omega_p^2 n \sum_{l=0}^{\infty} \sum_{m=-l}^{l} \sum_i \frac{l+1}{2l+1} Y_{lm}(\Omega) A_{lmi} \int_0^a \frac{r_1^{l+2} dr_1}{a^l} j_l(k_{li} r_1) \sin(\omega_{li} t). \quad (4.52)$$

Equation (4.47) thus attains the form

$$\frac{\partial^2 \sigma(\Omega, t)}{\partial t^2} = -\omega_p^2 a^2 \sum_{l=0}^{\infty} \sum_{m=-l}^{l} \frac{l}{2l+1} Y_{lm}(\Omega) \int d\Omega_1 \sigma(\Omega_1) Y_{lm}^*(\Omega_1)$$

$$+ \omega_p^2 n \sum_{l=0}^{\infty} \sum_{m=-l}^{l} \sum_i \frac{l+1}{2l+1} Y_{lm}(\Omega) A_{lmi} \times \int_0^a \frac{r_1^{l+2} dr_1}{a^l} j_l(k_{li} r_1) \sin(\omega_{li} t).$$

$$(4.53)$$

Assuming now that $\sigma(\Omega, t) = \sum_{l=0}^{\infty} \sum_{m=-l}^{l} q_{lm}(t) Y_{lm}(\Omega)$ and substituting it into the above equation, we obtain

$$\sum_{l=0}^{\infty} \sum_{m=-l}^{l} Y_{lm}(\Omega) \frac{\partial^2 q_{lm}(t)}{\partial t^2} = -\sum_{l=0}^{\infty} \sum_{m=-l}^{l} \frac{\omega_p^2 l}{2l+1} Y_{lm}(\Omega) q_{lm}(t)$$

$$+ \omega_p^2 \sum_{l=1}^{\infty} \sum_{m=-l}^{l} \sum_i \frac{l+1}{2l+1} Y_{lm}(\Omega) A_{lm}$$

$$\times \int_0^a \frac{r_1^{l+2} dr_1}{a^{l+2}} j_l(k_{li} r_1) \sin(\omega_{li} t). \quad (4.54)$$

From this equation we notice that for $l = 0$ we get $\partial^2 q_{00}/\partial t^2 = 0$ and thus $q_{00}(t) = 0$ (since $q(0) = 0$ and $\lim_{t \to \infty} q(t) < \infty$). For $l \geq 1$ we obtain

$$\frac{\partial^2 q_{lm}(t)}{\partial t^2} = -\frac{\omega_p^2 l}{2l+1} q_{lm}(t) + \sum_i \omega_p^2 \frac{l+1}{2l+1} A_{lm} n \int_0^a \frac{r_1^{l+2} dr_1}{a^{l+2}} j_l(k_{li} r_1) \sin(\omega_{li} t),$$

$$(4.55)$$

which induces the solution

$$q_{lm}(t) = \frac{B_{lm}}{a^2} \sin\left(\omega_p \sqrt{\frac{l}{2l+1}} t\right)$$

$$+ \sum_i A_{lm} \frac{(l+1)\omega_p^2}{\omega_p^2 - (2l+1)\omega_{li}^2} n \int_0^a \frac{r_1^{l+2} dr_1}{a^{l+2}} j_l(k_{li} r_1) \sin(\omega_{li} t)$$

$$(4.56)$$

and $\delta\rho_2(\mathbf{r},t) = \sum_{l=1}^{\infty} \sum_{m=-l}^{l} q_{lm}(t) Y_{lm}(\Omega) \delta(a-r)$. The first term in Eq. (4.56) describes the self-frequencies of surface plasmons, whereas the second term gives the surface plasmon oscillations induced by the volume plasmons. This induced part of the surface oscillations is nonzero only when the volume modes are excited and their amplitudes, A_{lmi}, are nonzero. The frequencies of self-oscillation of the surface plasmons are equal to $\omega_{l0} = \omega_p \sqrt{l/(2l+1)}$, corresponding to the various multipole modes (numbered by l). Note that these frequencies are lower than the bulk plasmon frequency ($\omega_p = \sqrt{ne^2 4\pi/m}$ in Gauss units or $\sqrt{ne^2/(\varepsilon_0 m)}$ in SI units), whereas the volume plasmon modes oscillate with frequencies higher than ω_p.

5
Damping of Plasmons in Metallic Nanoparticles

In this book our main attention is focused on plasmon damping, which is essential for energy transfer via plasmons. In this chapter various damping channels are discussed and compared. The attenuation rate for plasmons in metallic nanoparticles is derived in a far-reaching analytical form. The contributions to plasmon attenuation of electron scattering and radiation losses are presented. An accurate account of the Lorentz friction of plasmon oscillations in metallic nanoparticles is given. There is close consistency between theory and experiment, not achieved previously and revealing the nonharmonic character of plasmon oscillations. The related corrections to the popular conventional Mie approach to plasmons and to the commercial software COMSOL are defined and illustrated.

Plasmonic oscillations in metallic nanoparticles have recently received considerable attention owing to their prospective applications in photovoltaics. Solar cells modified by deposition on their surface of metallic nanoparticles are regarded as promising a new generation of solar cells with enhanced efficiencies due to plasmon mediation in the solar energy harvesting [33, 36, 2, 73, 74]. An increase in the efficiency of the photoeffect in semiconductors due to the mediation of plasmons in metallic compounds has been proved experimentally to exceed even 100%, which causes a few per cent growth in the overall efficiency of a solar cell (because the light absorption is only the initial step in a long chain of processes contributing to the overall solar cell functionality and its efficiency). The crucial role in the plasmon-aided photoeffect is played by the surface plasmons in the metallic nanoparticles, which can be easily excited by incident photons and can quickly transmit energy to the substrate semiconductor. This energy transit via plasmons is conditioned by various channels of plasmon damping, when the energy outflow reduces the plasmon oscillations. The higher the damping of plasmons, the more efficient is the energy transfer via plasmons, because the energy income from sunlight constantly supplements the losses through plasmon oscillations. Both the absorption and the emission of plasmons are linked by general quantum transition rules similar to those for ordinary optical transitions [75]. The latter are defined by the interaction of an electromagnetic wave with a quantum system (the electron in an atom or a band electron in a semiconductor) via the kinematic momentum shift caused by the vector potential of the electromagnetic field. In the case of plasmons, however, quite different and much more efficient coupling channels between plasmons and electrons occur. To recognize all these channels for energy transfer via plasmons, an understanding of plasmon damping is essential.

The energy of plasmon oscillations can be dissipated irreversibly into Joule heat by the scattering of electrons in a metallic nanoparticle. This unavoidable component of plasmon damping does not take part in the harvesting of solar energy in solar cells and causes parasite losses eventually to become phonon heat-like excitations in metallic nanoparticles (there are contributions from the scattering of electrons at metal imperfections and admixtures, from direct interaction with phonons and other electrons, and from scattering at the boundaries of a nanoparticle).

Plasmon oscillations correspond to a variation in time of the acceleration of electrons; thus the related Lorentz friction [55, 76] caused by radiation of an electromagnetic wave also contributes to the plasmon damping. The term Lorentz friction is conventionally applied to electromagnetic emission in the far-field radiation zone in the dielectric surroundings of a metallic nanoparticle (or in the vacuum). However, when another electric charge system is present in the vicinity of the metallic nanoparticle, the different type of coupling in the near-field zone may contribute to the plasmon energy losses. Such a situation occurs when the metallic nanoparticle is deposited on or embedded in a substrate semiconductor and the surface plasmons in the metallic nanoparticle couple with the semiconductor band electrons in the near-field zone of plasmon radiation. This photon-less energy transfer occurring at distances smaller than the plasmon resonance wavelength is especially large for dipole surface plasmons and causes a giant plasmon-mediated photoeffect. Noteworthy is an interesting situation when the energy receiver (the substrate semiconductor) influences the energy emitter (the plasmons in the metallic nanoparticles). This effect goes beyond classical electrodynamics, which does not describe such a feedback. In the case of metallic nanoparticles this feedback is, however, strong and causes a large damping of plasmons which is different from that occurring when plasmons freely emit radiation in the dielectric (or vacuum) surroundings. The feedback effect for plasmons is very sensitive to the nanoparticle size [77]. This significant effect is neglected in the conventional Mie approach or in numerical solutions of the Maxwell–Fresnel problem (e.g., by COMSOL). To address this phenomenon is the basic goal of the present text.

Radiation losses hamper the plasmon oscillations, but the higher the radiation losses, the better is the energy transmittance via plasmons. The radiation properties of plasmons also determine the subdiffraction plasmon–polariton weakly damped propagation in plasmonic wave guides consisting of metallic nano-chains or nanoarrays [3, 4, 39, 38, 78]. This phenomenon has prospective applications in nanophotonics and optoelectronics.

One should note that ultrasmall metal nanoparticles (clusters) of radius $a \in (1, 5)$ nm exhibit very weak radiation [19, 71, 20, 17] and that plasmon damping at this size scale is primarily due to the electron scattering resulting in ohmic losses and to electron collisions with the cluster boundary. The scattering of electrons results in Joule heat energy dissipation. Nevertheless, with increasing nanoparticle size, the plasmon radiation intensity grows rapidly, mostly because of the strengthening of the plasmon oscillation amplitude, which is proportional to the number of electrons in a nanoparticle. This grows as a^3 with the nanoparticle radius a.

In the case of the dielectric surroundings of a metallic nanoparticle the electron-plasma-oscillation radiative losses can be accounted for by the Lorentz friction [55, 76]. As mentioned above, the Lorentz friction force acts on nonuniformly accelerating charges (as in the case of oscillating charges), which radiate electromagnetic waves. The radiated energy can be treated as an effective kinetic energy loss, caused by a fictitious force – the Lorentz friction force – hampering the electron motion. In the case of an oscillating dipole, as is the situation for the dipole-type surface plasmons on a metallic nanosphere, the Lorentz friction is proportional to the third-order time derivative of this dipole [55, 76]. The Lorentz friction can be included in a description of the surface plasmons in metallic nanospheres via the typical damping term in the oscillatory equation. However, this term is proportional to the first-order time derivative of the oscillating dipole whereas, as stated above, the Lorentz friction is proportional to the third-order time derivative. The substitution of this third-order derivative by an appropriately lifted first-order derivative is the essence of the perturbative approximation [23, 77]. Such an approximation results in an approximately a^3 increase in the radiation damping rate for plasmons with nanosphere radius a. Such a fast growth of the radiation losses with nanoparticle radius suggests that the Lorentz friction dominates the other attenuation channels in the case of sufficiently large nanospheres. However, the rapid growth of a^3 quickly violates the necessary perturbation convergence limits, which precludes the use of this approximation for nanospheres with larger radii (in the case of Au nanoparticles in vacuum the perturbation approximation is correct only for $a \in (5, 25)$ nm). In order to describe the effect of the Lorentz friction beyond the scope of the perturbation approach regime, linearization of the third-order derivative must be avoided and the whole contribution of the third-order derivative has to be accounted for accurately. As will be shown below, these nonlinear corrections are essential for nanospheres of radii $a > 25$–30 nm (for Au in vacuum). Accurate accounting for the Lorentz friction (by using the third-order time derivative) can explain the anomalous size dependence in the light absorption that is observed in large metallic nanospheres (it is measured by the light extinction in water-colloidal solutions of metallic nanoparticles of varying size [77]). Exact inclusion of the Lorentz friction is critical for improving our understanding of the experimental data; this is unsatisfactory using the approximate perturbative approach, with a growing discrepancy occurring for the larger nanospheres ($a > 30$ nm for Au), when the perturbation approach fails.

As mentioned above, small metallic nanoparticles (clusters) of size 1–5 nm do not exhibit a radiation efficiency as high as nanospheres with larger radii (especially for $a > 10$ nm, for Au in vacuum).[1] Small metallic nanoparticles with radii of the order of a single nanometre were widely investigated in the 1980s, theoretically within the so-called local density approximation (LDA) or time-dependent local density approximation (TDLDA) [19, 71, 20, 79], and experimentally for clusters of alkali metals such as Na or K, which confirmed their poor radiation efficiency. The electromagnetic radiation of plasmons in these small systems turned out to be weak owing to the relatively small amount of electrons oscillating in such clusters, but other unique properties were important at this scale. Special attention was paid to the so-called spill-out of the electron liquid beyond the rim of the ion lattice and the influence of this effect on the plasmon oscillations, which is greater for smaller particles. Spill-out results in the dilution of the electron density inside the system and this is a reason for the redshift of the resonance frequency of surface plasmons (which is proportional to the electron density) observed experimentally for ultrasmall nanoparticles of K and Na [19]. The spill-out effect diminishes, however, with the increase of the nanoparticle size because of the lessening role of surface-related phenomena when the radius of the system grows. The spill-out takes place on the scale of the Thomas–Fermi length [16], which is of subnanometre value and is lower by at least one or two orders of magnitude in comparison to the nanosphere radius when $a > 5$ nm. In larger nanoparticles the role of the radiation phenomena rapidly grows, and for $a > 10$ nm (for Au in vacuum) Lorentz friction dominates the plasmon damping. In metallic nanoparticles with sizes of several tens of nanometres the radiative plasmon effects are of primary significance, and many quantum effects such as spill-out or Landau damping (corresponding to the decay of plasmons into electron–hole pairs that are of high energy with respect to the Fermi surface) [80] are completely negligible at this size scale.

To gain insight into plasmon behaviour in relatively large electron systems with 10^{5-8} electrons, i.e., with a nanosphere radius of several tens of nanometres, $a > 5$ nm, accurate numerical methods based on the Kohn–Sham approach (LDA

---

[1] The range is similar also for Ag or Cu, but generally it depends on the metal type and on the dielectric permittivity of the surroundings.

and TDLDA) are not useful as they are limited to a much smaller number of electrons, i.e., up to 200–300, as is the case for $a \sim 2$–3 nm nanoparticles. As the electron number increases above 300, these numerical methods are less effective because numerical-type constraints rapidly slow down calculations even at high-performance facilities. For large nanoparticles, various versions of the random phase approximation (RPA) turn out to be useful instead [19, 71, 21, 23]. Of particular interest are large nanoparticles of the noble metals (gold, silver, and also copper) because the surface plasmon resonances in nanoparticles of these metals lie within the visible light spectrum, which is important for photovoltaic applications.

The following sections contain a brief repetition of the analysis of the surface dipole-type plasmons in a single nanosphere within the RPA model [23, 77], as presented in the previous chapter, but with the inclusion of the damping effects caused both by scattering and radiation, the latter expressed by the fictitious Lorentz friction force acting on moving charges. A discussion of the difference between the perturbative approximation and the accurate treatment of the Lorentz friction will be presented in detail, together with a comparison with experiment. The last section focuses on some applications of the exact analysis of Lorentz friction to the widely developed Mie theory of the light properties of metallic nanoparticles, which are very useful and efficient for the interpretation of many experiments in the field of nanophotonics and plasmonics.

5.1 Damped Plasmonic Oscillations in Metallic Nanospheres in Dielectric Surroundings

Let us consider a metallic (Au) nanosphere of radius a immersed in a dielectric environment with relative permittivity ε. We assume the presence of a varying external electric field which can excite plasmon collective oscillations modes in the nanosphere. Such a situation could correspond to the so-called dipole limit for interaction with light, i.e., the case when the wavelength of the incident light well exceeds the nanosphere dimension. The surface plasmon resonance for metallic nanospheres in the noble metal case, for Au, Ag, and even Cu, with radius $a \in (5, 100)$ nm occurs for light of approximately 500 nm wavelength; thus, the conditions for the dipole approximation are fulfilled and the resonant incident electromagnetic signal can be treated as homogeneous over the entire nanoparticle in the range of radii considered. We will analyse the collective reaction of the electron liquid to this space-homogeneous but time-dependent field in the spherical metallic nanoparticle assuming the jellium model. This means that we neglect the dynamics of the ions, according to the scheme presented in the previous chapter.

As was pointed out, the ion dynamics is inertial owing to the much higher mass of the ions in comparison to the electron mass and can be neglected.[2] In the jellium model, the charge of the ion background is static and uniformly smeared over the whole nanosphere with density $n_e(r) = n_e \Theta(a - r)$, $n_e = N_e/V$, where N_e is the number of free electrons in the nanosphere, V is the volume of the nanosphere, Θ is the Heaviside step function, and $|e|n_e$ is the averaged density of the positive charge compensating the charge of the free electrons.

The plasmonic oscillations in a metallic nanosphere can be described by the local electron density fluctuation, defined as in the RPA approach for the bulk metal developed by Pines and Bohm [15]:

$$\rho(\mathbf{r},t) = \left\langle \Psi_e(t) \left| \sum_j \delta(\mathbf{r} - \mathbf{r}_j) \right| \Psi_e(t) \right\rangle, \tag{5.1}$$

where the $\mathbf{r}_j$ define the quasiclassically positions of the electrons and $|\Psi_e(t)\rangle$ is the multiparticle wave function of the electron system (for the Hamiltonian (5.4) below; see also (4.8)). The Fourier picture of (5.1) has the simple form

$$\tilde{\rho}(\mathbf{k},t) = \int \rho(\mathbf{r},t) e^{-i\mathbf{k}\cdot\mathbf{r}} d^3 r = \langle \Psi_e(t) | \hat{\rho}(\mathbf{k}) | \Psi_e(t) \rangle, \tag{5.2}$$

where the operator $\hat{\rho}(\mathbf{k}) = \sum_j e^{-i\mathbf{k}\cdot\mathbf{r}_j}$. The dynamic equation for this local electron density operator can be found by using the Heisenberg equation twice:

$$\frac{d^2 \hat{\rho}(\mathbf{k})}{dt^2} = \frac{1}{(i\hbar)^2} \left[\left[\hat{\rho}(\mathbf{k}), \hat{H}_e \right], \hat{H}_e \right], \tag{5.3}$$

where H_e is the Hamiltonian for electrons in the metallic nanosphere in the jellium model, but with the electron interactions expressed by local-electron-density operators that also include the interaction of the electrons with the jellium [23]:

$$\hat{H}_e = -\sum_{j=1}^{N_e} \frac{\hbar^2 \nabla_j^2}{2m} - \frac{e^2}{(2\pi)^3} \int d^3 k \tilde{n}_e(\mathbf{k}) \frac{2\pi}{k^2} \left(\hat{\rho}^+(\mathbf{k}) + \hat{\rho}(\mathbf{k}) \right)$$

$$+ \frac{e^2}{(2\pi)^3} \int d^3 k \frac{2\pi}{k^2} \left(\hat{\rho}^+(\mathbf{k}) \hat{\rho}(\mathbf{k}) - N_e \right), \tag{5.4}$$

where $\tilde{n}_e(\mathbf{k}) = \int d^3 r n_e(\mathbf{r}) e^{-i\mathbf{k}\cdot\mathbf{r}}$. Equation (5.3), after averaging over the multiparticle wave function and, according to the RPA scheme, neglecting sums of exponents with random phases, finally attains the form [23],

---

[2] The adiabatic approximation corresponds to neglecting the ion dynamics in comparison to the electron dynamics, since the ions are much heavier, though for some specific purposes one can consider dual-ion plasmons in metals. These, however, have a marginal significance for the electron plasmon response [19].

5.1 Damped Plasmonic Oscillations in Metallic Nanospheres

$$\frac{\partial^2 \delta\tilde{\rho}(\mathbf{r},t)}{\partial t^2} = \left[\frac{2}{3}\frac{\epsilon_F}{m}\nabla^2\delta\tilde{\rho}(\mathbf{r},t) - \omega_p^2\delta\tilde{\rho}(\mathbf{r},t)\right]\Theta(a-r)$$

$$-\frac{2}{3m}\nabla\left\{\left[\frac{3}{5}\epsilon_F n_e + \epsilon_F\delta\tilde{\rho}(\mathbf{r},t)\right]\frac{\mathbf{r}}{r}\delta(a-r)\right\}$$

$$-\left[\frac{2}{3}\frac{\epsilon_F}{m}\frac{\mathbf{r}}{r}\nabla\delta\tilde{\rho}(\mathbf{r},t) + \frac{\omega_p^2}{4\pi}\frac{\mathbf{r}}{r}\nabla\int d^3r_1 \frac{1}{|\mathbf{r}-\mathbf{r}_1|}\delta\tilde{\rho}(\mathbf{r}_1,t)\right]\delta(a-r).$$

(5.5)

In the above formula the Thomas–Fermi approximation for the kinetic energy has been used [16, 23] and ω_p is the bulk plasmon frequency, so that $\omega_p^2 = 4\pi n_e e^2/m$ in Gauss units (in SI $\omega_p^2 = e^2 n_e/\varepsilon_0 m$). We have used $\nabla\Theta(a-r) = -(\mathbf{r}/r)\delta(a-r)$, which gives rise to the Dirac delta distinguishing surface plasmon fluctuations in Eq. (5.5). Owing to the structure of Eq. (5.5), the general solution is thus decomposed into two parts related to the two distinct domains:

$$\delta\tilde{\rho}(\mathbf{r},t) = \begin{cases} \delta\tilde{\rho}_1(\mathbf{r},t) & \text{for } r < a, \\ \delta\tilde{\rho}_2(\mathbf{r},t) & \text{for } r \geq a \ (r \to a+), \end{cases}$$

(5.6)

corresponding respectively to the volume and surface charge fluctuations. These two parts of the local electron density fluctuations satisfy the following equations (for particularities of the relevant derivation see [23] and the previous chapter):

$$\frac{\partial^2 \delta\tilde{\rho}_1(\mathbf{r},t)}{\partial t^2} = \frac{2}{3}\frac{\epsilon_F}{m}\nabla^2\delta\tilde{\rho}_1(\mathbf{r},t) - \omega_p^2\delta\tilde{\rho}_1(\mathbf{r},t),$$

(5.7)

and (in the next equation $\epsilon = 0+$ is introduced formally to ensure the correct definition of the Dirac delta),

$$\frac{\partial^2 \delta\tilde{\rho}_2(\mathbf{r},t)}{\partial t^2} = -\frac{2}{3m}\nabla\left\{\left[\frac{3}{5}\epsilon_F n_e + \epsilon_F\delta\tilde{\rho}_2(\mathbf{r},t)\right]\frac{\mathbf{r}}{r}\delta(a+\epsilon-r)\right\}$$

$$-\left[\frac{2}{3}\frac{\epsilon_F}{m}\frac{\mathbf{r}}{r}\nabla\delta\tilde{\rho}_2(\mathbf{r},t) + \frac{\omega_p^2}{4\pi}\frac{\mathbf{r}}{r}\nabla\int d^3r_1 \frac{1}{|\mathbf{r}-\mathbf{r}_1|}(\delta\tilde{\rho}_1(\mathbf{r}_1,t)\Theta(a-r_1)\right.$$

$$\left. + \delta\tilde{\rho}_2(\mathbf{r}_1,t)\Theta(r_1-a))\right]\delta(a+\epsilon-r).$$

(5.8)

For the solution of Eqs. (5.7) and (5.8) we represent the two parts of the electron fluctuation, as described in the previous chapter, in the following manner:

$$\begin{aligned}\delta\tilde{\rho}_1(\mathbf{r},t) &= n_e F(\mathbf{r},t) & \text{for } r < a, \\ \delta\tilde{\rho}_2(\mathbf{r},t) &= \sigma(\Omega,t)\delta(r+\epsilon-a), \ \epsilon = 0+ & \text{for } r \geq a \ (r \to a+),\end{aligned}$$

(5.9)

supplemented with the initial conditions $F(\mathbf{r},t)|_{t=0} = 0$, $\sigma(\Omega,t)|_{t=0} = 0$ (Ω is the spherical angle, representing the spherical coordinate, θ and φ), with the neutrality condition $F(\mathbf{r},t)|_{r\to a} = 0$, and with $\int \rho(\mathbf{r},t)d^3r = N_e$.

After some algebra (as presented in detail in [23] and in the previous chapter) one finds for a nanoparticle with spherical symmetry (i.e., a nanosphere) that the volume plasmons are described by

$$F(\mathbf{r},t) = \sum_{l=1}^{\infty} \sum_{m=-l}^{l} \sum_{n=1}^{\infty} A_{lmn} j_l(k_{nl}r) Y_{lm}(\Omega) \sin(\omega_{nl} t) \tag{5.10}$$

and the surface plasmons are described by

$$\sigma(\Omega,t) = \sum_{l=1}^{\infty} \sum_{m=-l}^{l} \frac{B_{lm}}{a^2} Y_{lm}(\Omega) \sin(\omega_{0l} t)$$

$$+ \sum_{l=1}^{\infty} \sum_{m=-l}^{l} \sum_{n=1}^{\infty} A_{lmn} \frac{(l+1)\omega_p^2}{l\omega_p^2 - (2l+1)\omega_{nl}^2} Y_{lm}(\Omega) n_e$$

$$\times \int_0^a dr_1 \frac{r_1^{l+2}}{a^{l+2}} j_l(k_{nl} r_1) \sin(\omega_{nl} t), \tag{5.11}$$

where $j_l(\xi) = \sqrt{\pi/2\xi}\, I_{l+1/2}(\xi)$ is a spherical Bessel function, $Y_{lm}(\Omega)$ is a spherical harmonic, $\omega_{nl} = \omega_p\sqrt{1 + x_{nl}^2/(k_T^2 a^2)}$ are the frequencies of the volume plasmon self-oscillations, $k_T = \sqrt{6n_e e^2/\epsilon_F}$ is the Thomas–Fermi length (ϵ_F is the Fermi energy), and x_{nl} are the nodes of the Bessel function $j_l(\xi)$; $k_{nl} = x_{nl}/a$ and $\omega_{0l} = \omega_p\sqrt{l/(2l+1)}$ are the frequencies of the electron surface self-oscillations (the surface plasmon frequencies). In the formula for $\sigma(\Omega,t)$, the first term corresponds to the surface self-oscillations while the second term describes the surface oscillations induced by the volume plasmons. The frequencies of the surface self-oscillations corresponding to multipole modes enumerated by l have the form

$$\omega_{0l} = \omega_p \sqrt{\frac{l}{2l+1}}, \tag{5.12}$$

which, for $l = 1$, gives the dipole-type surface oscillation frequency $\omega_{01} = \omega_p/\sqrt{3}$ originally found by Mie [18] (for simplicity, hereafter we will set $\omega_1 = \omega_{01}$).

The advantage of this RPA approach consists in the oscillatory form of both Eqs. (5.7) and (5.8). Plasmon damping can thus be included in a phenomenological manner, by the addition of an attenuation term to the plasmon dynamic equations, i.e.,

$$-\frac{2}{\tau_0} \frac{\partial \delta \rho_{1(2)}(\mathbf{r},t)}{\partial t},$$

is added to the right-hand sides of Eqs. (5.7) and (5.8). The damping ratio $1/\tau_0$ accounts for electron scattering losses and can be approximated by the formula [24]

5.1 Damped Plasmonic Oscillations in Metallic Nanospheres

$$\frac{1}{\tau_0} \simeq \frac{v_F}{2\lambda_B} + \frac{Cv_F}{2a}, \tag{5.13}$$

where C is a constant of order unity, a is the nanosphere radius, v_F is the Fermi velocity in the metal, and λ_B is the electron mean free path in the bulk metal (including the scattering of electrons from other electrons, from impurities, and from phonons [24]); for example, for Au, $v_F = 1.396 \times 10^6$ m/s and $\lambda_B \simeq 53$ nm (at room temperature); the latter term in formula (5.13) accounts for the scattering of electrons at the boundary of the nanoparticle, whereas the former term corresponds to scattering processes similar to those in the bulk metal (leading to irreversible ohmic energy losses). Other quantum effects, as for example Landau damping (especially important in small nanoparticles [80, 79]), corresponding to the decay of a plasmon to a high-energy particle–hole pair, are of less significance for nanosphere radii larger than 2–3 nm [80] and are completely negligible for radii larger than 5 nm.

The electron response to the driving field $\mathbf{E}(t)$ (which is homogeneous over the nanosphere, corresponding to the dipole approximation) resolves to a single dipole-type mode, effectively described by a three-component function denoted here as $Q_{1m}(t)$ ($l = 1$ and $m = -1, 0, 1$). The dynamical equation (5.8) reduces in this case to the following:

$$\frac{\partial^2 Q_{1m}(t)}{\partial t^2} + \frac{2}{\tau_0} \frac{\partial Q_{1m}(t)}{\partial t} + \omega_1^2 Q_{1m}(t)$$
$$= \sqrt{\frac{4\pi}{3} \frac{en_e}{m}} \left[E_z(t)\delta_{m,0} + \sqrt{2} \left(E_x(t)\delta_{m,1} + E_y(t)\delta_{m,-1} \right) \right], \tag{5.14}$$

where $\omega_1 = \omega_p/\sqrt{3\varepsilon}$ is the dipole surface plasmon frequency found with the semiclassical RPA approach [23] ($\omega_p = 4\pi ne^2/m$ is the plasmon resonance frequency in the bulk metal and ε is the relative permittivity of the material surrounding the nanosphere). For $\varepsilon = 1$ this frequency coincides with the classical Mie frequency, $\omega_{Mie} = \omega_p/\sqrt{2\varepsilon + 1}$ [17]. The semiclassical RPA dipole frequency ω_1 is somewhat smaller than ω_{Mie} for $\varepsilon > 1$, as illustrated in Fig. 5.1; this result is supported by a more exact numerical TDLDA calculus indicating some overestimation of the plasmon resonance frequency by the Mie formula [81]. Note that the classical Mie frequency is based on a greatly simplified dielectric function for the metal [18, 17].

The electron density fluctuations described by Eq. (5.14) are as follows:

$$\delta\rho(\mathbf{r},t) = \begin{cases} 0, & \text{for } r < a, \\ \sum_{m=-1}^{1} Q_{1m}(t) Y_{1m}(\Omega) & \text{for } r \geq a,\ r \to a+. \end{cases} \tag{5.15}$$

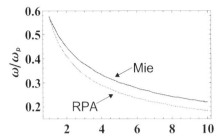

Figure 5.1 For comparison, the RPA semiclassical formula for the dipole resonance frequency of surface plasmons in a metallic nanosphere, $\omega_1 = \omega_p/\sqrt{3\varepsilon}$, and the classical formula of Mie for this frequency, $\omega_{Mie} = \omega_p/\sqrt{2\varepsilon+1}$.

For the dipole-type plasmonic oscillations presented by Eq. (5.15), the dipole $\mathbf{d}(t)$ of the electron system can be written as follows:

$$\mathbf{d}(t) = e \int d^3r \mathbf{r} \delta\rho(\mathbf{r},t) = \frac{\sqrt{2\pi}}{\sqrt{3}} ea^3 \left[Q_{1,1}(t) \quad Q_{1,-1}(t) \quad \sqrt{2} Q_{1,0}(t) \right] \quad (5.16)$$

and the $\mathbf{d}(t)$ satisfies the following equation (obtained via rewriting Eq. (5.14)),

$$\left(\frac{\partial^2}{\partial t^2} + \frac{2}{\tau_0} \frac{\partial}{\partial t} + \omega_1^2 \right) \mathbf{d}(t) = \frac{a^3 4\pi e^2 n_e}{3m} \mathbf{E}(t) = \varepsilon a^3 \omega_1^2 \mathbf{E}(t). \quad (5.17)$$

According to Eq. (5.16) the plasmon oscillation dipole scales with the nanosphere radius a as $\sim a^3$, which indicates that all the electrons in the sphere contribute to the surface plasmon excitations. This can be referred to the fact that the surface plasmon modes correspond to uniform translation-type oscillations of the whole electron liquid. Inside the sphere the uniformly shifted charge of the electrons is still exactly compensated by the static and uniform positive jellium, while an unbalanced charge density fluctuation occurs only on the surface. One can note that in the case of the volume plasmons the noncompensated charge density fluctuations are present inside the sphere, because the volume plasmon modes are related to compressional modes resulting in volume charge fluctuations not balanced by the jellium along the radial direction inside the nanosphere.

The damping term $1/\tau_0$ accounts for dissipation of the plasmon energy by electron scattering, similarly to that for the ohmic losses. This channel of energy dissipation includes all types of scattering phenomena: electron–electron, electron–phonon, and electron–admixture collisions, as is usually the case for ohmic losses, but additionally a contribution related to electron scattering on the nanoparticle boundary [24]. All these scattering processes cause the attenuation of plasmons, and their energy is dissipated irreversibly and finally converted into heat.

5.1 Damped Plasmonic Oscillations in Metallic Nanospheres

By reducing structural imperfections in the metal one can diminish these losses to some limiting value. The electron scattering channel of plasmon attenuation turns out to be especially important for small metallic nanoparticles, for which it is the predominant mechanism of plasmon damping, together with Landau damping [80, 79]. In small metallic nanoparticles radiation effects have a negligible role because of the low total number of electrons ($\sim 100-300$ for radius $a \sim 2-3$ nm). Instead, for ultrasmall nanoparticles, nanosphere-edge scattering plays a significant role as its contribution to the damping-time ratio is proportional to $1/a$ and thus is large for small a but quickly diminishes when the radius increases. The scattering of electrons by the boundary of the nanosphere may undergo different regimes depending on the microscopic details of the surface. One can correct these regimes to two limiting model scattering types, reflection and diffusive, which differ in their attenuation rate. In order to model various types of boundary scattering of electrons one can introduce in Eq. (5.13) an effective constant C of the order of unity which can be helpful in accommodating the theory to the experimental data, utilizing the $1/a$ scaling of this attenuation channel. Below we will show that radiation losses, which also contribute to the overall attenuation of plasmons, start to be important in the case of larger metallic nanoparticles with a sufficiently large number of electrons. Radiation losses scale as a^3 (for nanospheres that are not too large, e.g., for Au up to about 30 nm) and therefore radiative losses quickly dominate the plasmon damping at the range of 10 nm (for Au, Ag, and Cu). Therefore, at some particular nanosphere size we encounter a cross-over in the size dependence of the plasmon damping: for lower radii the size dependence scales as $1/a$ whereas for larger radii it scales as a^3 [77]. This pronounced cross-over in the size dependence of surface plasmon damping is illustrated in Fig. 5.3.

The radiation energy losses of oscillating electrons related to the surface plasmon time-dependent dipole can be expressed by the Lorentz friction [55], i.e., by an

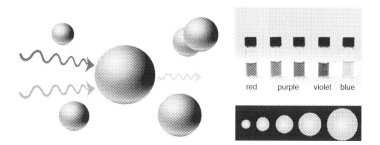

Figure 5.2 In an experiment [77] with light passing through a colloidal solution of Au nanoparticles one can measure the extinction of light for various wavelengths; that depends strongly on the nanosphere size, resulting in a variation in the colour for colloidal solutions with different sized particles.

effective electric field hampering the movement of charges and thus lowering the amplitude of dipole oscillations:

$$\mathbf{E}_L = \frac{2}{3c^3}\frac{\partial^3 \mathbf{d}(t)}{\partial t^3}. \tag{5.18}$$

Hence, one can rewrite Eq. (5.17) including the Lorentz friction term as

$$\left(\frac{\partial^2}{\partial t^2} + \frac{2}{\tau_0}\frac{\partial}{\partial t} + \omega_1^2\right)\mathbf{d}(t) = \varepsilon a^3 \omega_1^2 \mathbf{E}(t) + \varepsilon a^3 \omega_1^2 \mathbf{E}_L, \tag{5.19}$$

or, in a more explicit form, for the case $\mathbf{E} = 0$ we have

$$\left(\frac{\partial^2}{\partial t^2} + \omega_1^2\right)\mathbf{d}(t) = \frac{\partial}{\partial t}\left(-\frac{2}{\tau_0}\mathbf{d}(t) + \frac{2}{3\omega_1\sqrt{\varepsilon}}\left(\frac{\omega_p a}{c\sqrt{3}}\right)^3 \frac{\partial^2}{\partial t^2}\mathbf{d}(t)\right). \tag{5.20}$$

One can solve Eq. (5.20) by application of the perturbation calculus, i.e., one can treat the right-hand side of this equation as a small perturbation. Therefore, in the zeroth step of the perturbation calculus we get $\left(\frac{\partial^2}{\partial t^2} + \omega_1^2\right)\mathbf{d}(t) = 0$ and thus $(\partial^2/\partial t^2)\mathbf{d}(t) = -\omega_1^2\mathbf{d}(t)$. Next, in the first-order step of the perturbation method, we substitute the latter formula into the right-hand side of Eq. (5.20), obtaining,

$$\left(\frac{\partial^2}{\partial t^2} + \frac{2}{\tau}\frac{\partial}{\partial t} + \omega_1^2\right)\mathbf{d}(t) = 0, \tag{5.21}$$

where the effective attenuation rate is given by

$$\frac{1}{\tau} = \frac{1}{\tau_0} + \frac{\omega_1}{3\sqrt{\varepsilon}}\left(\frac{\omega_p a}{c\sqrt{3}}\right)^3. \tag{5.22}$$

In these steps the Lorentz friction is accounted for in the total attenuation rate $1/\tau$ utilizing the approximate perturbative linearization of the third-order derivative in Eq. (5.18). Such an approximation leads to an effective model of radiation losses within the scheme of the damped harmonic oscillator. Nevertheless, as mentioned previously, this approximation is justified only for small perturbations, i.e., when the radiation term, the second term in Eq. (5.22), which is proportional to a^3, is sufficiently small, as required by the perturbation calculus limits. For nanospheres of radii $a < 30$ nm (e.g., for Au in vacuum) the perturbation constraints are not violated; this has been independently confirmed experimentally for Au and Ag nanospheres [77]. For nanosphere radii 5–30 nm the perturbation approximation explains the experimentally observed redshift of the resonance surface plasmon frequency with increasing radius a, recordable from the measurement of light extinction spectra [77]. In this case the solution of Eq. (5.21) has the following form: $\mathbf{d}(t) = \mathbf{d}_0 e^{-t/\tau}\cos(\omega_1' t + \phi)$, where $\omega_1' = \omega_1\sqrt{1 - 1/(\omega_1\tau)^2}$ gives the experimentally observed redshift of the plasmon resonance owing to the $\sim a^3$

5.2 Attenuation of Dipole Surface Plasmons with Exact Inclusion of Lorentz Friction

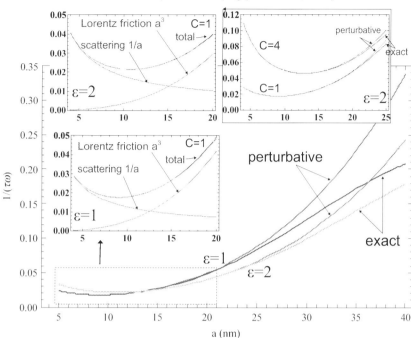

Figure 5.3 For comparison, the different contributions to plasmon oscillation damping (Au nanoparticles in vacuum and in water) in the size region $a \sim 10\text{–}20$ nm (in insets), where the cross-over in the size dependence of the damping rate takes place. In the upper right inset the effect of changing the constant C in Eq. (5.13) is illustrated; this models two scattering regimes for electrons on a nanosphere boundary.

increase in damping caused by radiation losses in this size window. The Lorentz friction term in Eq. (5.22) starts to dominate plasmon damping for approximately $a \geq 12$ nm (for Au and Ag it is almost independent of the relative permittivity of the surrounding medium ε, as is illustrated in the insets to Fig. 5.3) owing to its strong a^3 dependence; see Fig. 5.3. The plasmonic oscillation damping grows sharply with a, causing a redshift of the surface dipole plasmon resonance that is in very good agreement with the experimental data for $10 < a < 30$ nm (Au and Ag) [77].

5.2 Attenuation of Dipole Surface Plasmons with Exact Inclusion of the Lorentz Friction

Let us consider the dynamic equation (5.20) for surface plasmons in a metallic nanosphere, now without substituting the Lorentz friction term

$$\frac{2}{3\omega_1\sqrt{\varepsilon}}\left(\frac{\omega_p a}{c\sqrt{3}}\right)^3 \frac{\partial^3 \mathbf{d}(t)}{\partial t^3}$$

by the approximate perturbative formula

$$-\frac{2\omega_1}{3\sqrt{\varepsilon}}\left(\frac{\omega_p a}{c\sqrt{3}}\right)^3 \frac{\partial \mathbf{d}(t)}{\partial t}.$$

This approximate linearization of the third-order derivative of $\mathbf{d}(t)$ is justified only when the total attenuation term

$$\frac{\partial}{\partial t}\left[-\frac{2}{\tau_0} + \frac{2}{3\omega_1\sqrt{\varepsilon}}\left(\frac{\omega_p a}{c\sqrt{3}}\right)^3 \frac{\partial^2}{\partial t^2}\right]\mathbf{d}(t)$$

is sufficiently small. To compare the different contributions to Eq. (5.20) one must change to a dimensionless time variable $t' = \omega_1 t$. Then Eq. (5.20) attains the form

$$\frac{\partial^2 \mathbf{d}(t')}{\partial t'^2} + \frac{2}{\tau_0 \omega_1}\frac{\partial \mathbf{d}(t')}{\partial t'} + \mathbf{d}(t') = \frac{2}{3\sqrt{\varepsilon}}\left(\frac{\omega_p a}{c\sqrt{3}}\right)^3 \frac{\partial^3 \mathbf{d}(t')}{\partial t'^3}. \tag{5.23}$$

In Eq. (5.23) the dimensionless damping ratio caused by scattering is given by

$$\frac{2}{\tau_0 \omega_1} = \frac{2}{\omega_1}\left(\frac{v_F}{2\lambda_b} + \frac{v_F}{2a}\right) \simeq 0.027$$

for $a = 10$ nm (values for the other parameters are those for Au and Ag, as collected in Table 5.1) and diminishes with increasing radius a; see the Fig. 5.3 insets. One can also estimate the dimensionless coefficient of the Lorentz friction term in Eq. (5.23). It equals to

$$\frac{2}{3\sqrt{\varepsilon}}\left(\frac{\omega_p a}{c\sqrt{3}}\right)^3$$

Table 5.1 *Nanosphere parameters assumed for calculation of the damping rates for surface plasmons*

Material	Au	Ag
bulk plasmon energy, $\hbar\omega_p$	8.57 eV	8.56 eV
bulk plasmon frequency, ω_p	1.302×10^{16}/s	1.3×10^{16}/s
Mie dipole plasmon energy, $\hbar\omega_1$	4.94 eV	4.93 eV
Mie frequency, $\omega_1 = \omega_p/\sqrt{3}$	0.752×10^{16}/s	0.75×10^{16}/s
constant in Eq. (5.13), C	1, 4	1, 4
Fermi velocity, v_F	1.396×10^6 m/s	1.4×10^6 m/s
bulk mean free path (room temp.), λ_B	53 nm	57 nm

5.2 Attenuation with Exact Inclusion of Lorentz Friction

and rapidly increases with the radius a because it is proportional to a^3 (for $a = 10$ nm it is rather small, $\simeq 0.0104$ but for $a = 25$ nm it increases more than ten times to $\simeq 0.1638$; the example is taken for Au nanospheres in vacuum).

Utilizing the perturbation method to solve Eq. (5.23) one obtains the renormalized attenuation rate for the effective damping term in the form

$$\frac{1}{\omega_1 \tau_0} + \frac{1}{3\sqrt{\varepsilon}} \left(\frac{\omega_p a}{c\sqrt{3}} \right)^3.$$

Owing to the rapid increase of the factor a^3 with the radius a, the effective damping term obtained by the perturbation approach quickly reaches the value 1, for which the oscillator enters the overdamped regime. The value 1 of this renormalized attenuation rate is attained at $a \simeq 57$ nm (for Au in vacuum). Nevertheless this is an artefact of the perturbation calculus, and one can check that the exact solution of Eq. (5.23) agrees sufficiently well with the solution obtained using the perturbation approach only up to $a \sim 20$–30 nm. For higher values of a the rapid increase in the effective attenuation rate causes an unacceptable discrepancy between the perturbation approximation and the accurate solution, as we illustrate in Figs. 5.4 and 5.5.

In Fig. 5.5 the self-frequency and the damping rate for dipole-type surface plasmons are plotted as functions of the nanosphere radius a, for the perturbation approximation, for which

$$\omega = \omega_1 \sqrt{1 - \left[\frac{1}{\omega_1 \tau_0} + \frac{1}{3\sqrt{\varepsilon}} \left(\frac{\omega_p a}{c\sqrt{3}} \right)^3 \right]^2},$$

and for the exact solution of Eq. (5.23). The lines corresponding to the approximate solution end at $a \simeq 57, 64, 69, 72, 75$ nm (for Au with the relative dielectric permittivity of the surroundings $\varepsilon = 1, 2, 3, 4, 5$, respectively). At these values of a the effective attenuation rate found from the perturbation calculus reaches the limiting value 1 (when the corresponding resonance frequency vanishes and the resonance wavelength $\lambda = c 2\pi/\omega \to \infty$). This limiting behaviour corresponds to the termination of oscillating solutions of the perturbative oscillatory equation. In this case we are in the overdamped regime, with an aperiodic solution. This is, however, an artefact caused by the perturbation solution method, which cannot be applied for too high a value of the perturbation term. For an exact solution of Eq. (5.23) this singular behaviour disappears and an oscillating solution exists for larger a as well. Thus, one can see that the plasmon resonance redshift is strongly overestimated within the framework of the perturbative approach to the

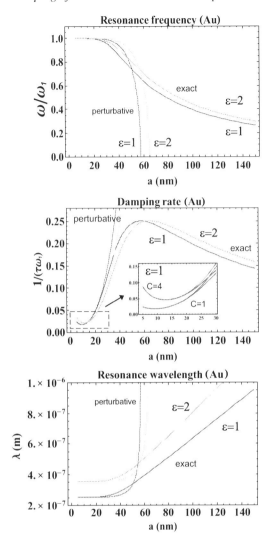

Figure 5.4 (Left and centre) The real and imaginary parts of the exact oscillating solution of Eq. (5.23): the resonance frequency and the damping rate given by Eqs. (5.24) are plotted as functions of the nanosphere radius a for two distinct dielectric surrounding media (vacuum and water); the lines corresponding to the perturbative solutions are presented for comparison. In the right-hand panel the resonance wavelength is depicted versus the radius a for vacuum ($\varepsilon = 1$) and water surroundings ($\varepsilon = 2$); in the inset in the central panel the impact of the constant C in Eq. (5.13) is illustrated.

Lorentz friction unless the nanosphere radius is lower than ~ 20 nm (for Au in vacuum).

The dynamical equation for dipole surface plasmon excitations including the exact form for the Lorentz friction, i.e., Eq. (5.23), is a third-order linear differential

5.2 Attenuation with Exact Inclusion of Lorentz Friction

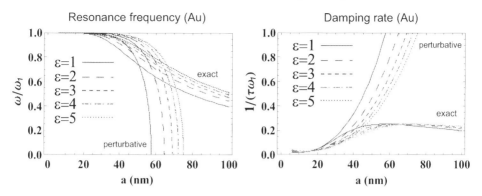

Figure 5.5 The resonance frequency and the damping rate for the oscillating solutions of Eq. (5.23), for five different values of ε. The exact and the approximate perturbative approach dependences are presented. The termination of the perturbative solutions displays an artefact of the perturbation method.

equation. This equation can be solved exactly and the exponent coefficients Ω of its solutions $\sim e^{i\Omega t'}$ can be listed in the following analytical form:

$$\Omega_1 = -\frac{i}{3l} - \frac{i 2^{1/3}(1+6lq)}{3l\mathcal{A}} - \frac{i\mathcal{A}}{2^{1/3} 3l} = i\alpha \in \text{Im},$$

$$\Omega_2 = -\frac{i}{3l} + \frac{i(1+i\sqrt{3})(1+6lq)}{2^{2/3} 3l \mathcal{A}} + \frac{i(1-i\sqrt{3})\mathcal{A}}{2^{1/3} 6l} = \omega + i\frac{1}{\tau}, \quad (5.24)$$

$$\Omega_3 = -\omega + i\frac{1}{\tau} = -\Omega_2^*,$$

where $\mathcal{A} = \left(2 + 27l^2 + 18lq + \sqrt{4(-1-6lq)^3 + (2+27l^2+18lq)^2}\right)^{1/3}$, $q = 1/\tau_0\omega_1$ and $l = (2/3\sqrt{\varepsilon})\left(a\omega_p/c\sqrt{3}\right)^3$. Here Im stands for the set of imaginary numbers. The functions ω and $1/\tau$ (in dimensionless units, i.e., divided by ω_1) are plotted in Fig. 5.4 versus the nanosphere radius a.

We note that Eq. (5.23) has two types of particular solutions $e^{i\Omega t'}$, with complex self-frequencies Ω as given in Eq. (5.24). The solutions given by Ω_2 and Ω_3 describe oscillating behaviour with frequency ω and positive damping (note that $i\Omega_2$ and $i\Omega_3$ are mutually conjugate; thus Ω_2 and Ω_3 have real parts of opposite sign, but the same imaginary parts, these latter, being positive, correspond to the ordinary damping rate for positive time). The self-frequency given by Ω_1 turns out to correspond to an unstable exponentially rising function, as Ω_1 is a purely negative imaginary solution. Such an unstable solution is a well-known artefact in Maxwell electrodynamics (see e.g. [55] for the particularities) and corresponds to infinite self-acceleration of a free charge exposed to the action of the Lorentz friction force. This solution in turn corresponds to a singular solution of the equation $m\dot{v} = \text{const.} \times \ddot{v}$, which is associated with the formal renormalization of the so-called field mass of the charge. The field mass in Maxwell electrodynamics is infinite for a pointlike charge and is assumed to be cancelled by an arbitrarily assumed negative infinite non-field mass, resulting in the ordinary mass of the charge (this is, however, not defined mathematically in a proper way, which links with some inconsistency in the classical electrodynamics of pointlike charges). Therefore, this unphysical singular particular solution should be discarded. The other, oscillatory-type, solutions describe damped oscillations which resemble the solution of the ordinary damped harmonic oscillator. It must be emphasized, however, that Eq. (5.23) does not describe a damped harmonic oscillator and therefore the attenuation rate and frequency ($1/\tau$ and ω as given by Eq. (5.24)) do not agree with corresponding quantities for an ordinary damped harmonic oscillator. They are expressed by the analytical formulae for Ω_2 or Ω_3 in Eq. (5.24). They can be easily evaluated for various values of a and compared with corresponding quantities found within the perturbation-calculus-based approach. Such a comparison is illustrated in Figs. 5.4 and 5.5. From the comparison presented in Fig. 5.4 one can note that application of the perturbation approach leads to a highly overestimated damping rate for $a > 30$ nm (for Au in vacuum). For $30 < a < 50$ nm, for gold nanoparticles, a small underestimation of the resonance wavelength occurs whereas for $a > 50$ nm there is a rapid and very strong overestimation of the resonance wavelength, as illustrated in the left-hand panel in Fig. 5.4. Moreover, as has already been mentioned above, the approximate solution terminates at $a = 57, 64, 69, 72, 75$ nm (for Au; $\varepsilon = 1, 2, 3, 4, 5$ respectively), above which an approximate oscillatory-type solution does not exist within the perturbative calculus approach. One can finally conclude that utilization of the approximate formula to account for Lorentz friction damping in the form given by (5.22) is justified up to $a \simeq 20$ nm (for Au in vacuum); it causes only a small discrepancy (a practically negligible error in the damping rate and only a small underestimation of the resonance wavelength) at $20 < a < 30$ nm. However, for $a > 30$ nm, for Au, these approximate perturbative parameters of the oscillatory plasmon excitations deviate strongly from the exact values. Also illustrated in Figs. 5.4 and 5.5 are the dependences of the limiting values for the radii listed above, which relate the accuracy of the perturbation approach for the radiative losses of plasmons to the relative permittivity of the surroundings of the metallic nanoparticles, ε.

The accurate value of the damping rate as plotted in Fig. 5.4 reaches its minimum at a certain value of the nanosphere radius,

$$a^* = \left(\frac{9}{2}\varepsilon \frac{v_F c^3}{\omega_p^4}\right)^{1/4}, \qquad (5.25)$$

and reaches its maximum at a^{**}. We note that the latter value might be defined as the zero point of the appropriate derivative expressed analytically using Eq. (5.24), though the related formula for this derivative is cumbersome owing to the

Table 5.2 *Radius of the nanosphere corresponding to the minimal (a^*) and maximal (a^{**}) values of surface plasmon damping*

	a^*, a^{**} (nm)	
$n = \sqrt{\varepsilon}$	Au	Ag
$n = 1$, vacuum	8.77, 57.4	8.78, 57.42
$n = 1.4$	10.4, 64.3	10.4, 64.4
$n = 2$	12.4, 72	12.4, 72.1

complicated analytical expression for the function $1/\tau$ given by Eq. (5.24)). Both a^* and a^{**} are listed in Table 5.2 for $\varepsilon = 1, 2, 4$ (a^{**} depends weakly on the material, Au or Ag, whereas it depends strongly on the relative permittivity of the dielectric surroundings and is located approximately at 57 nm (for Au in vacuum)).

5.3 Comparison of Surface Plasmon Oscillation Features Including Lorentz Friction with the Experimental Data and Simplified Mie Approach

For comparizon of the model results for surface plasmon damping in a metallic nanosphere with the experimentally observed behaviour, one can use published experimental data [77] for the extinction of light after passing through water-colloidal solutions of Au nanospheres with varying radii. Extinction peaks that correspond to plasmon excitations in metal nanospheres are plotted in Fig. 5.8, in the left-hand panel. In the right-hand panel of this figure analogous data for Ag nanoparticles are presented for comparison, though the fit is not so good as for Au, most probably owing to surface imperfections of Ag nanospheres and their possible reshaping. For the Au nanoparticles, the coincidence of the peak positions, their widths, and the evolution with increase of the nanosphere radius is visible on comparison of Fig. 5.6 with Fig. 5.8.

From the experimentally collected extinction peaks presented in Fig. 5.8 (left) one can extract their centre positions with respect to the nanosphere radius a. These points perfectly fit the resonance self-frequencies defined by the exact solution of Eq. (5.23) obtained by including the Lorentz friction. This coincidence of the exact solution with the experimental data points is visualized in Fig. 5.7 (left). In this figure the approximate perturbative curve is also plotted for comparison. The increasing discrepancy between the approximate perturbative solution and the experimental data for larger nanosphere radii is clearly visible, in contrast with the satisfactory agreement with the exact solution.

Better fitting is obtained after inclusion of the skin effect renormalization for the Lorentz friction term by slightly reducing the number of electrons accessed by the

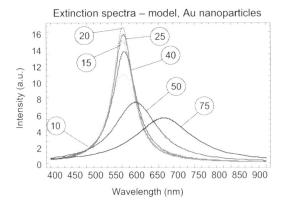

Figure 5.6 Extinction spectra for nanoparticles of Au calculated as Lorentzian functions with centre positions and widths given by the resonance frequencies and damping rates of Eq. (5.24), for nanosphere radii as indicated. The similarity to the attenuation curves obtained by measurements [77] of extinction spectra for light passing through a small glass container filled with a water-colloidal solution of metallic nanospheres with various radii and with the same volume density of metallic components is noticeable; see the left-hand panel in Fig. 5.8.

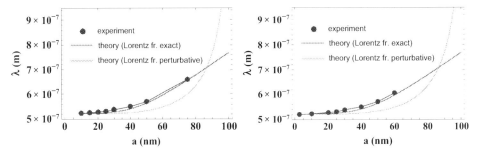

Figure 5.7 (Left) For comparison, resonance self-energies (expressed in terms of wavelength) for radiatively damped surface plasmons and experimental data from extinction measurements (data after [77]) for Au nanospheres in colloidal-water solution with radii 10, 15, 20, 25, 30, 40, 50, 75 nm: points, experiment; grey line, perturbative approximation; black line, exact inclusion of the Lorentz friction ($\varepsilon = 2.6$ and ω_p renormalized by a factor 0.78, the same as in Fig. 5.6). (Right) A similar comparison for the surface plasmon resonances in colloidal Au particles of different radii (experimental data for radii 2.6, 10, 20, 25, 30, 40, 50, 60 nm, after Fig. 12.17 in [17]).

incident light wave to those in an activation layer of thickness h, i.e., by the factor $(a^3 - (a-h)^3)/a^3$, where $h < a$, and by the accompanying reduction in plasmon frequency due to the lowering of the effective density. For the fittings presented in Fig. 5.7 we assumed a realistic skin-effect layer thickness with corresponding reduction in the plasmon frequency.

5.3 Comparison with Data and Simplified Mie Approach

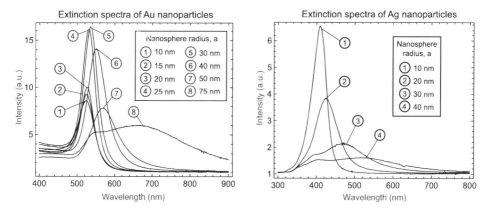

Figure 5.8 Attenuation curves for nanoparticles of Au (left) and Ag (right) obtained by measurement of extinction spectra for light passing through a water-colloidal solution of metallic nanospheres with various radii (after [77]); the deformation of the regular Lorentzian extinction feature shape is noticeable at $a = 75$ nm (Au) and at $a = 30, 40$ nm (Ag).

The size dependence of the redshift of the surface plasmon resonances observed experimentally [82, 22, 77] can also be interpreted using the Mie approach [17, 83, 84]. To explain the cross-over in the size dependence of the experimentally observed redshift of the plasmon resonance with increasing nanosphere radius, within the Mie theory two regimes are considered [83]: an intrinsic size effect (for $a < 20$ nm, for Au) and an extrinsic size effect (for $a > 20$ nm). The intrinsic size effect is associated with the dipole approximation of the Mie response, which is justified for nanosphere radii considerably smaller than the surface plasmon resonance wavelength. To account for absorption in the Mie approach, the dielectric function of the metallic nanosphere is accounted for phenomenologically. For relatively small particles, as in the case of the intrinsic size effect, this dielectric function includes electron scattering (as given by Eq. (5.13)), which results in a $\sim 1/a$ redshift of the dipole Mie resonance. Nevertheless, for larger nanospheres (with a greater than about 20 nm, for Au), for which it is observed experimentally that the resonance redshift rises with increasing a, the second mechanism is suggested in the Mie theory and referred to as the extrinsic size effect [83, 22]. The extrinsic regime in the Mie theory amounts to the inclusion of multipole mixing in the electromagnetic response, which leads to a related shift of the resonance energy with growth in radius. To obtain agreement with the experimentally observed size effect in the redshift of the plasmon resonance for larger nanospheres, the radiation correction to the dielectric function is introduced in the form of a model damping term proportional to the number of electrons and thus $\sim a^3$ [84]. This overestimates, however, the radiative damping and in the extrinsic limit for the size effect in the Mie theory this too-high redshift is reduced by inclusion of the multipole components in the Mie response; this lowers the redshift for larger nanosphere radii. This lowering of the redshift is due to the increasing contribution of multipole plasmon excitations to the overall Mie response when the radius grows beyond the dipole approximation limit and to the related increase in the resonance energy [17], mitigating the redshift. A similar growth in the multipole surface plasmon resonance frequency is also observed in the framework of the RPA approach [23]; cf. Eqs. (5.11) and (5.12), where the multipole-mode frequencies are $\omega_{0l} = \omega_p \sqrt{l/(2l+1)}$ and they increase with l.

To explain the Lorentz-friction-induced corrections, let us first address the mutual links between the Mie theory and the microscopic RPA method. Note from [22] that the major drawback of the Mie theory and its modifications is that the underlying electron relaxation mechanisms responsible for the absorption of incident light are all included in the macroscopic material dielectric function, which does not distinguish between several possible energy dissipation processes. A microscopic picture for the plasmon absorption is therefore lacking within the Mie theory. It means that the Mie theory, which is a classical solution of the Maxwell equations for an incident planar electromagnetic wave with imposed spherical boundary conditions, does not include the internal electron dynamics in the metal except in the phenomenologically modelled dielectric function. The electron scattering processes, i.e., the dissipation due to electron–electron scattering, electron–phonon collisions, electron scattering from admixtures or crystal imperfections, and the nanoparticle boundary scattering, can be in principle calculated using the many-body framework typical for solids (such as the collision term in the Boltzmann equation or more generally by the Green function approach). All these contributions lead to the scattering time rate given by Eq. (5.13). This dissipation-channel time rate is used also in the Mie theory for small nanoparticle radii, for which in experiments a redshift proportional to $1/a$ is observed (note that Landau damping and the redshift scale similarly owing to spill-out, which is important in ultrasmall nanoparticles (1–3 nm) [19] but negligible for radii of order 10 nm). For this range of radii ($\sim$ 10 nm for Au) the dipole approximation holds, and this simplifies the Mie-type response. Thus, the inclusion of the scattering using Eq. (5.13) gives reasonable agreement with experimental observations both in the Mie theory and in other approaches, including RPA. Within the Mie approach this is conveyed by the intrinsic size effect. Nevertheless, for larger radii (greater than about 20 nm for Au) the experimentally observed resonance redshift is not proportional to $1/a$ but, rather, is proportional to a or to some power of a, e.g., a^3 (for a not too large). As an intrinsic mechanism including only electron scattering does not offer such a behaviour, in the Mie response radiation losses are also taken into account [84] and simultaneously the mixing of multipole components is proposed. One can thus explain the experimentally observed inverted redshift because it rises with a, though irregularly for larger nanospheres as it is due to an increase in the radiation-induced damping. This is reduced, however, by multipole resonances which are shifted with l-multipole mode number towards higher energies. Nevertheless, the frontier between the dipole and multipole limits is not sharply defined and, rather, can be accommodated to the experimental data. The reason is that the resonance surface plasmon wavelength is around 500 nm (for Au), and even shifts with growth in radius to around 650 nm. Thus radii of 20 or 30 nm do not differ significantly on the wavelength scale and indeed the important multipole contribution occurs rather for larger nanospheres, at approximately $a > 60$ nm, as $2a/650 = 0.18$ for $a = 60$ nm is much lower than one-quarter of a wavelength and thus the electric field of an incident electromagnetic wave is still sufficiently homogeneous over the whole nanosphere; this justifies the dipole approximation.

The Lorentz-friction-induced radiation losses, which, for radii $a < 10$ nm (for Au), are much lower than the scattering term given by Eq. (5.13), for $a > 12$ nm strongly contribute to the plasmon energy dissipation (for Au in vacuum and for radii that are a little larger in the case of water surroundings), as visualized in Fig. 5.3. At these radiis the inverse time for dissipation caused by the radiation of energy (Lorentz friction) is similar to the inverse time for dissipation caused by scattering; indeed, if one estimates it from Eq. (5.13) for $a = 12$ nm, one obtains 7×10^{13}/s, and the second term in Eq. (5.22), the Lorentz friction term, is of the same magnitude. For larger a the Lorentz friction losses quickly dominate the plasmon damping, as depicted in Fig. 5.3. Thus for nanospheres with radii of the order of 5–10 nm (and even up to 15 nm, see Fig. 5.3) the Mie approach, neglecting the Lorentz friction losses, gives a reasonable intrinsic size effect. This is, however, not so when we are well past the cross-over point and arrive at $a \sim 15$–25 nm or more. For such radii, radiation losses dominate plasmon damping by a factor greater than 1 (it is about 5 at $a = 20$ nm). This was not included in the conventional Mie theory for the intrinsic size effect. Thus, the radiative damping of plasmons (i.e., the Lorentz friction) must be included (for $a > 12$ nm, at least) in the modelled (even by a simplified Lorentzian) dielectric function of the metallic nanosphere. This will give the required redshift in the corrected intrinsic-size-effect regime even using the dipole approximation only. The irregular size effect of plasmon damping caused by the Lorenz friction, if accounted for accurately, explains well the experimentally observed irregular (i.e., not proportional to a^3) size effect for the redshift. This Lorentz-friction-induced correction mixes, however, with the extrinsic size effect due to the multipole contributions, but it does this for radii $a > 60$ nm (for Au), significantly exceeding the previously suggested limit of 20 nm [83, 22, 84]. The quadrupole contribution, and the higher multipoles at larger radii, result in the deformation and further broadening of the extinction features, no longer allowing a Lorentzian form (the higher-energy quadrupole secondary broad peak occurs first at a smaller wavelength in association with the dipole peak, which is broadened and red shifted by the Lorentz friction). In experiments it is visible for Au at 75 nm and Ag at 30–40 nm [77, 82]. Remarkably, for Ag this deformation of the extinction features occurs at smaller radii than for Au [77, 82], see Fig. 5.8; this is probably due to the lower resonance wavelength for Ag nanospheres (about 400 nm) and moreover to scattering by the nanosphere facets, charges, or molecular contamination [82]. Reference [83] raises the question of the role of stabilizing molecules attached to the surface of nanoparticles to prevent them from aggregation or precipitation in solution and also other contaminations of the nanoparticle surface which could affect the scattering and absorption of light in an extinction experiment with colloidal solutions of the more chemically reactive metals (such as Ag, in comparison to Au).

By using an accurate treatment for the Lorentz friction term it was shown that the $\sim a^3$ growth of the radiation-induced damping of plasmons accounted for phenomenologically in the Mie theory quickly saturates, which is quite reasonable and, what is more important, it agrees well with the experimental observations; it is clear that cubic growth cannot continue and must saturate and, actually, the curve even starts to drop for larger a, as shown in Figs. 5.4 and 5.5. Thus, inclusion of the exact form of the Lorentz friction damping rate, as given by Eqs. (5.24), in the modelled dielectric function for a nanosystem (needed in the Mie theory and especially important in the intrinsic-size-effect regime) is a significant correction to the Mie approach and contributes to the interpretation of light extinction measurements for Au nanospheres with radii in the region of 10–80 nm. Thus the intrinsic regime with the inclusion of Lorentz friction can be extended to larger radii (up to about 60 nm for Au), where the extrinsic multipole effect starts to contribute with gradually growing intensity for higher radii; however, up to 60 nm the extrinsic effect mixes rather weakly with the effects of saturation and lowering of the damping rate due to Lorentz friction.

Multipole corrections cannot be avoided for radii $a > 60$ nm, as is evident from the deformation of the Lorentzian shape of the extinction features, accompanied by their significant broadening, for Au observed at $a = 75$ nm [77, 82]. The accurate microscopic analysis of the Lorentz friction carried out within the RPA approach supports the modelling of the dielectric function needed in the Mie theory and thus can be utilized to prolong the intrinsic size effect in the Mie approach beyond the 20 nm size limit, to larger radii for which the dipole approximation still holds (up to approximately 60 nm for Au). One should note however, that since the dielectric function simultaneously represents all energy dissipation channels in a combined manner, details of the size effect caused by the exact form of the Lorentz friction (beyond the scope of the Mie theory at present [84]) must be included in the modelled dielectric function. Thus, the specific size dependence of the damping time rate for plasmons due to Lorentz friction, derived from an exact solution of the relevant RPA equation for the electron density dynamics in nanospheres (as depicted in Figs. 5.4 and 5.5), may be employed to explain the observed irregular size effect of plasmon resonance in the radius window 10−60 nm for Au.

We can conclude that the damping of surface plasmons in large metallic nanospheres is overwhelmed by the radiation energy losses, which strongly exceed the electron scattering attenuation of plasmons for nanosphere radii larger than a critical value (approximately 12 nm for Au in vacuum). The radiation losses can be expressed by the so-called Lorentz friction, which hampers plasmon oscillations in a metallic nanoparticle. The related damping rate is proportional initially to the volume of the nanosphere; this results from the perturbative approach to the Lorentz friction contribution to plasmon damping. This rapid increase of the radiation losses saturates, however, for larger nanospheres and then slowly drops down. The deviation of the perturbative approximation with respect to the exact Lorentz friction term grows quickly for $a > 30$ nm (for Au in vacuum) and means that one has to dismiss the harmonic damped oscillator approximation as a model of plasmon damping. The solution of the dynamical equation for the local electron density including the exact form of the Lorentz friction determines both the resonance frequency of plasmons and their damping beyond the damped harmonic oscillator model. The related irregular size-dependent redshift of the plasmon dipole resonance agrees very well with the experimental data for metallic nanospheres with large radii (10−75 nm for Au).

5.4 Numerical Modelling of Plasmon Resonances in Metallic Nanoparticles

The ability to manipulate light at a sub wavelength scale is attributed to high-energy collective excitations in metals called plasmons. The properties of these excitations in metallic nanoparticles depend on their size and shape, on the surrounding medium, and on the light frequency [85–87]. The unique properties of plasmons in metallic nanostructures have found applications in photonics, optoelectronics, and photovoltaics [2, 88, 89, 3].

The classical solution of the metal–insulator Fresnel problem with spherical boundaries was found by Mie [18], and was soon generalized to the spheroidal case [90]. Mie theory utilizes the decomposition of a planar electromagnetic wave in a basis of spherical functions, as is appropriate for the spherical boundary conditions imposed by the geometry of small metallic particles that are scattering and absorbing incident light. This problem can be solved analytically within classical electrodynamics, but the metal characterization must be made phenomenologically in the form of a model dielectric function defining the metallic material. The assumed dielectric function is modelled from experimental data for bulk samples, and the strong size dependence at the nanoscale is typically neglected. The aim was to improve the predefined dielectric function by utilizing the knowledge of plasmons in metallic nanoparticles gained from the quantum RPA method, as presented in previous sections. This approach allows for precise modelling of the dielectric function for a nanoparticle, but only of spherical shape. For an analysis of plasmons in nanoparticles of an arbitrary shape, numerical methods must be employed. For the solution of such problems many methods have been developed based on the finite element approximation for the solution of the Maxwell–Fresnel differential equations, including, e.g., the finite difference time domain method [91, 92] and the boundary element method [93]. The main idea of such numerical calculations of the electromagnetic field distribution is a simulation of the space–time field dynamics on a discrete predefined grid and the achievement of convergence when the scale of the grid is shortened.

There have been reported extensive numerical studies of the optical properties of metallic nanoparticles in a variety of particle shapes [29, 84, 95, 95]. Also, shell-type particles [94] including those covered with a dielectric [96] have been studied. Easy access in numerical simulations to the variation of both the shape and size of the metallic elements allows for the recognition of plasmon resonances in the case when multipole modes contribute to the electromagnetic response [29, 84].

Nevertheless, all numerical simulations of the electromagnetic response of metallic nanostructures are limited by a serious drawback relating to the predefined dielectric function of a metal component, which is a prerequisite for a numerical procedure. This dielectric function is usually taken from experiment, but in bulk or thin-film geometries, and is much different from that for a nanostructure, which has, however, not been measured. The most significant error is related to the neglect of the strong plasmon damping in nanostructures, which is not present in the bulk substance. Simplified phenomenological trials to model the plasmon damping in nanoparticles are also insufficient. As shown above, the perturbation approach, resulting in an a^3 growth in the plasmon damping is, however, not justified for radii $a > 30$ nm (for Au in vacuum) [26]. To enhance the accuracy an exact formula for the plasmon damping must to be utilized, which gives the proper behaviour of the Lorentz friction in metallic nanoparticles. Instead of the $\sim a^3$ rapid increase, saturation of damping growth occurs, as depicted in Fig. 5.9.

The perturbation approach may be used only for particles with radii lower than about 30 nm (Au); for larger nanoparticles the exact solution of the dynamical equation for plasmons differs significantly. This has a pronounced consequence that plasmon oscillations in larger nanoparticles are not of the harmonic type. In particular, the relation between the damping rate and the frequency shift, being of damped oscillator-type form,

$$\omega'_1 = \sqrt{\omega_1^2 - \frac{1}{\tau^2}}, \qquad (5.26)$$

5.4 Numerical Modelling of Plasmon Resonances in Metallic Nanoparticles

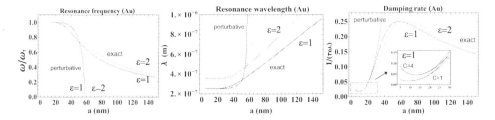

Figure 5.9 The real (left-hand panel) and imaginary parts (right-hand panel) of the exact oscillating solution given by Eq. (5.27) versus nanosphere radius a, for two distinct dielectric surrounding media (vacuum and water). The resonance frequency expressed as the resonance wavelength (central panel). The lines corresponding to the perturbative solutions are presented for comparison; in the inset in the right-hand panel the effect of changes in the constant C in Eq. (5.13) is illustrated.

where ω_1' is the shifted resonance frequency, does not hold for larger nanoparticles. The exact solutions for the damping rate and the self-frequency including Lorentz friction are as follows [26] (the exponents of the solutions depend on $e^{i\Omega t}$ for the self-modes of Eq. (5.23) (cf. Eq. (5.24)), in apparent disagreement with Eq. (5.26):

$$\Omega_1 = -\frac{i}{3l} - \frac{i 2^{1/3}(1 + 6lq)}{3l\mathcal{A}} - \frac{i\mathcal{A}}{2^{1/3}3l} = i\alpha \in \text{Im},$$

$$\Omega_2 = -\frac{i}{3l} + \frac{i(1 + i\sqrt{3})(1 + 6lq)}{2^{2/3}3l\mathcal{A}} + \frac{i(1 - i\sqrt{3})\mathcal{A}}{2^{1/3}6l} = \omega + i\frac{1}{\tau}, \qquad (5.27)$$

$$\Omega_3 = -\omega + i\frac{1}{\tau} = -\Omega_2^*,$$

where $\mathcal{A} = \left(\mathcal{B} + \sqrt{4(-1 - 6lq)^3 + \mathcal{B}^2}\right)^{1/3}$, $q = 1/\tau_0\omega_1$, $\mathcal{B} = 2 + 27l^2 + 18lq$, and $l = \frac{2}{3\sqrt{\varepsilon}}\left(\frac{a\omega_p}{c\sqrt{3}}\right)^3$. To better visualize the difference between the exact and perturbation solutions one can illustrate the functions for ω and $1/\tau$ (in dimensionless units, i.e., divided by ω_1) in Fig. 5.9 versus the nanosphere radius a. The Lorentz friction terms representing the radiative losses of surface dipole plasmons and the self-frequency of these plasmons are complicated functions with respect to the particle radius, as given by Eq. (5.27) (expressed by imaginary and real parts of Ω, correspondingly). A perturbative-type approximation of the plasmon damping by a simple $\sim a^3$ dependence is erroneous outside a relatively narrow size window, e.g., 12–30 nm for Au in vacuum [27, 26].

The numerical study of plasmon resonances in metallic nanoparticles normally utilizes the finite element method to solve the Maxwell–Fresnel equations with the dielectric function for the metal assumed as a prerequisite. Such an approach is realized by the commercial system COMSOL. Alternatively, the Mie-type approach is very popular, being numerically quick as it is based on a semi-analytical formulation [18, 29, 84, 17]. The Mie approach utilizes an analytical solution of the Maxwell equations for an incident planar wave scattered and absorbed by a metallic sphere including all multipole terms, which is an advantage of the method. The disadvantage of this approach consists in the phenomenological assumption of the dielectric function for a metallic nanoparticle, usually modelled upon the Drude form, typically using experimental data for the bulk material. None of the microscopic electron dynamics details are directly described in this way in which it is assumed that all the electron dynamics is included in the dielectric function taken from experiment. The problem is, however, that this dielectric function is not only dependent on the material but is also strongly sensitive to the nanoparticle size and shape, which cannot be experimentally expressed in a single factor, as is needed for the Mie-type calculus. Therefore, the modelling of the electron liquid response, as presented previously, with the quantum RPA method, may serve as an important supplementation to the numerical Mie algorithm. The same holds for other methods of solution of the Maxwell–Fresnel equations utilizing the finite element method in both the space and time domains, such as are implemented in the numerical system COMSOL. In the case of spherical geometry COMSOL re-derives the analytical Mie solution for the scattering and absorption cross sections. The COMSOL system is especially convenient for the study of nanoparticle geometries other than spherical. Practically any shape of nanoparticle may be assumed without any difference in efficiency of the finite element numerical procedure. The drawback is, however, the same here as in the Mie-type approach – the complete classical electrodynamical model has no access to specific plasmon behaviour at the nanoscale. The predefined dielectric function used in COMSOL is also of Drude form, with parameters taken from experiment for the bulk material, thus neglecting the strong size dependence of the plasmon resonance and damping at the nanoscale.

To apply the finite element method practically for a solution of the Maxwell–Fresnel equation for a metallic nanoparticle, one can use the commercial software package COMSOL Multiphysics 5.0, including the Wave Optics module [97]. The usefulness of COMSOL in plasmonic structure modelling has been verified by comparison with other calculation methods such as the Mie theory or the multiple expansion method [98]. The COMSOL system consumes considerable information resources; however, it is quick and efficient if applied to relatively simple three-dimensional models only [94] or more complicated two- or one-dimensional models (including, e.g., plasmon–polariton propagation along nanoparticle chains [99]). The most important advantage of the COMSOL simulation is its flexibility with regard to the shape and material composition of the analysed system. The drawback of such plasmon simulation is, as in the Mie approach, the predefined phenomenological dielectric function in the Drude–Lorentz form [17], which is not specific to the nanoscale corrections. Often a linear interpolation of the experimental results

5.4 Numerical Modelling of Plasmon Resonances in Metallic Nanoparticles

from bulk or thin-film measurements is used [100] with some phenomenological size corrections for small particles ($a < 20$ nm) [25], in the following form:

$$\epsilon(\omega, a) = \epsilon(\omega, a = \infty) + \omega_p^2 \left[\frac{1}{\omega^2 + \gamma_\infty^2} - \frac{1}{\omega^2 + \gamma(a)^2} \right]$$
$$+ i \frac{\omega_p^2}{\omega} \left[\frac{\gamma(a)}{\omega^2 + \gamma(a)^2} - \frac{\gamma_\infty}{\omega^2 + \gamma_\infty^2} \right], \tag{5.28}$$

where $\gamma(a) = \gamma_\infty + Av_F/a$ and $\gamma_\infty = v_F/\lambda_B$. We propose to improve this approach via inclusion of the exact attenuation term as given by Eqs. (5.27), obtained by the RPA method presented in previous chapters.

For an illustrative model we will consider a system of domains: a metallic nanosphere with radius a with its centre at the origin, and with dielectric medium surrounding the particle, conventionally assumed as perfectly matched layers (PMLs) which quench back-reflections from boundaries. The restriction to a spherical symmetry allows us to reduce the model to one-quarter of the sphere, but the generalization to an arbitrary shape is obvious. The nanoparticle is assumed to be illuminated by a plane electromagnetic wave polarized in the z direction and propagating along the x direction. The thickness of the PMLs of the surrounding medium is accommodated to the incident radiation wavelength. The discretization essential for COMSOL simulation is achieved by a built-in free meshing algorithm, which selects a suitable space division into tetrahedral finite elements in the domain of the nanoparticle and the surrounding medium. The PMLs are divided into five spherical layers by using a mesh with spherical symmetry. The maximum element size inside the metal is chosen to suit the nanosphere radius, as $a/5$ for $a < 60$ nm and 13.5 nm for $a > 60$ nm. For the surrounding medium, the maximum element size is chosen as $\lambda/6$. The extinction cross section is estimated via the sum of the scattering and the absorption, given by the formulae

$$Q_{sc} = \frac{4}{\sqrt{\epsilon_0/\mu_0} E_0^2} \int_\Sigma \mathbf{S} \cdot \hat{\mathbf{n}} d\Omega, \tag{5.29}$$

$$Q_{abs} = \frac{4}{\sqrt{\epsilon_0/\mu_0} E_0^2} \int_V U_v dV, \tag{5.30}$$

where Σ is the boundary between the PMLs and surrounding medium, $\hat{\mathbf{n}}$ is a vector normal to the Σ surface, E_0 is the amplitude of the electric component of the incident electromagnetic wave, $\mathbf{S}$ is the Poynting vector, U_v indicates ohmic heat losses in the nanoparticle and V is the nanoparticle volume.

The surface plasmon resonance simulated by the commercial numerical COMSOL system[3] includes simultaneously all multipole components of the

---
[3] The COMSOL system utilizes a numerical solution of the Maxwell–Fresnel differential equations by the method of finite elements; it uses as prerequisites the material parameters, including the predefined dielectric

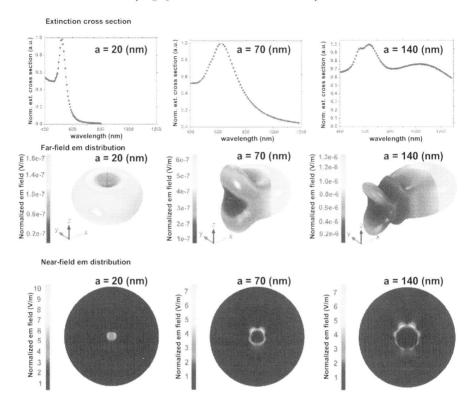

Figure 5.10 The extinction spectra, far-field scattering radiation, and near-field electric field distribution for three Au nanospheres with radii equal to 20 nm, 70 nm, and 140 nm. The field distributions correspond to illumination by an electromagnetic wave with wavelength 400 nm. The emergence of higher harmonics (multipole) modes with increasing particle radius (density maps in colours are given in [28]) can be seen.

electromagnetic response. It is noticeable that, for particles larger than ~ 0.2 of the wavelength of the incident light, the quadrupole mode contributes. Figure 5.10 shows for comparison the optical response of three Au nanospheres with radii 20 nm, 70 nm, and 140 nm. The particles are embedded in water and illuminated with an electromagnetic wave with wavelength 400–1200 nm. The influence of the nanosphere radius on the optical response of the system is visible in the extinction spectra (top), the far-field scattering radiation distribution (middle), and the near-field electric field distribution (bottom). In the case of a relatively small

---

function of a metal (in the form of a Drude model [17]); the advantage of the COMSOL system consists in its ability to solve arbitrary asymmetric nanoparticle-shape problems including various types of deposition of a metallic nanoparticle on arbitrary substrates; the disadvantage is the neglect of quantum plasmon effects, important in the nanoscale.

5.4 Numerical Modelling of Plasmon Resonances in Metallic Nanoparticles

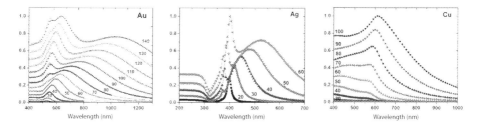

Figure 5.11 Normalized extinction spectra calculated for Au, Ag, and Cu nanospheres surrounded by water ($\varepsilon = 1.77$) with COMSOL (points) and spectra obtained with the Mie model (solid lines) for various size of nanospheres of these metals: Au (30–140 nm), Ag (20–60 nm), Cu (20–100 nm). The Mie–COMSOL agreement corresponds to the same dielectric function being assumed as the prerequisite for both fully equivalent solutions of the Maxwell–Fresnel equations.

particle with radius 20 nm and illuminated with a 400 nm wavelength planar wave ($a/\lambda = 0.05$), the dipole approximation is justified and only a dipole-type response is visible. However, for 70 nm ($a/\lambda = 0.175$) the quadrupole mode occurs, whereas for particles with radius 140 nm ($a/\lambda = 0.35$) the octupole mode also contributes.

Some extinction spectra for Au, Ag, and Cu nanospheres obtained by a COMSOL simulation are presented in Fig. 5.11. From the extinction spectra we can determined the surface plasmon resonance wavelength as a function of the nanosphere radius. The resonance wavelength redshift with increase in the nanosphere radius is visible. The COMSOL resonances were compared with experimental data taken from the extinction spectra measured in colloidal solutions of metallic nanoparticles with various sizes. The resonances were compared also with the theoretical microscopic quantum predictions within the RPA approach including the exact Lorentz friction. This comparison for Au (presented in Fig. 5.12) shows good agreement of the exact RPA calculations with the experimental data and the numerical estimations lifted by the exact Lorentz friction. By virtue of the agreement of the simulation with the experimental data one can conclude that the observed redshift in plasmon resonance is caused mostly by the radiative damping of plasmon oscillations, i.e., by the Lorentz friction.

For plasmons in metallic nanoparticles as described in the RPA approach one should take into account also the skin effect, due to which effectively only electrons from an outer layer of thickness h of the nanoparticle are excited by external radiation. These electrons contribute also to the Lorentz friction term, which describes the plasmon damping; hence this term must then be reduced by the factor $(a^3 - (a-h)^3)/a^3$ when $h < a$.

Note that in the case of Ag nanoparticles the experimental data are shifted to longer wavelengths in comparison to both the numerical and theoretical predictions,

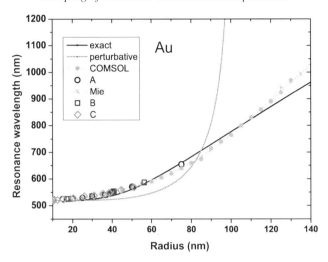

Figure 5.12 The COMSOL (grey dotted line with points), Mie (grey dotted line with stars), and RPA predictions (black line shows exact RPA calculations of Lorentz friction, grey solid line shows approximate Lorentz friction calculation) and the experimental data: A, [77]; B, [101]; C, [102]. The plots are for the plasmon resonance redshift as a function of nanosphere radius for Au nanospheres. For small radii (10–20 nm) the change in the resonance wavelength is relatively small; then we observe an increase in the redshift 'speed' (for 20–60 nm), and finally stabilization (> 60 nm).

as is visualized in Fig. 5.13. This discrepancy is probably caused by the silver nanoparticle synthesis method, which could leave on the silver surface a residual amount of organic surfactant (citrate) or other optically active molecules utilized in the synthesis. This effect can be illustrated by COMSOL simulation via the addition of a dielectric coating to the model of the Ag nanoparticle. In Fig. 5.14 one can observe a relatively large shift in the plasmon resonance due to the additional dielectric shell (with refractive index $n = 1.5$ in the figure). Also, the dependence on the refractive index of the dielectric shell with thickness 10 nm is visualized. Both an increase in the shell thickness and an increase in the refractive index of the shell material cause a redshift of the resonance wavelength. This shift in the first case is relatively small, from 407 nm at $d = 4$ nm to 419 nm at $d = 20$ nm in comparison to 423 nm for an infinite layer thickness. The optimal fit between the theory and experimental data for Ag is obtained using a refractive index value 1.5 for the dielectric outer layer.

The behaviour described above at the infinite limit for the thickness of the outer dielectric layer shows that the plasmon resonance wavelength can be strongly affected by the surroundings of the nanoparticle (which can be utilized in, e.g., refractive index detectors). In the case of Ag (see Fig. 5.15) the resonance wavelength could be shifted from 400 nm in air to 1000 nm in a medium with

5.4 Numerical Modelling of Plasmon Resonances in Metallic Nanoparticles 91

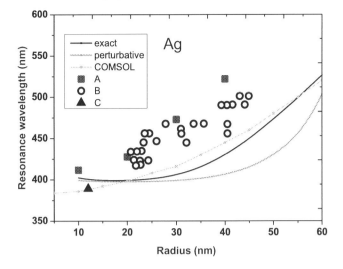

Figure 5.13 For comparison, COMSOL predictions (grey line with small points) and RPA predictions (black line, exact inclusion of the Lorentz friction; solid grey line for approximate Lorentz friction) and experimental data (A, [77]; B, [103]; C, [85] for the plasmon resonance redshift as a function of nanosphere radius for Ag nanospheres.

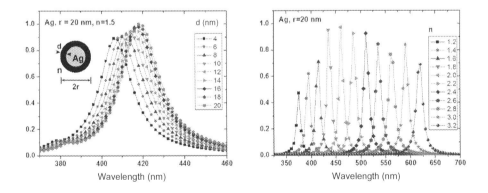

Figure 5.14 Normalized extinction cross sections for Ag nanospheres covered with dielectric shells of various thicknesses d with fixed refractive index $n = 1.5$ (left) and with fixed thickness $d = 20$ nm and different refractive indices n (right).

refractive index equal to 4. One can also notice that the higher multipole resonances accompanying the dipole resonance are proportionally shifted (see Fig. 5.15).

In COMSOL simulation one can easy consider the case of elongated nanoparticles such as spheroids or rods. In such a geometry with cylindrical symmetry, two different surface plasmon modes, longitudinal (oscillations along the long axis) and transverse (oscillations along the short axis), exist. These modes can be

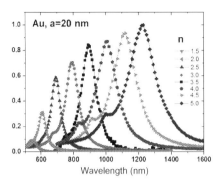
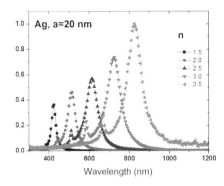

Figure 5.15 Normalized extinction spectra for Au (left) and Ag (right) nanospheres for dielectric surroundings with various refractive indices. The resonance peak shifts towards longer wavelengths with increasing refractive index of the surrounding medium.

observed, when illuminating the nanoparticle with light of different polarizations. In the case of a spheroid-shaped particle, the Ganss development of the Mie theory in the dipole limit is analytically available [25], allowing the following condition for the frequencies of both modes:

$$\text{Re}\,\epsilon(\omega_{res}) = \frac{\epsilon_m(L_i - 1)}{L_i}, \qquad (5.31)$$

where $\epsilon(\omega_{res})$ is the value of the dielectric function of the metal at the resonance frequency, ϵ_m is the dielectric permittivity of the surrounding medium, L_i are the so-called geometrical depolarization factors determined by the particle geometry,

$$L_x = \frac{1-e^2}{e^2}\left(-1 + \frac{1}{2e}\ln\frac{1+e}{1-e}\right), \qquad L_{y,z} = \frac{1 - L_x}{2}, \qquad (5.32)$$

where $e^2 = 1 - \chi^2$ (χ is the aspect ratio). In Fig. 5.18 we illustrate the extinction cross sections for a nano-rod with aspect ratio (length/width) equal to 3 for different polarizations of the incident light; these were calculated using COMSOL. The angle α in Fig. 5.16 varies from 0° to 90°, i.e., from an electric field perpendicular to the long rod axis to one parallel to the long rod axis. The related electric field distributions (on the sections along the long and short rod axes) are illustrated in Fig. 5.17. By varying the light polarization, two modes can be excited in a selective manner. The longitudinal one dominates. For a polarization angle larger than 30° (see Fig. 5.18) only one mode is visible. Additionally, in Figs. 5.18 and 5.19 the dependence of resonance wavelength on particle ratio is presented.

The Mie approach in its original version [18] gives an analytical formulation for the scattering and absorption cross sections in the case of spherical geometry and

5.4 *Numerical Modelling of Plasmon Resonances in Metallic Nanoparticles* 93

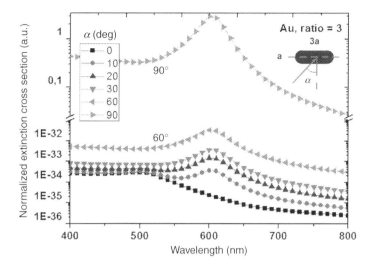

Figure 5.16 Extinction spectra for nano-rods for various incident light polarizations relative to the rod axis.

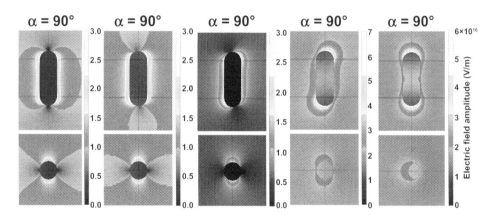

Figure 5.17 The electric field distribution near metallic nano-rods for different incident light polarizations (wavelength 600 nm) relative to the rod axis. One can observe two plasmon excitations: longitudinal, excited when the light is polarized parallel to the rod axis, and transverse, when the light is polarized perpendicularly to the rod axis (density maps are given in colour in [28]).

the easiest form for the dielectric function of the metallic nanoparticle [17]. It can be developed for ellipsoid symmetry also [17]. Because of this analytical formulation, it is much quicker to use in calculations than a numerical solution of the Maxwell–Fresnel equations. To meet the experimental data the Mie approach has been improved, mostly via phenomenological modelling of the dielectric function of the metallic particle. The original formulation [17] gives the Mie frequency as

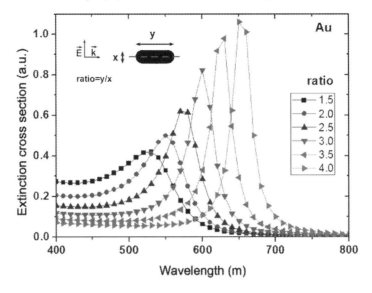

Figure 5.18 Extinction spectra for Au nanorods with various aspect ratios y/x. The incident light is polarized parallel to the rod axis.

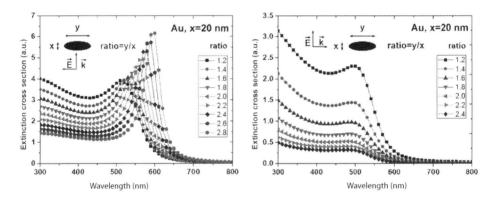

Figure 5.19 Extinction spectra for Au prolate spheroids with different aspect ratios. (Left) The incident light is polarized parallel to the larger particle axis; (right) the incident light is polarized perpendicularly to that axis.

independent of the sphere size, $\omega_{Mie} = \omega_p/\sqrt{3}$ in the simplest case, which does not agree with experimental observations. For ultrasmall nanoparticles some radius dependence can be introduced by the dependence of the volume plasmon frequency ω_p on the electron density n ($\omega_p \sim \sqrt{n}$) and by accounting for the dilution of electrons caused by spill-out. The resulting redshift of the Mie resonance agrees with observations for extremely small nanoparticles, where spill-out is important [19]. For larger nanospheres this correction disappears together with the spill-out contribution $\sim 1/a$. Nevertheless, for the range $a \in (5, 75)$ nm (for Au in vacuum) a strong size dependence of the surface plasmon resonance is observed that is clearly beyond the simplest version of the Mie approach. Moreover, at $a \sim 12$ nm (Au in vacuum) a cross-over in the size dependence of the resonance frequency is observed. To cope with this problem, some artificial corrections to the Mie approach have been applied. To explain this cross-over in the size dependence of the experimentally observed redshift of the plasmon resonance in metallic nanoparticles with growing nanosphere radius, two regimes of the Mie theory were introduced in [83]: an intrinsic size effect (for $a < 20$ nm for Au) and an extrinsic size effect (for $a > 20$ nm). The intrinsic size effect refers to the dipole approximation of the Mie response, when other multipole contributions are neglected. Simultaneously, the electron scattering energy dissipation is included in the dielectric function (with the damping rate defined by Eq. (5.13)), which gives a $\sim 1/a$ redshift of the dipole Mie resonance due to the $\sim 1/a$ contribution of scattering from the nanoparticle boundary. For larger nanospheres (approximately for $a > 20$ nm for Au), when experimentally the resonance redshift rises with increase in a, an extrinsic mechanism is introduced into the Mie theory [83, 22]. The latter mechanism corresponds to the inclusion of the multipole response and corrections for the pole of the dipole response component. To obtain agreement with the experimentally observed size effect in the redshift of the plasmon resonance in larger nanospheres, radiation corrections to the dielectric function have been introduced that are proportional to the number of electrons and thus $\sim a^3$ [84]. This overestimates, however, the radiative damping for radii > 30 nm (for Au).

This phenomenological modelling in the Mie approach gives the cross-over in the surface plasmon resonance from $1/a$ to a^3 using the dipole approximation only (by modifying the metal's dielectric function). The frontier between the dipole and multipole limits is not sharply defined and multipole corrections occur gradually with stronger intensity for larger nanoparticles. As we have demonstrated, the Lorentz-friction-induced radiation losses strongly contribute to plasmon energy dissipation for $a > 12$ nm (for Au in vacuum). Thus, for nanospheres with radii of order of 5–10 nm (and even up to 15 nm, see Fig. 5.3) the Mie approach gives a reasonabe intrinsic size effect neglecting the Lorentz friction losses. This does not hold, however, at $a \sim 15$–25 nm or at larger scales for the radius. For such radii radiation losses dominate the plasmon damping. This is not included in the conventional Mie theory for the intrinsic size effect. The accurate irregular size effect of plasmon damping caused by Lorentz friction (as presented previously, see formula (5.27)) explains well the experimentally observed irregular (i.e., not proportional to a^3) size effect for the redshift. Therefore the Mie approach must be modified by Eq. (5.27) instead of the incorrect a^3 contribution to the dielectric function. This strongly modifies the dipole contribution in the Mie response in consistency with observations up to $a < 60$ (for Au in vacuum) without any multipole contributions. The latter start to be important for radii $a > 60$ nm (Au) significantly exceeding the previously suggested limit of 20 nm as the onset of the extrinsic regime [83, 22, 84]. This agrees better with observations – the multipole contribution for $a > 60$ nm (Au) results in deformation and greater broadening of the extinction features, no longer allowing a Lorentzian form (the higher-energy quadrupole secondary broad peak occurs first at smaller wavelength in association with the dipole peak broadened and redshifted by the Lorentz friction). In the experimental results it is visible for Au at 75 nm [77, 82].

As previously discussed [27, 26], the $\sim a^3$ growth of the radiation-induced damping of plasmons, phenomenologically accounted for in the Mie theory, must quickly saturate. Thus, the inclusion of the exact form of the Lorentz friction damping rate (as given by Eq. (5.27)) in the modelled dielectric function for nanosystems (needed in the Mie theory and especially important in its intrinsic size effect regime) gives a significant correction of the Mie approach. The intrinsic Mie regime with inclusion of the Lorentz friction can be extended to larger radii (up to abuot 60 nm for Au), where the extrinsic multipole effect starts to contribute with gradually growing intensity for higher radii; however, up to 60 nm the extrinsic effect mixes rather weakly with the saturation and lowering of the damping rate due to Lorentz friction [27].

The multipole corrections cannot be avoided for radii $a > 60$ nm. The deformation of the Lorentzian shape describing the extinction features is accompanied by their significant broadening, for Au observed at $a = 75$ nm [77, 82]. However, in the radius window 10–60 nm (for Au) the dipole approximation is sufficient to explain the observed irregular size effect of plasmon resonance, provided the accurate Lorentz friction contribution (5.27) to dipole plasmon damping is included.

The analysis presented above agrees well with the experimental data for Au. As mentioned above, some underestimation for Ag of the resonance wavelength is probably connected with the modification of the surface of Ag nanoparticles caused by the chemical manufacturing method. Similar properties are confirmed and verified by a microscopic RPA approach to plasmons, using a COMSOL finite element numerical solution of Maxwell equations and application of the conventional Mie approach. The Mie and COMSOL methods must be, however, suitably lifted to incorporate the size effect at the nanoscale by appropriate modelling of the dielectric function. Agreement is clear, owing to the common geometry of the considered nanoparticles for any approach. It is important, however, that the form of the true dielectric function is greatly different from that of the bulk metal supported by experimental data for the bulk metal or thin films at best. Thus, it has been demonstrated that the corrections required at the nanoscale can be taken from the microscopic RPA model of plasmons, including plasmon damping due to the Lorentz friction, which is found to be strongly size-sensitive.

6
Plasmon Photovoltaic Effect

One of the most important applications of nano-plasmonics is the mediation of light capture in a semiconductor by surface plasmons in metallic nanoparticles deposited on the semiconductor substrate. The increase in efficiency of solar cells thus metallically modified on the nanoscale has been experimentally demonstrated. In this chapter we provide a microscopic quantum explanation of the plasmon-induced photovoltaic phenomenon. Using Fermi's golden rule the efficiency of the plasmon-mediated channel for solar energy transfer is derived in analytical form. A comparison with experiments is given. Moreover, an experiment demonstrating the depth range of the plasmonic photoeffect in the substrate semiconductor of a metallized solar cell is described.

Considerable interest in the so-called plasmonic photovoltaic effect has emerged within the past decade. This is a result of the experimental demonstration of a significant increase in efficiency of the photoeffect when the incident-photon energy transfer to electrons is mediated by surface plasmons in metallic nanoparticles deposited on the photoactive surface of the photodiode or semiconductor solar cell [32–36, 104–106]. The planar density of the metallic nanoparticle coverage that causes a strong effect is rather small – in the range 10^{8-10} per cm^2 for Au or Ag nanoparticles randomly distributed on the surface. It has been demonstrated that more dense coverings do not lead to strengthening of the effect but, in contrast, may even diminish the photovoltaic efficiency. The covering densities mentioned above and the typical sizes of nanoparticles, with radius $a \in (10, 50)$ nm, make these coverings inexpensive even though noble metals may be used (the best is Au, superior over Ag or Cu in lifting the efficiency of the photoeffect). The problem is, however, the technique of nanoparticle deposition in large scales and the durability of the metallic nanocoverings of solar cells in commercial use.

The discovery of the plasmon-mediated photoeffect is of great prospective importance for the development of a new generation of solar cells with enhanced efficiency, especially for improvement via the metallic nanomodifications of thin-film cells, and even organic cells and perovskite cells (the latter are especially interesting owing to their simple and cheap chemical methods of manufacture) [36, 2, 105, 107–112, 73, 74]. The feasibility of plasmonic improvements has been supported by the successful application of metallic nanocomponents to about a factor of ten increase in Raman spectrometer resolution via the concentration of local laser illumination near the curvature of metallic nanoconcentrators [113, 114]. A similar improvement of scanning tunnelling microscopy by the utilization of surface plasmons has been proposed [115]. Moreover, another phenomenon that is worthy of consideration is plasmon-assisted transport in metallic nanostructures, using hybridized collective modes of surface plasmons and photons, i.e., plasmon–polaritons [3, 116], with properties that are highly attractive for applications in optoelectronics [3, 4] with subdiffraction arrangements of light–plasmon circuits. In such circuits, light signals are converted into plasmon–polaritons with the same energy but with a wavelength one or two orders of magnitude lower [116, 39, 38, 78]. This direction of research meets also the rapidly developing, field of metamaterials.

As was demonstrated in Chapter 4, in finite nanometre-scale metallic systems new types of plasmon oscillation arise that do not occur in the bulk case. These new types of oscillation are called surface plasmons and correspond to a translational movement of the electron liquid with respect to the ion jellium, resulting in unbalanced-charge fluctuations at the surfaces of the finite nanoparticles. Oscillations that have an inhomogeneous compressional character along the system radius are referred to as volume plasmons in the finite system, in close analogy to plasmons in the bulk case. Remarkably, the energies of the surface plasmon modes are lower than the bulk plasmon energy $\hbar\omega_p$ whereas the energies of the volume plasmon modes in nanoparticles are larger than $\hbar\omega_p$, as was shown in Chapter 4.

Finite metallic systems have been widely investigated by a thorough quantum analysis, including the application of the shell model for a small number of electrons in analogy to multi-electron atoms, and by the development of numerical quasi-exact studies employing the Kohn–Scham approximation (which is efficient up to about 300 electrons only) [19–21, 71, 79]. For larger nanoparticles the random phase approximation (RPA) approach proves to be effective [21, 19, 71]. As a rule, in this treatment, the jellium model is utilized. In this model the ion dynamics are considered negligible because in metals the ions have a much greater inertia than the electron fluid owing to the much greater mass of the ions compared with the electron mass [19–21].

In bulk metals, the plasmon energy is $\hbar\omega_p = \sqrt{e^2 n/(\varepsilon_0 m)} \simeq 10$ eV, which is larger than the Fermi energy. This fact indicates that plasmons are not excitations near the Fermi surface and that they interact only weakly with the Landau quasiparticles (the electrons near the Fermi surface) because of the large energy incommensurability. The origin of such high-energy plasmon excitations is the

Coulomb interaction between the electrons and between the electrons and the jellium, described in the Pines–Bohm RPA theory of plasmons [16, 14, 15], as presented in Chapter 3.

Progress in the RPA formulation of plasmons in metallic nanospheres [23] has enabled the description of both volume and surface plasmons in spheres with radii ranging between approximately 5 nm and 100 nm especially interesting for Au, Ag or Cu particles, because at this particle-size range the surface plasmon resonances of these metals are in the visible part of the electromagnetic spectrum, conveniently overlapping with the sun's spectrum at the Earth's surface. In this size range of metallic nanoparticles, the quasiclassical RPA approach proves to be sufficiently accurate. For smaller clusters more precise quantum models are needed, especially for radii of 1–2 nm, for which the pronounced surface effects are of primary importance. One such effect is the so-called spill-out of the electron liquid beyond the jellium rim. The spill-out occurs on the scale of the Thomas–Fermi length [16] and gradually loses its significance as the sphere radius increases. Similarly, another quantum phenomenon, the Landau damping of plasmons as a result of their decay into high-energy particle–hole excitations far from the Fermi surface [80], is of decreasing significance as the radius increases. These quantum phenomena, although of importance for small metallic nanoparticles clusters, are negligible for radius $a > 5$ nm. In smaller nanoparticles, the spill-out effect is considered to be the reason for the redshift of the dipole surface-oscillation frequency (the Mie frequency) as a result of the decrease in the electron density caused by spill-out. For ultrasmall nanoparticles, this redshift property has been confirmed experimentally in K, Au, and Ag [71]. In larger metallic nanospheres a strong redshift of the plasmon resonance is also observed but is apparently caused by another mechanism because, for $a > 5$ nm, all surface-induced effects are negligible. The most important factor in larger metallic nanospheres, in the radius range $a \in (5, 100)$ nm, is the rapid enhancement of radiation losses of plasmon oscillations whereas radiation is rather insignificant in small metallic nanoparticles. The radiation-loss-induced damping of plasmons causes a redshift of the resonance, and this satisfactorily explains the plasmonic frequency shift in large metallic nanospheres [77], as was shown in Chapter 5.

Plasmon damping via the radiation phenomenon in large nanospheres explains why such metallic particles are so efficient in light-energy transfer. The large radiation effect for plasmons in metallic nanoparticles means that they also have large absorption. So they quickly gather energy from incident light photons. When metallic nanoparticles are deposited on the photoactive surfaces of a semiconductor (as in solar cells) the coupling of plasmons to the substrate semiconductor band electrons opens an especially efficient channel for energy transfer [32–34, 117–120].

Thus metallic nanospheres (or nanoparticles of other shapes [121]) can act as light converters because of their capability for high absorption and radiation of energy, which originates from the mutual mirror relationship between the two directions of quantum transitions [75].

The incident photon energy that is absorbed in surface plasmon oscillations in metallic nanoparticles can thus be transferred to a semiconductor substrate in a very efficient manner, highly exceeding the efficiency of the ordinary photoeffect. We can identify the origin of this efficiency enhancement as being due to two competing factors. The first is the local concentration of the electric field of the surface plasmon oscillations near the curvature of a metallic nanoparticle and, as a result, collecting the energy of many photons, increasing step by step the plasmon oscillation amplitude. The latter effect is proportional to a^3 (a is the nanosphere radius) and reflects the fact that all electrons in the nanosphere participate in the oscillations even though charge fluctuations occur only on the surface because of the screening by the jellium of translational oscillations in the internal volume. The second factor is the possibility of indirect interband electron transitions in the semiconductor substrate, which are allowed for a system with broken translational symmetry. This occurs for the dipole interactions of the surface plasmon oscillations in the local nanoscale system. The momentum conservation rule is not restrictive here, in contrast with the ordinary photoeffect. In the ordinary photoeffect the momentum (or quasi-momentum) of an electron undergoing a photon-induced transition between the valence and conduction bands does not change because of the extremely low momentum of photons that is required for consistency in energy with the forbidden gap in the semiconductor (because of the high velocity of light). This results in so-called vertical optical transitions, significantly reducing the efficiency of the conventional photoeffect. However, when there are mediating interactions with plasmons, all indirect interband excitations are allowed, which enhances the efficiency of this channel. It is clear that this effect favours nanospheres of smaller radius, for which the breaking of translation invariance is more important, whereas the effect weakens with increasing radius. This concept was originally formulated in [23] and was recently utilized in another publication [122]. It was also discussed in an earlier paper [109].

In the next section we summarize briefly the model constructed previously to describe plasmon oscillations in large metallic nanospheres in the framework of the RPA approach [23], in analogy to the Pines–Bohm RPA theory of plasmons in a bulk metal [16, 14]. We recall that an advantage of this approach [23, 77] is its fully analytical formulation, which enables its straightforward and controllable application to a description of plasmon damping; numerically, the problem is beyond the capability of even large simulations.

6.1 Semiclassical RPA Approach to Plasmons in Large Metallic Nanospheres

The RPA theory of plasmon excitations in metallic nanospheres [23] is effective and sufficiently accurate for large nanoparticles of radius $a \in (5, 100)$ nm. For such large nanoparticles, the spill-out of electrons beyond the jellium rim is negligible, which allows for the separation of the volume and surface types of charge oscillation. Otherwise, when the Thomas–Fermi length, i.e., the scale of the spill-out, is significant in comparison to the radius, the surface becomes fuzzy, which leads to the mixing of the surface and volume charge fluctuations. It has been proven via TDLDA numerical calculations employing the Kohn–Sham approach to multi-electron clusters [19, 71] that the separation of the volume and surface modes becomes clear for more than approximately 60 electrons, whereas for smaller nanoparticles the mixing of modes occurs in accordance with the shell model. The Thomas–Fermi length in a metal is of the order of the interparticle separation; therefore, for $a > 5$ nm, we can safely assume that the mixing of modes disappears, similarly to other surface-type effects [80]. Although spill-out was included in the original formulation of the RPA approach [23], in the following description we will completely neglect it because the Thomas–Fermi length is on the subnanometre scale, more than one order of magnitude smaller than the nanosphere radii considered.

For completeness we briefly summarize again the most important steps in the description of plasmons in nanoparticles. Similarly to the treatment in the RPA approach for a bulk metal as formulated by Pines and Bohm [14–16], the local electron density can be represented by the quasiclassical formula [16, 15],

$$\rho(\mathbf{r},t) = \left\langle \Psi_e(t) \left| \sum_j \delta(\mathbf{r} - \mathbf{r}_j) \right| \Psi_e(t) \right\rangle, \tag{6.1}$$

where the Dirac deltas quasiclassically define the positions of the electrons. The Fourier picture of Eq. (6.1) takes the convenient form

$$\tilde{\rho}(\mathbf{k},t) = \int \rho(\mathbf{r},t) e^{-i\mathbf{k}\cdot\mathbf{r}} d^3r = \langle \Psi_e(t) | \hat{\rho}(\mathbf{k}) | \Psi_e(t) \rangle, \tag{6.2}$$

where the operator $\hat{\rho}(\mathbf{k})$ is given by $\sum_j e^{-i\mathbf{k}\cdot\mathbf{r}_j}$. The dynamic equation for this local electron density operator can be determined using the Heisenberg equation in the form

$$\frac{d^2 \hat{\rho}(\mathbf{k})}{dt^2} = \frac{1}{(i\hbar)^2} \left[\left[\hat{\rho}(\mathbf{k}), \hat{H}_e \right], \hat{H}_e \right], \tag{6.3}$$

where the Hamiltonian H_e for the electrons in a metallic nanosphere based on the jellium model (cf. Eq. (4.8)) is given by

$$\hat{H}_e = \sum_{j=1}^{N_e} \left(-\frac{\hbar^2 \nabla_j^2}{2m}\right) - \frac{e^2}{(2\pi)^3} \int d^3k \tilde{n}_e(\mathbf{k}) \frac{2\pi}{k^2} \left(\hat{\rho}^+(\mathbf{k}) + \hat{\rho}(\mathbf{k})\right)$$
$$+ \frac{e^2}{(2\pi)^3} \int d^3k \frac{2\pi}{k^2} \left(\hat{\rho}^+(\mathbf{k})\hat{\rho}(\mathbf{k}) - N_e\right), \quad (6.4)$$

where $\tilde{n}_e(\mathbf{k}) = \int d^3 r n_e(\mathbf{r}) e^{-i\mathbf{k}\cdot\mathbf{r}}$ and $4\pi/k^2 = \int d^3 r r^{-1} 1 e^{-i\mathbf{k}\cdot\mathbf{r}}$: The jellium [19, 17, 123] consists of the background ion charge uniformly distributed over the sphere; $n_e(\mathbf{r}) = n_e \Theta(a - r)$, where $n_e = N_e/V$ and $n_e|e|$ is the average positive charge density, $V = 4\pi a^3/3$ is the sphere volume, and Θ is the Heaviside step function.

Equation (6.3), after averaging over the multiparticle wave function and, according to the RPA scheme, neglecting the multiplications of density fluctuations with different phases, ultimately attains the following form [23]:

$$\frac{\partial^2 \delta\tilde{\rho}(\mathbf{r},t)}{\partial t^2} = \left[\frac{2\epsilon_F}{3m}\nabla^2 \delta\tilde{\rho}(\mathbf{r},t) - \omega_p^2 \delta\tilde{\rho}(\mathbf{r},t)\right]\Theta(a-r)$$
$$- \frac{2}{3m}\nabla\left\{\left[\frac{3}{5}\epsilon_F n_e + \epsilon_F \delta\tilde{\rho}(\mathbf{r},t)\right]\frac{\mathbf{r}}{r}\delta(a-r)\right\}$$
$$- \left[\frac{2\epsilon_F}{3m}\frac{\mathbf{r}}{r}\nabla\delta\tilde{\rho}(\mathbf{r},t) + \frac{\omega_p^2}{4\pi}\frac{\mathbf{r}}{r}\nabla\int d^3 r_1 \frac{1}{|\mathbf{r}-\mathbf{r}_1|}\delta\tilde{\rho}(\mathbf{r}_1,t)\right]\delta(a-r). \quad (6.5)$$

In the above formula the Thomas–Fermi approximation for the kinetic energy has been used [16, 23] and ω_p is the bulk plasmon frequency, $\omega_p^2 = 4\pi n_e e^2/m$ (in SI units, $\omega_p^2 = e^2 n_e/(\varepsilon_0 m)$). Here we have used $\nabla\Theta(a-r) = (-\mathbf{r}/r)\delta(a-r)$), which gives rise to the surface plasmon fluctuations indicated by the Dirac delta in Eq. (6.5). Because of the structure of Eq. (6.5), its general solution can be decomposed into two parts related to two distinct domains, as was discussed in Chapter 4, and we rewrite them here:

$$\delta\tilde{\rho}(\mathbf{r},t) = \begin{cases} \delta\tilde{\rho}_1(\mathbf{r},t) & \text{for } r < a, \\ \delta\tilde{\rho}_2(\mathbf{r},t) & \text{for } r \geq a \ (r \to a+), \end{cases} \quad (6.6)$$

corresponding to the volume and surface charge fluctuations, respectively. These two components of the local electron density fluctuations satisfy the following equations (see Eq. (5.7)):

$$\frac{\partial^2 \delta\tilde{\rho}_1(\mathbf{r},t)}{\partial t^2} = \frac{2\epsilon_F}{3m}\nabla^2 \delta\tilde{\rho}_1(\mathbf{r},t) - \omega_p^2 \delta\tilde{\rho}_1(\mathbf{r},t) \quad (6.7)$$

6.1 Semiclassical RPA Approach to Plasmons in Large Metallic Nanospheres   103

and (see Eq. (5.8), here, $\epsilon = 0+$,

$$\frac{\partial^2 \delta\tilde{\rho}_2(\mathbf{r},t)}{\partial t^2} = -\frac{2}{3m}\nabla\left\{\left[\frac{3}{5}\epsilon_F n_e + \epsilon_F \delta\tilde{\rho}_2(\mathbf{r},t)\right]\frac{\mathbf{r}}{r}\delta(a+\epsilon-r)\right\}$$

$$-\left[\frac{2\epsilon_F}{3m}\frac{\mathbf{r}}{r}\nabla\delta\tilde{\rho}_2(\mathbf{r},t) + \frac{\omega_p^2}{4\pi}\frac{\mathbf{r}}{r}\nabla\int d^3r_1 \frac{1}{|\mathbf{r}-\mathbf{r}_1|}(\delta\tilde{\rho}_1(\mathbf{r}_1,t)\Theta(a-r_1)\right.$$

$$+ \delta\tilde{\rho}_2(\mathbf{r}_1,t)\Theta(r_1-a))\bigg]\delta(a+\epsilon-r). \tag{6.8}$$

To solve Eqs. (6.7) and (6.8), let us represent the two components of the electron fluctuation in the following manner:

$$\begin{aligned}\delta\tilde{\rho}_1(\mathbf{r},t) &= n_e F(\mathbf{r},t) && \text{for } r < a, \\ \delta\tilde{\rho}_2(\mathbf{r},t) &= \sigma(\Omega,t)\delta(r+\epsilon-a), \epsilon = 0+, && \text{for } r \geq a \ (r \to a+),\end{aligned} \tag{6.9}$$

with initial conditions $F(\mathbf{r},t)|_{t=0} = 0$ and $\sigma(\Omega,t)|_{t=0} = 0$ (Ω is the spherical (solid) angle); moreover, $F(\mathbf{r},t)|_{r\to a} = 0$ and $\int \rho(\mathbf{r},t)d^3r = N_e$ is the neutrality condition. After some algebra [23] one can find, given the spherical symmetry of the system,

$$F(\mathbf{r},t) = \sum_{l=1}^{\infty}\sum_{m=-l}^{l}\sum_{n=1}^{\infty} A_{lmn} j_l(k_{nl}r) Y_{lm}(\Omega) \sin(\omega_{nl}t) \tag{6.10}$$

and

$$\sigma(\Omega,t) = \sum_{l=1}^{\infty}\sum_{m=-l}^{l} \frac{B_{lm}}{a^2} Y_{lm}(\Omega) \sin(\omega_{0l}t)$$

$$+ \sum_{l=1}^{\infty}\sum_{m=-l}^{l}\sum_{n=1}^{\infty} A_{lmn} \frac{(l+1)\omega_p^2}{l\omega_p^2 - (2l+1)\omega_{nl}^2} Y_{lm}(\Omega) n_e$$

$$\times \int_0^a dr_1 \frac{r_1^{l+2}}{a^{l+2}} j_l(k_{nl}r_1) \sin(\omega_{nl}t), \tag{6.11}$$

where the $j_l(\xi) = \sqrt{\pi/(2\xi)}I_{l+1/2}(\xi)$ are the spherical Bessel functions, the $Y_{lm}(\Omega)$ are spherical harmonics, the $\omega_{nl} = \omega_p\sqrt{1 + x_{nl}^2/(k_T^2 a^2)}$ are the frequencies of the volume plasmon self-oscillations, $k_T = \sqrt{6n_e e^2/\epsilon_F}$ is the Thomas–Fermi length (ϵ_F is the Fermi energy), the x_{nl} are the nodes of the Bessel function $j_l(\xi)$, $k_{nl} = x_{nl}/a$, and the $\omega_{0l} = \omega_p\sqrt{l/(2l+1)}$ are the frequencies of the electron surface self-oscillations (the surface plasmon frequencies). The function $F(\mathbf{r},t)$ describes the volume plasmon oscillations whereas the function $\sigma(\Omega,t)$ describes the surface plasmon oscillations. In the formula for $\sigma(\Omega,t)$, Eq. (6.11), the first term corresponds to surface self-oscillations, whereas the second term describes surface oscillations induced by the volume plasmons. The frequencies of the surface self-oscillations are

$$\omega_{0l} = \omega_p \sqrt{\frac{l}{2l+1}}, \tag{6.12}$$

which, for $l = 1$, gives the dipole-type surface oscillation frequency originally found by Mie [18, 17], $\omega_{01} = \omega_p/\sqrt{3}$ (for simplicity, we denote it as $\omega_1 = \omega_{01}$).

6.2 Damping of Plasmons in Large Nanospheres

Now we again summarize the most important aspects of plasmon damping. The advantage of the RPA Eqs. (6.7) and (6.8) for the volume and surface plasmons in a metallic nanoparticle is their oscillatory form. This property allows phenomenological inclusion of the damping of the plasmons, in analogy to oscillator damping, via the addition to the right-hand sides of Eqs. (6.7) and (6.8) the term

$$-\frac{2}{\tau_0} \frac{\partial \delta \rho_{1(2)}(\mathbf{r},t)}{\partial t},$$

with indices 1 or 2, respectively. The damping of the plasmons that is included in τ_0 accounts for the scattering of electrons from other electrons, from defects, from phonons, or from nanoparticle boundaries; all these scattering channels lead to a damping rate expressed by the simplified formula [24]

$$\frac{1}{\tau_0} \simeq \frac{v_F}{2\lambda_B} + \frac{C v_F}{2a}, \tag{6.13}$$

where C is a constant of order unity, a is the nanosphere radius, v_F is the Fermi velocity in the metal, and λ_B is the electron mean free path in the bulk (including effects due to the scattering of electrons on other electrons, on impurities and on phonons [24]). As was discussed previously (see Eq. (5.13)), the latter term in Eq. (6.13) accounts for the scattering of electrons from the boundary of the nanoparticle, whereas the former term corresponds to scattering processes similar to those in the bulk. The other effects, such as Landau damping (especially important in small clusters [80, 79]), which corresponds to the decay of a plasmon into a high-energy particle–hole pair, are of less significance for nanosphere radii larger than 2–3 nm [80] and are completely negligible for radii larger than 5 nm.

Irreversible dissipation of the plasmon energy due to scattering, as expressed by (6.13), affects each mode of both the volume and surface plasmons. In the case of interest here, i.e., that of plasmons induced by incident sunlight, one can observe that only dipole surface plasmons will contribute to the electromagnetic response of a nanosphere of radius $a \in (5, 100)$ nm. This follows from the restriction imposed by the dipole approximation, which holds in the case of such metallic nanospheres; in this case the resonance wavelength (corresponding to photons of energy $\hbar\omega_1$, which, for Au and Ag, is of the order of 500 nm) is far larger than the nanosphere radius and the electric field of the incident light is essentially homogeneous over the entire nanoparticle. Such a forcing electric field can excite only the dipole mode of the surface plasmons, which can be represented by the function $Q_{1m}(t)$, $m = -1, 0, 1$. This function satisfies the following equation (which is Eq. (6.8) for just the $l = 1$ spherical component, with damping and a homogeneous forcing field):

$$\frac{\partial^2 Q_{1m}(t)}{\partial t^2} + \frac{2}{\tau_0}\frac{\partial Q_{1m}(t)}{\partial t} + \omega_1^2 Q_{1m}(t)$$
$$= \sqrt{\frac{4\pi}{3}}\frac{en_e}{m}\left[E_z(t)\delta_{m0} + \sqrt{2}\left(E_x(t)\delta_{m1} + E_y(t)\delta_{m-1}\right)\right]. \quad (6.14)$$

The corresponding electron density fluctuations are given by

$$\delta\rho(\mathbf{r},t) = \begin{cases} 0 & \text{for } r < a, \\ \sum_{m=-1}^{1} Q_{1m}(t)Y_{1m}(\Omega) & \text{for } r \geq a, \ r \to a+. \end{cases} \quad (6.15)$$

Because the $l = 1$ case describes dipole-type fluctuations, one can calculate the dipole of the plasmon oscillations given by Eq. (6.15):

$$\mathbf{D}(t) = e\int d^3r\,\mathbf{r}\,\delta\rho(\mathbf{r},t) = \frac{4\pi}{3}e\mathbf{q}(t)a^3. \quad (6.16)$$

This dipole can be represented by the vector $\mathbf{q}(t)$, which is related to the functions $Q_{1m}(t)$ as follows: $Q_{11}(t) = \sqrt{8\pi/3}q_x(t)$, $Q_{1-1}(t) = \sqrt{8\pi/3}q_y(t)$, and $Q_{10}(t) = \sqrt{4\pi/3}q_x(t)$. The vector $\mathbf{q}(t)$ satisfies the following equation (cf. Eq. (6.14)):

$$\left(\frac{\partial^2}{\partial t^2} + \frac{2}{\tau_0}\frac{\partial}{\partial t} + \omega_1^2\right)\mathbf{q}(t) = \frac{en_e}{m}\mathbf{E}(t). \quad (6.17)$$

6.2.1 Radiation from Dipole Surface Plasmons – Lorentz Friction for Plasmons

Plasmon oscillations are themselves a source of electromagnetic radiation. This radiation depletes the energy of the plasmons, resulting in hampering of their oscillations, which can be regarded in terms of the Lorentz friction in the case of an oscillating and radiating dipole [55]:

$$\mathbf{E}_L = \frac{2}{3c^3}\frac{\partial^3 \mathbf{D}(t)}{\partial t^3}, \quad (6.18)$$

where $\mathbf{E}_L$ represents an additional electric field hampering the charge oscillations (the Lorentz friction), c is the speed of light, and $\mathbf{D}(t)$ is surface plasmon oscillation dipole. According to Eq. (6.16), we arrive at the following form for $\mathbf{E}_L$:

$$\mathbf{E}_L = \frac{2e}{3c^3} \frac{4\pi}{3} a^3 \frac{\partial^3 \mathbf{q}(t)}{\partial t^3}. \quad (6.19)$$

Substituting this expression into Eq. (6.17), we obtain

$$\left(\frac{\partial^2}{\partial t^2} + \frac{2}{\tau_0}\frac{\partial}{\partial t} + \omega_1^2\right)\mathbf{q}(t) = \frac{en_e}{m}\mathbf{E}(t) + \frac{2}{3\omega_1}\left(\frac{\omega_1 a}{c}\right)^3 \frac{\partial^3 \mathbf{q}(t)}{\partial t^3}. \quad (6.20)$$

In the zeroth-order approximation (neglecting attenuation and for $\mathbf{E} = 0$),

$$\left(\frac{\partial^2}{\partial t^2} + \omega_1^2\right)\mathbf{q}(t) = 0. \quad (6.21)$$

In the next step of the perturbation approach, which is iteratively applied to solve Eq. (6.20), one substitutes $-\omega_1^2 \mathbf{q}(t)$ for $\partial^2 \mathbf{q}(t)/\partial t^2$ (according to Eq. (6.21)) in the second term on the right-hand side of that equation. Thus the Lorentz friction term is included in a renormalized damping rate τ:

$$\left(\frac{\partial^2}{\partial t^2} + \frac{2}{\tau}\frac{\partial}{\partial t} + \omega_1^2\right)\mathbf{q}(t) = \frac{en_e}{m}\mathbf{E}(t), \quad (6.22)$$

where

$$\frac{1}{\tau} = \frac{1}{\tau_0} + \frac{\omega_1}{3}\left(\frac{\omega_1 a}{c}\right)^3 \simeq \frac{v_F}{2\lambda_B} + \frac{Cv_F}{2a} + \frac{\omega_1}{3}\left(\frac{\omega_1 a}{c}\right)^3. \quad (6.23)$$

The contributions of scattering and radiation to the overall damping of the plasmons, in the framework of the perturbation approach presented above, are given for comparison in Fig. 6.1 (left). The renormalized damping increases rapidly with increasing radius a and causes a significant redshift of the self-frequency of the free-dipole-type surface plasmon oscillations, $\omega_1' = \sqrt{\omega_1^2 - 1/\tau^2}$. The radius-dependent

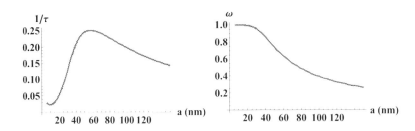

Figure 6.1 Real and imaginary parts of the solution of Eq. (6.20), i.e., the self-frequency and the damping rate, given by Eq. (6.24), versus a ($1/\tau$ and ω in units of ω_1).

6.2 Damping of Plasmons in Large Nanospheres

shift in the resonance caused by strong radiation-induced plasmon damping has been confirmed experimentally [77] via measurements of light extinction in colloidal solutions of nanoparticles of different sizes (in [77], such measurements are reported for Au nanospheres of 10–80 nm radius and Ag nanospheres of 10–60 nm radius). These measurements support the a^3 increase in plasmon damping described above. However, for $a > 30$ nm, some deviation from this a^3 increase is evident in experimental results [77]. This deviation is related to the constraints of the applied perturbative solution procedure. With increasing a, the attenuation rapidly increases as a^3 and can no longer be treated as a small perturbation. Thus, an exact solution of Eq. (6.20) is necessary. This equation is a third-order linear differential equation and can be solved exactly. The related self-frequency and its damping rate are given by the following expression:

$$\omega + i\frac{1}{\tau}$$
$$= -\frac{i}{3l} + \frac{i(1 + i\sqrt{3})(1 + 6lq)}{3 \times 2^{2/3}l \left(2 + 27l^2 + 18lq + \sqrt{4(-1-6lq)^3 + (2+27l^2+18lq)^2}\right)^{1/3}}$$
$$+ \frac{i(1 - i\sqrt{3})\left(2 + 27l^2 + 18lq + \sqrt{4(-1-6lq)^3 + (2+27l^2+18lq)^2}\right)^{1/3}}{6 \times 2^{1/3}l},$$
(6.24)

where $q = 1/\tau_0\omega_1$ and $l = (2/3)(a\omega_1/c)^3$. The functions ω and $1/\tau$ (in dimensionless units, i.e., divided by ω_1) are plotted in Fig. 6.2 versus nanosphere radius a. Strong deviation from the perturbatively obtained behaviour, $1/\tau \sim a^3$, is apparent for $a > 30$ nm. For $a < 25$ nm, the approximate perturbative expressions for $1/\tau$ and $\omega = \omega_1\sqrt{1 - (1/\omega_1\tau)^2}$ agree with the exact solution in Eq. (6.24). The Lorentz friction term is given in Eq. (6.23) (and also in Eq. (6.24) for a lower than approximately 11 nm, for Au and Ag). In this size regime for metallic particles, we encounter a conspicuous cross-over of the attenuation rate with respect to a. The minimum damping is achieved at

$$a^* = \left(\frac{9}{2}\varepsilon \frac{v_F c^3}{\omega_p^4}\right)^{1/4}$$
(6.25)

and for $a < a^*$ the damping ratio increases with decreasing a approximately as $1/a$, whereas for $a > a^*$ this ratio increases with increasing a in proportion to a^3 (if $a < 25$ nm, for Au in vacuum). The value of a^* can be estimated for Au and Ag; see Table 6.1. As is apparent in Fig. 6.1, the attenuation rate $1/\tau$ reaches a maximum at a certain value of a. This value, denoted by a^{**}, is also listed in

Table 6.1 *Radii of a nanosphere corresponding to the minimal value of surface plasmon damping (a^*) and to the maximal value (a^{**})*

	a^*, a^{**} (nm)	
$\sqrt{\varepsilon}$	Au	Ag
1, vacuum	8.77, 57.4	8.78, 57.42
1.41	10.4, 64.3	10.4, 64.4
1.73	11.5, 68.7	11.6, 68.8

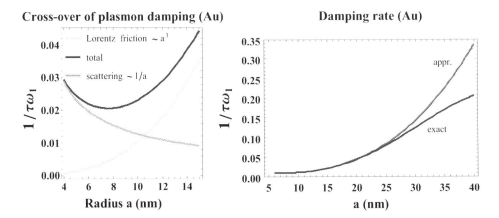

Figure 6.2 (Left) Cross-over in the size dependence of the damping rate: the contributions to the damping rate of surface plasmon oscillations in a nanosphere as a function of the nanosphere radius, including the scattering attenuation ($\sim v_F/(2\lambda_B) + C v_F/(2a)$) (mid-grey line) and the Lorentz friction damping ($\sim a^3$) (light grey line). For radii greater than approximately 11 nm (for Au), the second channel dominates the overall damping (black line). (Right) The damping rate calculated with the perturbative treatment of the Lorentz friction, as in Eq. (6.23), with, for comparison, the damping rate calculated using the exact formula (6.24) for the Lorentz friction.

Table 6.1 (one can observe that a^* and a^{**} depend weakly on the material but more strongly on the dielectric forming the surroundings).

In the region of the cross-over in the size dependence of the plasmon damping, the perturbative approximation agrees with the exact solution, as illustrated in Fig. 6.2 (right). However, the perturbatively calculated damping is heavily overestimated for $a > 30$ nm (for Au in vacuum). At $a \simeq 57$ nm (for Au in vacuum), the perturbative value of $1/\tau \omega_1$ equals 1, which corresponds to the overdamped regime for the oscillatory equation. This is an artefact of the perturbation approach and has been removed in the exact solution of Eq. (6.20) presented in Eq. (6.24). The exact oscillating solution also exists for $a > 57$ nm, although the relation $\omega = \omega_1\sqrt{1 - 1/(\tau\omega_1)^2}$ no longer holds (for $a > 25$ nm). The resonance self-frequency and the attenuation rate are given by the exact formula (6.24) and are illustrated in Figs. 6.1 and 6.2 (right). The exact solution in Eq. (6.24) improves the fit with experiment of the resonance frequency of dipole surface plasmons.

The radiative damping of plasmons in metallic nanospheres, as described above, corresponds to the free oscillations of a dipole in vacuum or in dielectric surroundings (in the latter case, ω_1 must be renormalized by the factor $1/\sqrt{\varepsilon}$, where ε is the relative permittivity of the surrounding dielectric medium). If, however, another electrical system is present in the vicinity of the nanosphere, the situation changes significantly. The coupling of plasmons to other charged systems causes resonant energy transfer, which results in an increase in the plasmon damping. For the case in which a nanosphere is deposited on the surface of a semiconductor, the near-field coupling of the plasmons with the band electrons of the semiconductor must be taken into account. In such a case, a quantum mechanical estimation of the related attenuation rate can be found by applying the Fermi golden rule to the semiconductor interband transitions induced by dipole near-field coupling with the plasmons [23, 124]; we will analyse this in the following section for nanospheres of various radii deposited on a semiconductor substrate.

6.3 Transfer of Sunlight Energy to a Semiconductor Mediated by Surface Plasmons through the Channel of Dipole Coupling in the Near-Field Regime

6.3.1 Fermi Golden Rule for Band Electron Transitions due to Coupling with Plasmons

The presence of an energy receiver in the vicinity of metallic nanospheres with plasmons, as in the case of a semiconductor layer on which nanoparticles are deposited, provides a channel for energy transfer from the plasmons to the band electrons. The plasmon damping in this case is highly modified in comparison with free oscillations in a metallic nanoparticle placed in dielectric surroundings and radiating energy in the far-field zone. In the latter case the locally plane-wave photons take away energy via the conventional Lorentz friction mechanism. These photons have no contact with the source after being emitted. In the near-field zone when plasmons couple with the band electrons in the substrate semiconductors the situation is different. In the near-field zone photons have not yet been formed and the coupling of the emitter and the receiver causes a strong mutual influence. Beyond this zone the classical Maxwell electrodynamics properties of the emitter depend on the receiver, which must be taken into account quantum mechanically using the quantum transition scheme [75]. To assess the efficiency of energy transfer from the plasmons to the semiconductor substrate via the near-field coupling of plasmons and electrons, we will apply the Fermi golden rule. This allows for the calculation of the transition probability per time unit for an electron to hop between the valence and conduction bands as a result of the coupling of the electrons to the dipole surface plasmons in a metallic nanosphere.

The Fourier components of the magnetic field ($\mathbf{B}_\omega$) and electric field ($\mathbf{E}_\omega$) produced at a distant point $\mathbf{R}$ by a monochromatic plasmonic dipole oscillation at the origin, $\mathbf{D} = \mathbf{D}_0 e^{-i\omega t}$, can be represented by the equations [55]

$$\mathbf{B}_\omega = \frac{ik}{\sqrt{\varepsilon}} [\mathbf{D}_0 \times \hat{\mathbf{n}}] \left(\frac{ik}{R} - \frac{1}{R^2} \right) e^{ikR} \tag{6.26}$$

and

$$\mathbf{E}_\omega = \frac{1}{\varepsilon} \left\{ \mathbf{D}_0 \left(\frac{k^2}{R} + \frac{ik}{R^2} - \frac{1}{R^3} \right) + \hat{\mathbf{n}}(\hat{\mathbf{n}} \cdot \mathbf{D}_0) \left(-\frac{k^2}{R} - \frac{3ik}{R^2} + \frac{3}{R^3} \right) \right\} e^{ikR}, \tag{6.27}$$

where we use the following notation for the retarded argument: $i\omega(t - R/c) = i\omega t - ikR$ and $\hat{\mathbf{n}} = \mathbf{R}/R$. The contributions with denominators R^3, R^2, and R are referred to as the near-, medium- and far-field zones of the dipole radiation, respectively. For interactions with the adjacent layer of the semiconductor, the near-field zone is of the most significance. One can therefore neglect the terms containing $1/R$ and $1/R^2$. Under the further assumption that $e^{ikR} = 1$ for the near field, one obtains $\mathbf{B}_\omega = 0$ and $\mathbf{E}_\omega = (1/\varepsilon R^3)[3\hat{\mathbf{n}}(\hat{\mathbf{n}} \cdot \mathbf{D}_0) - \mathbf{D}_0]$, which corresponds to the dipole electric field. The magnetic field disappears in the near-field zone. Photons are not created yet at these small distances much less than the resonance wavelength. Thus, the related potential for the near-field interaction of the Mie-type surface plasmons with the band electrons can be written as follows [55]:

$$w = e\psi(\mathbf{R}, t) = \frac{e}{\varepsilon R^2} \hat{\mathbf{n}} \cdot \mathbf{D}_0 \sin(\omega t + \alpha) = w^+ e^{i\omega t} + w^- e^{-i\omega t},$$

$$w^+ = (w^-)^* = \frac{e}{\varepsilon R^2} \frac{e^{i\alpha}}{2i} \hat{\mathbf{n}} \cdot \mathbf{D}_0. \tag{6.28}$$

The terms in w^+ and w^- correspond to emission and absorption, respectively ($\hat{\mathbf{n}} = \mathbf{R}/R$ and $\mathbf{D}_0$ is the plasmon dipole amplitude). We will consider the emission of energy from the plasmon oscillations and the transfer of this energy to the electron system in the semiconductor substrate.

The semiconductor band system is modelled here in the simplest form, the single-band parabolic effective-mass approximation. According to the Fermi golden rule, the interband transition probability is given by the expression

$$w(\mathbf{k}_1, \mathbf{k}_2) = \frac{2\pi}{\hbar} \left| \langle \mathbf{k}_1 | w^+ | \mathbf{k}_2 \rangle \right|^2 \delta(E_p(\mathbf{k}_1) - E_n(\mathbf{k}_2) + \hbar\omega), \tag{6.29}$$

6.3 Transfer of Sunlight Energy Mediated by Surface Plasmons

where the Bloch states in the conduction and valence bands are assumed to be plane waves (for the sake of simplicity):

$$\Psi_{\mathbf{k}_1} = \frac{1}{(2\pi)^{3/2}} e^{i\mathbf{k}_1 \cdot \mathbf{R} - i E_p(\mathbf{k}_1) t/\hbar}, \qquad \Psi_{\mathbf{k}_2} = \frac{1}{(2\pi)^{3/2}} e^{i\mathbf{k}_2 \cdot \mathbf{R} - i E_n(\mathbf{k}_2) t/\hbar},$$

$$E_p(\mathbf{k}_1) = -\frac{\hbar^2 \mathbf{k}_1^2}{2m_p^*} - E_g, \qquad E_n(\mathbf{k}_2) = \frac{\hbar^2 \mathbf{k}_2^2}{2m_n^*},$$

(6.30)

where the indices n and p refer to electrons from the conduction and valence bands, respectively, and E_g is the forbidden gap.

Here, let us make a technical comment. The electron wave functions as given above are assumed to be normalized to the Dirac delta, which corresponds to infinite particle movement and a continuous energy spectrum. The modulus squared of the wave function does not, in this case, carry a probabilistic interpretation, for which it must be normalized to unity. To restore proper normalization, the expression

$$w(\mathbf{k}_1, \mathbf{k}_2) = \frac{2\pi}{\hbar} |\langle \mathbf{k}_1 | w^+ | \mathbf{k}_2 \rangle|^2 \delta(E_p(\mathbf{k}_1) - E_n(\mathbf{k}_2) + \hbar \omega)$$

must be divided by the squared Dirac delta function, which can be expressed by the factor $(V/(2\pi)^3)^2$, taking into account that

$$\frac{1}{(2\pi)^3} \int d^3 r e^{i\mathbf{k} \cdot \mathbf{r}} = \delta(\mathbf{k}) \simeq \text{(for } \mathbf{k} = 0\text{)} \frac{V}{(2\pi)^3} \quad (V \to \infty).$$

(6.31)

The same factor occurs, however, from the density of states, when one integrates over all initial and final states $\mathbf{k}_1, \mathbf{k}_2$, which results in the factor $(2V/(2\pi)^3)^2$ (the factor 2 arises because of spin degeneracy). Thus, the renormalization factors mutually cancel. Taking into account the renormalization described above, we find the matrix element

$$\langle \mathbf{k}_1 | w^+ | \mathbf{k}_2 \rangle = \frac{1}{(2\pi)^3} \int d^3 R \frac{e}{\varepsilon 2i} e^{i\alpha} \hat{\mathbf{n}} \cdot \mathbf{D}_0 \frac{1}{R^2} e^{-i(\mathbf{k}_1 - \mathbf{k}_2) \cdot \mathbf{R}}.$$

(6.32)

Here, the integrals can all be computed analytically. After some algebra, one arrives at the expression (see also the derivation of Eq. (6.75) below)

$$\langle \mathbf{k}_1 | w^+ | \mathbf{k}_2 \rangle = \frac{-1}{(2\pi)^3} \frac{e e^{i\alpha}}{\varepsilon} \mathbf{D}_0 \cos \Theta \, (2\pi) \int_a^\infty dR \frac{1}{q} \frac{d}{dR} \frac{\sin(qR)}{qR}$$

$$= \frac{1}{(2\pi)^2} \frac{e e^{i\alpha}}{\varepsilon} \frac{\mathbf{D}_0 \cdot \mathbf{q}}{q^2} \frac{\sin(qa)}{qa} \xrightarrow{a \to 0} \frac{1}{(2\pi)^2} \frac{e e^{i\alpha}}{\varepsilon} \frac{\mathbf{D}_0 \cdot \mathbf{q}}{q^2}.$$

(6.33)

To estimate the absorption probability per time unit, a summation over all initial and final states in both bands must be performed. Thus, we arrive at the following formula for the transition probability:

$$\delta w = \int d^3 k_1 \int d^3 k_2 \left[f_1 (1 - f_2) w(\mathbf{k}_1, \mathbf{k}_2) - f_2 (1 - f_1) w(\mathbf{k}_2, \mathbf{k}_1) \right],$$

(6.34)

where f_1 and f_2 represent temperature-dependent Fermi–Dirac distribution functions for the initial and final states, respectively. Emission and absorption are included but, for room temperature, one can assume that $f_2 \simeq 0$ and $f_1 \simeq 1$, which leads to

$$\delta w = \int d^3 k_1 \int d^3 k_2 \cdot w(\mathbf{k}_1, \mathbf{k}_2). \tag{6.35}$$

In the above formula, we have eliminated the state density factors through probability normalization, as mentioned above. After integrating we arrive at the expression (see Eq. (6.85) below)

$$\delta w = \frac{4}{3} \frac{\mu^2 (m_n^* + m_p^*) 2(\hbar\omega - E_g) e^2 D_0^2}{\sqrt{m_n^* m_p^*} 2\pi \hbar^5 \varepsilon^2} \int_0^1 dx \frac{\sin^2(xa\xi)}{(xa\xi)^2} \sqrt{1-x^2}$$
$$= \frac{4}{3} \frac{\mu^2}{\sqrt{m_n^* m_p^*}} \frac{e^2 D_0^2}{2\pi \hbar^3 \varepsilon^2} \xi^2 \int_0^1 dx \frac{\sin^2(xa\xi)}{(xa\xi)^2} \sqrt{1-x^2}. \tag{6.36}$$

In the limiting cases, we obtain (see the derivation of Eq. (6.87) below)

$$\delta w = \begin{cases} \dfrac{4}{3} \dfrac{\mu \sqrt{m_n^* m_p^*} (\hbar\omega - E_g) e^2 D_0^2}{\hbar^5 \varepsilon^2} & \text{for } a\xi \ll 1, \\[2ex] \dfrac{4}{3} \dfrac{\mu^{3/2} \sqrt{2} \sqrt{\hbar\omega - E_g} e^2 D_0^2}{a \hbar^4 \varepsilon^2} & \text{for } a\xi \gg 1, \end{cases} \tag{6.37}$$

where

$$\xi = \frac{\sqrt{2(\hbar\omega - E_g)(m_n^* + m_p^*)}}{\hbar}.$$

In the latter case represented in Eq. (6.37) we have applied the approximation

$$\int_0^1 dx \frac{\sin^2(xa\xi)}{(xa\xi)^2} \sqrt{1-x^2} \approx \text{(for } a\xi \gg 1) \frac{1}{a\xi} \int_0^\infty d(xa\xi) \frac{\sin^2(xa\xi)}{(xa\xi)^2} = \frac{\pi}{2a\xi},$$

whereas in the former case $\int_0^1 dx \sqrt{1-x^2} = \pi/4$.

Note that the above formulae are fundamentally distinct from those for the ordinary photoeffect, for which (see Section 6.5, the derivation of Eq. (6.84))

$$\delta w_0 = \frac{4\sqrt{2}}{3} \frac{\mu^{5/2} e^2}{m_p^{*2} \omega \varepsilon \hbar^3} \left(\frac{\varepsilon E_0^2 V}{8\pi \hbar \omega} \right) (\hbar\omega - E_g)^{3/2}. \tag{6.38}$$

Taking into account that the number of photons in the volume V is equal to $(\varepsilon E_0^2 V / (8\pi \hbar \omega))$, the probability of single-photon absorption by the semiconductor per unit time takes the following form for the ordinary photoeffect:

$$q_0 = \delta w_0 \left(\frac{\varepsilon E_0^2 V}{8\pi \hbar \omega} \right)^{-1} = \frac{4(4)\sqrt{2}}{3} \frac{\mu^{5/2} e^2}{m_p^{*2} \omega \varepsilon \hbar^3} (\hbar\omega - E_g)^{3/2}, \tag{6.39}$$

where the factor 4 corresponds to the spin degeneracy of the band electrons.

6.3.2 Plasmon Damping Rate due to Near-Field Coupling with Semiconductor-Band Electrons

The probability per time unit of a transition between states of the band electrons in the semiconductor substrate is given by Eq. (6.37), as derived in Section 6.5 below in accordance with the Fermi golden rule. Let us emphasize that because of the absence of momentum conservation for the near-field dipole coupling in the vicinity of a nanosphere, all interband transitions contribute, not only the direct ones (as in the case of interaction with a plane wave). This results in the enhancement of the transition probability for near-field coupling in comparison with the photon (plane-wave) absorption rate in a semiconductor in the ordinary photoeffect. However, this enhancement is gradually quenched as the radius a increases, as is expressed in Eq. (6.37).

Assuming now that dipole plasmon oscillations are excited by the rapid switching off of the uniform electric field, so that $E(t) = E_0(1 - \Theta(t))$, we consider the dipole-type solution for surface plasmons [23],

$$\mathbf{D}(t) = [0, 0, D_0 e^{-t/\tau'} \cos(\omega_1' t)] \tag{6.40}$$

with

$$D_0 = \frac{e^2 n_e}{m \omega_1^2} E_0 \frac{4\pi}{3} a^3. \tag{6.41}$$

In such a case, one can estimate the total energy transfer $\mathcal{A}$ to the semiconductor (assuming that the dominant channel of dissipation is the near-field interaction with the semiconductor substrate, neglecting the shift of ω_1' with respect to ω_1 caused by dissipation, and using Eq. (6.37) for δw):

$$\mathcal{A} = \beta \int_0^\infty \delta w \hbar \omega_1 dt = \beta \hbar \omega_1 \delta w \tau'/2$$

$$= \begin{cases} \dfrac{2}{3} \dfrac{\beta \omega_1 \tau' \mu \sqrt{m_n^* m_p^*} (\hbar \omega_1 - E_g) e^2 D_0^2}{\hbar^4 \varepsilon^2} & \text{for } a\xi \ll 1, \\[2mm] \dfrac{2}{3} \dfrac{\beta \omega_1 \tau' \mu^{3/2} \sqrt{2} \sqrt{\hbar \omega_1 - E_g} e^2 D_0^2}{a \hbar^3 \varepsilon^2} & \text{for } a\xi \gg 1, \end{cases} \tag{6.42}$$

where β accounts for the proximity constraints, not included microscopically in the model, that reduce the near-field contact of the sphere with the semiconductor medium. For the case of nanospheres deposited on the surface of a semiconductor layer, $\beta \sim h^2/a^2 \sim 10^{-3}$ for $a \sim 50$ nm (h is the effective range of the near-field coupling), whereas for nanospheres that are entirely embedded in their semiconductor surroundings, β would be significantly enhanced. Comparing the value given by the formula (6.42) with the energy loss estimated in [23] (the total energy of the plasmon oscillations), one finds that

$$\frac{1}{\tau'\omega_1} = \begin{cases} \dfrac{4\beta\mu\sqrt{m_n^* m_p^*}(\hbar\omega_1 - E_g)e^2 a^3}{3\hbar^4 \varepsilon} & \text{for } a\xi \ll 1, \\ \dfrac{4\beta\mu^{3/2}\sqrt{2}\sqrt{\hbar\omega_1 - E_g}\, e^2 a^2}{3\hbar^3 \varepsilon} & \text{for } a\xi \gg 1. \end{cases} \quad (6.43)$$

For nanospheres of Au deposited on an Si layer, we obtain for a in nm

$$\frac{1}{\tau'\omega_1} = \begin{cases} 44.092\beta a^3 \dfrac{\mu}{m}\dfrac{\sqrt{m_n^* m_p^*}}{m} & \text{for } a \ll 0.15\sqrt{m/(m_n^* + m_p^*)}, \\ 13.648\beta a^2 \left(\dfrac{\mu}{m}\right)^{3/2} & \text{for } a \gg 0.15\sqrt{m/(m_n^* + m_p^*)}, \end{cases} \quad (6.44)$$

where for light (heavy) carriers in Si, $m_n^* = 0.19m(0.98m)$ and $m_p^* = 0.16m(0.52m)$, with m the bare electron mass, $\mu = m_n^* m_p^*/(m_n^* + m_p^*)$, $E_g = 1.14$ eV, $\varepsilon = 12$, and $\hbar\omega_1 = 2.72$ eV. For these parameters and for nanospheres with radius a in the range 5–50 nm, the second case of Eq. (6.44) applies. The value of the parameter β obtained from a fit to the experimental data is equal to approximately 0.002 (see Table 6.2 for the details of this fitting).

One can also notice that for a very high value of β ($\beta > \beta^*$ in Fig. 6.3), the plasmon oscillator enters the overdamped regime. The energy transfer to the semiconductor substrate via the near-field coupling channel is so efficient that it quenches free plasmon oscillations. In this case the estimation of the damping rate presented above is not applicable because free oscillations are thus not possible in the case of such strong attenuation. Nevertheless, it is possible for oscillations to continue but in the forced-oscillator regime. In this case, the energy transfer to the substrate system (calculated on the basis of on the Fermi golden rule and given by Eq. (6.42)) is equal to the output power of the forcing field. According to the standard formula for a harmonically forced and damped oscillator, this output power (averaged over the oscillation period) is equal to

$$\langle P \rangle = \frac{E_0^2 a^3 \omega_p^2}{3} \frac{\omega^2 \tau}{(\omega_1^2 - \omega^2)^2 \tau^2 + 4\omega^2}. \quad (6.45)$$

Table 6.2 *Fitting the parameter β to the experimental data presented in [33] for Au nanospheres on an Si layer*

a (nm)	n_s (10^8/cm²)	$\hbar\omega = \hbar\omega_1\sqrt{1 - 2(\tau'\omega_1)^{-2}}$ (eV) (theoretical)	$\hbar\omega$ (eV) (exp.)	I'/I
50	0.8	2.37 ($\beta = 0.0016$), 2.14 ($\beta = 0.002$), 2.25 ($\beta = 0.0018$)	2.25	1.55
40	1.6	2.58 ($\beta = 0.0016$), 2.48 ($\beta = 0.002$), 2.53 ($\beta = 0.0018$)	2.48	1.9
25	6.6	2.70 ($\beta = 0.0016$), 2.68 ($\beta = 0.002$), 2.69 ($\beta = 0.0018$)	2.70	1.75

6.3 Transfer of Sunlight Energy Mediated by Surface Plasmons

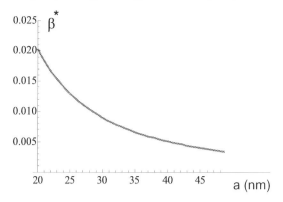

Figure 6.3 Critical value β^* of the β parameter for a plasmon oscillator in the overdamped regime $\beta > \beta^*$ as a function of the nanosphere radius a.

We also have, for the power of the plasmonic energy output, $\langle P \rangle = \beta \hbar \omega \delta w$. A comparison of the two expressions yields

$$\frac{1}{\tau} = \frac{\omega_p^2 a^3}{3\omega^2} \beta \hbar \omega \frac{\delta w}{D_0^2}. \tag{6.46}$$

Note that, in such an approach, one obtains the damping rate as a function of the photon frequency. Moreover, it should be noted here that a phenomenological approach analogous to that for a driven and damped oscillator will not correspond exactly to a real quantum system of plasmons, which are quantum quasiparticles that are given by the poles of an appropriate response function (i.e., a retarded Green function), where the real and imaginary parts of the pole are well defined. However, in the case of strong coupling with a nearby receiver of energy when the damping is strong, the similarity to the quasiparticle model lessens (quasiparticles do not exhibit frequency and damping rates identical to those of a forced and damped oscillator). Thus, for the strong damping case, and especially when the corresponding oscillator model lies in the overdamped regime, the quasiparticle approach to these forced plasmon oscillations does not hold. For the damped and driven oscillator approach, one can consider the driven oscillations of a plasmon with a frequency far lower than the free plasmon resonance frequency and with a damping rate that is dependent on the frequency of the forcing field. A comparison of the results for the damping rates obtained using the two approaches is presented in Fig. 6.4, where it is evident that, for wavelengths close to the dipole surface plasmonic resonance, the damping rate for the damped and driven oscillator model is slightly greater than the corresponding damping rate obtained using the quasiparticle approach to the plasmons.

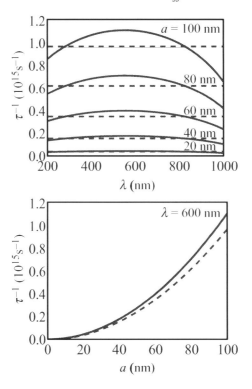

Figure 6.4 (Upper panel) The damping rate τ^{-1} versus the wavelength λ and the nanoparticle radius a, estimated in the phenomenological approach to plasmon oscillations using the damped and driven oscillator model, in which the damping is dependent on $\omega = 2\pi c/\lambda$ (solid line), and also in the quasiparticle plasmon model, in which the damping rate is independent of ω (broken lines). In the lower panel the damping rate for $\lambda = 600$ nm versus a is presented.

6.3.3 Transfer of Light Energy via the Plasmon Channel

To assess the efficiency of the plasmon near-field coupling channel, one can estimate the ratio of the probability of energy absorption in the semiconductor via the mediation of the surface plasmons (per single photon incident on a metallic nanosphere, q_m) to the probability of energy absorption in the semiconductor directly from plane-wave illumination (also per single photon, q_0, as given by Eq. (6.39)). The probability q_m is equal to the product of δw, given by Eq. (6.37), and the number of metallic nanospheres N_m divided by the photon density, with an additional phenomenological factor β that is responsible for all effects that are not directly taken into account (such as deposition separation and surface properties reducing the coupling strength):

$$q_m = \beta N_m \delta w \left(\frac{\varepsilon E_0^2 V}{8\pi \hbar \omega} \right)^{-1}. \tag{6.47}$$

In the balanced state, when incoming light energy is transferred to the semiconductor via plasmon near-field coupling, we can consider the stationary solution

6.3 Transfer of Sunlight Energy Mediated by Surface Plasmons

for a driven and damped plasmon oscillator. The driving force is the electric field of the incident plane wave, and the damping force is the near-field energy transfer described by $1/\tau$, calculated using formula (6.37) in the manner described in the previous subsection. The resulting redshifted resonance, with a simultaneously reduced amplitude, allows the energy transfer to the semiconductor to accommodate to balance the incident photon energy. Within this model, the amplitude of the resonant plasmon oscillations $D_0(\omega)$ is thus determined by

$$f(\omega) = \frac{1}{\sqrt{(\omega_1^2 - \omega^2)^2 + 4\omega^2/\tau^2}}.$$

The extremum of the redshifted resonance is attained at $\omega_m = \omega_1\sqrt{1 - 2(\omega_1\tau)^{-2}}$, with the corresponding amplitude $\sim \tau/(2\sqrt{\omega_1^2 - \tau^{-2}})$. The shift is proportional to $1/(\omega_1\tau^2)$ and scales with the nanosphere radius a, in a similar manner, i.e., diminishing with decreasing a, to that observed in experiments [33] (note again that for scattering-induced $1/\tau_0$, the dependence on a is the opposite, i.e., it increases with decreasing a). In the case of this energy transfer balance, according to Eq. (6.37) one obtains

$$q_m = \begin{cases} \beta C_0 \dfrac{128}{9}\pi^2 a^3 \dfrac{\mu\sqrt{m_n^* m_p^*}}{m^2}(\hbar\omega - E_g)\dfrac{e^6 n_e^2 \omega}{\hbar^4 \varepsilon^3} f^2(\omega) & \text{for } a\xi \ll 1, \\[2mm] \beta C_0 \dfrac{128}{9}\sqrt{2}\pi^2 a^2 \dfrac{\mu^{3/2}}{m^2}\sqrt{\hbar\omega - E_g}\dfrac{e^6 n_e^2 \omega}{\hbar^3 \varepsilon^3} f^2(\omega) & \text{for } a\xi \gg 1, \end{cases}$$
(6.48)

where $f(\omega)$ is given above and corresponds to the amplitude factor for the forced oscillator and $D_0 = (e^2 n_e E_0 4\pi a^3/(3m)) f(\omega)$; furthermore, $C_0 = N_m 4/3\pi a^3/V$, V is the volume of the semiconductor, and $\xi = \sqrt{2(\hbar\omega - E_g)(m_n^* + m_p^*)/\hbar}$.

Given that q_0 is the direct photoeffect transition probability (see (6.39)), the ratio q_m/q_0 can be expressed as follows:

$$\frac{q_m}{q_0} = \begin{cases} \dfrac{8\pi^2 a^3 \beta C_0 \sqrt{m_n^*}(m_p^*)^{5/2} e^4 n_e^2 \omega^2 f^2(\omega)}{3\sqrt{2}\mu^{3/2}m^2\sqrt{\hbar\omega - E_g}\hbar^3\varepsilon^2} & \text{for } a\xi \ll 1, \\[2mm] \dfrac{8\pi^2 a^2 \beta C_0 (m_p^*)^2 e^4 n_e^2 \omega^2 f^2(\omega)}{3\mu m^2(\hbar\omega - E_g)\hbar^2\varepsilon^2} & \text{for } a\xi \gg 1. \end{cases}$$
(6.49)

This ratio proves to be of order $10^4 \times 40\beta/H(\text{nm})$ (note that, at a typical surface density of nanoparticles, e.g., $n_s \sim 10^8/\text{cm}^2$, $C_0 = n_s 4\pi a^3/(3H)$ where H is the depth of the semiconductor layer). This, including the phenomenological factor β and the depth H, is sufficient to explain the scale of the experimentally observed strong enhancement of the absorption and emission rates. The strong enhancement of the transition probability is related to the fact that momentum-non-conserved transitions are allowed; however, it rapidly decreases with increasing radius a in accordance with formula (6.37). For example, for $a \sim 10-60$ nm this results in a reduction in the transition probability, which can be included [23] by means of the effective phenomenological factor β (with a fitted experimental value of $\beta \simeq 28 \times 10^{-3} (50/a(\text{nm}))^2$) in the case of the atomic limit, $a = 0$. The enhancement of the near-field-induced interband transition in the case of large nanospheres, however, is still significant because the diminished role of the size-related quenching of the transitions is partially compensated by the $\sim a^3$ growth of the dipole amplitude of the plasmon oscillations.

The high (although decreasing with increasing a) efficiency of the near-field energy transfer from the surface plasmons to the semiconductor substrate is primarily attributable to the contribution of all interband transitions, not just the direct (vertical) transitions as in the case of the ordinary photoeffect, because of the absence of the momentum conservation constraints on the nanosystem. The strengthening of the probability transition as a result of the contribution of all indirect interband paths of excitation in the semiconductor is probably responsible for the experimentally observed strong enhancement of the light absorption and emission in diode systems mediated by the surface plasmons in nanoparticle surface coverings [32–34, 117–119].

For comparison with experiment, we can estimate the photocurrent in the case of an Si photodiode with a metallically modified photoactive surface. This photocurrent is given by $I' = |e|N(q_0 + q_m)A$, where N is the number of incident photons, q_0 and q_m are the probabilities of single-photon absorption in the ordinary photoeffect [59] and of single-photon absorption mediated by the presence of metallic nanospheres, respectively, and $A = \tau_f^n/t_n + \tau_f^p/t_p$ is the amplification factor; here $\tau_f^{n(p)}$ are the recombination times of carriers of both signs and $t_{n(p)}$ are the drive times, i.e., the times required for the carriers to traverse the distance between electrodes. From the above formulae, it follows that if $I = I'(q_m = 0)$, is the photocurrent without metallic modifications then

$$\frac{I'}{I} = 1 + \frac{q_m}{q_0}, \qquad (6.50)$$

where the ratio q_m/q_0 is given by Eq. (6.49).

The results are given in Fig. 6.5 and reproduce the experimental behaviour well [33]. Both channels of photon absorption that result in a photocurrent in the semiconductor sample are included, namely, direct ordinary photoeffect absorption, with the transition probability q_0, and plasmon-mediated absorption, with transition probability q_m. Note also that certain additional effects, such as the reflection of incident photons or destructive interference on the metallic net, may contribute; these effects are accounted for phenomenologically in the plasmon-mediated channel by the fitted experimental factor β. However, these corrections are rather

6.3 *Transfer of Sunlight Energy Mediated by Surface Plasmons* 119

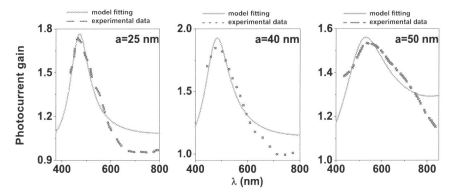

Figure 6.5 Dependence of the normalized photocurrent $I'(\lambda)/I$: for comparison, the experimental data [33] for (left) $a = 25$ nm and $n_s = 6.6 \times 10^8/\text{cm}^2$, (centre) $a = 40$ nm and $n_s = 1.6 \times 10^8/\text{cm}^2$, and (right) $a = 50$ nm and $n_s = 0.8 \times 10^8/\text{cm}^2$ ($H = 3$ μm).

weak for the low densities of metallic coverings considered here, of the order of $10^8/\text{cm}^2$, and for nanosphere sizes much lower than the resonance wavelength, although they would play a stronger reducing role for larger concentrations and larger nanosphere sizes [105, 111]. The resonance threshold is accounted for by the damped resonance envelope function in Eq. (6.50), which also includes the semiconductor band-gap limit.

As indicated in Fig. 6.5, the relatively high value of $q_m/q_0 \sim 10^4 40\beta/H(\text{nm})$ enables, by an increase in β or a reduction in H (at constant n_s), a significant enhancement in the efficiency of photoenergy transfer to the semiconductor that is mediated by the surface plasmons in the nanoparticles deposited on the active layer. However, because an enhancement of β could easily push the system into the overdamped regime, the more favourable prospect is to lower H, as in a thin-film solar cell with a smaller active-layer depth.

Using an Si photodiode, a photocurrent enhancement close to the factor 2 was achieved (see the solid lines in Fig. 6.5). The structure of the diode in the experiment was optimized to reveal the plasmonic effect and was not the structure used in high-efficiency silicon-based solar cells. The estimated junction depth was approximately 80 nm from the device surface, i.e., sufficiently close to justify the near-field approximation for dipole coupling with plasmons. It is worth noting that a strong photocurrent enhancement was observed at relatively low surface concentrations of the metallic nanocomponents. However, in many efficiency-optimized devices used in real solar cells, experimentally observed enhancements due to metallic nanocomponents are not so pronounced [36]. The details of the experimental setup utilized for the fitting presented in Fig. 6.5 are described in [33]. The plasmon effect is related only to the shallowest layer of the photodiode, at a depth of a few hundred nanometres at most, and plays an important role in the particular diode setup with suitably located active layer only. If the active layer is located at a larger distance from the metal nanoparticle plasmonic light converters then the strengthening of the efficiency would disappear.

The overall behaviour of $I'/I(\omega) = 1 + q_m/q_0$, as calculated in accordance with Eq. (6.49) and depicted in Fig. 6.5, is quite consistent with the experimental observations [33] in terms of the positions, heights, and shapes of the photocurrent curves for the different samples (the strongest enhancement is achieved for $a = 40$ nm).

Plasmon mediation in solar energy harvesting also opens the possibility of nonlinear photovoltaic (PV) effects related to high concentration of the electric field in the local surroundings of the metallic nanoparticles [125]. In the case of such PV effects, two-photon absorption is considered as an important nonlinear channel [126] but the inclusion of metallic nanocomponents would cause another effect, that of local modification of the band structure in the substrate semiconductor, which would interfere with the absorption of energy via the dipole channel from surface plasmons. Remarkable is the presence of nonlinear effects even at relatively small intensities of the incident photon beam [125, 127]. These nonlinear phenomena lie outside our present consideration but ought to be taken into account in a more realistic analysis together with other competing factors; the result would finally be an increase in the efficiency of solar cells due to the plasmon effect being less than the strengthening factors obtained within the idealized and simplified models.

6.4 Experimental Demonstration of the Proximity Constraints of the Plasmon Effect

The problem arises of how large is the range of the plasmon photovoltaic effect. An experimental verification of the micrometre scale of this range has been performed [30]. The idea of this experiment consists in the preparation of a double-layered semiconductor substrate structure with a spacer layer separating the metallic nanoparticles from the bottom layer. The bottom layer and the spacer layer are photoactive, but their contributions to the photocurrent can be distinguished because of the different spectral characteristics of both materials. Measurement of the photocurrent with and without metallic nanoparticles deposited on the top of the spacer layer enables a determination of the range of the plasmonic effect in comparison with the spacer thickness.

The spacer material in the experimental setup was ZnO, with a direct band gap of 3.34 eV [128]. The spacer layer consisted of vertically aligned nano-rods of ZnO [129] with length of about 1 μm and diameter of order 300 nm. Such a structure is manufactured by a deposition method [130]. In the experiment with the layered

6.4 *Proximity Constraints of the Plasmon Effect* 121

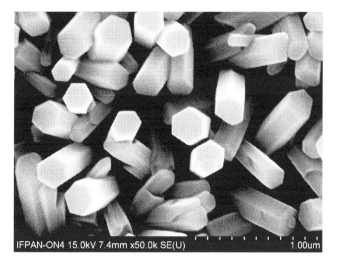

Figure 6.6 Micrographs of the ZnO nano-rod structure deposited on the Si substrate.

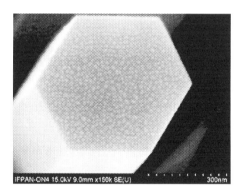 

Figure 6.7 (Left) Magnified SEM image showing the Ag nanoparticles covering the ZnO nano-rods. (Right) AZO layer is deposited on the ZnO nano-rods with their Ag covering, to provide a transparent conductive oxide layer, with thickness 350 nm.

structure described above, n-type ZnO nano-rods were used that had been grown via a low-temperature hydrothermal method on the Si substrate, see Fig. 6.6. The tops of the ZnO nano-rods were then covered with silver nanoparticles with radius approximately 5 nm.

Images obtained using SEM of the structure before deposition of a transparent AZO[1] film are shown in Fig. 6.6, and at a higher magnification in Fig. 6.7. The sample with Ag nanoparticles and the related reference sample without

---

[1] ZnO doped with aluminum to produce a low-resistivity but highly transparent layer.

nanoparticles are shown in Figs. 6.6 and 6.7. The ZnO nano-rods in Fig. 6.6 have a hexagonal shape, indicating the wurtzite crystalline structure. The surface density of the Ag nanoparticles was about $5 \times 10^{10}/cm^2$.

As the bottom substrate, p-type Si(100) wafer was used. The substrate with ZnO seeds on its surface was next placed in a reaction mixture. Here the growth of the ZnO nano-rods proceeded at atmospheric pressure and a temperature of approximately 60 °C. The hexagonal final nano-rods were of $\sim$ 1 μm in length and 300 nm in diameter.

On the top of ZnO nano-rods, Ag nanoparticles with radius ca. 5 nm were distributed via short-duration sputtering (Fig. 6.7 (left)). The total system was finally covered by an AZO film of thickness 350 nm by means of low-temperature atomic layer deposition. The AZO layer serves as a transparent conductive layer (Fig. 6.7 (right)).

The photoresponsivity was measured using a PV quantum efficiency measurement system with a temperature control. No bias was applied during the measurement, so that the zero-bias photocurrent was measured. The result of the photoresponsivity measurement is illustrated in Fig. 6.8. Notice that the pronounced increase (by a factor 2) in PV efficiency due to the plasmonic effect is observed only as a result of the ZnO layer of nano-rods covered with Ag nanoparticles, whereas the efficiency of the Si substrate at distances from the Ag nanoparticles of more than 1 μm was much more weakly affected. This observation provides direct evidence that the plasmonic PV efficiency enhancement is a short-range effect not exceeding 1 μm distance between the metallic nanoparticles and the substrate.

In Fig. 6.8 the photoresponsivity of the p-Si/n-ZnO/Ag NPs/AZO system under study and that of a reference diode without nanoparticles, p-Si/n-ZnO/AZO, as well as that for a sole Si photodiode are shown for comparison. The spectral responsivity for the tested system with and without Ag nanoparticles was lower than that of the commercial Si photodiode within the whole wavelength range because the diodes under study were not optimized to reach the best performance (unlike the reference Si diode). Within the 400–600 nm wavelength range, the responsivity does not follow the linear dependence expected for Si. Instead, the measured responsivity grows and reaches a local maximum plateau. Silver nanoparticles deposited on top of the whole structure increase the effect, with an additionally hump around 500 nm.

Other studies of the p-Si/n-ZnO [131, 132] and n-Si/p-ZnO [133] systems suggest that the enhancement of the responsivity within the 400−600 nm wavelength range is due to the subband-gap trap states in ZnO and at the ZnO-Si interface [131, 132]. All these arguments support the interpretation that the signal within 400−600 nm is related to the ZnO layer [134]. Some authors [135] have also highlighted the role of surface defects in ZnO nanostructures, which could enhance the

6.4 Proximity Constraints of the Plasmon Effect

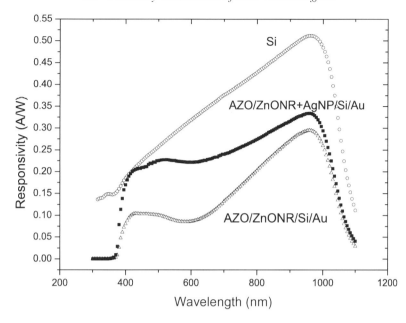

Figure 6.8 Spectral responsivity of the calibrated Si photodiode (open circles) and the studied p-Si/n-ZnO(nano-rods)/AZO photodiode, without silver nanoparticles (triangles) and with silver nanoparticles (squares). NR, nano-rods as in Fig. 6.6; NP, nanoparticles as in Fig. 6.7; Au, ohmic bottom contact. There is a strengthening, by a factor ~ 2, of the PV response of the two-layer system consisting of ZnO nano-rods (with length approximately 1 µm) deposited on the Si substrate and covered with Ag nanoparticles of approximately 5 nm in radius; the band structure of ZnO allows one to distinguish the related PV response from that of the Si component of the system and to observe that the plasmon effect applies only to the nearby nano-rods with respect to the Ag nanoparticles.

range of the photoresponsivity especially effectively for a nano-rod geometry with a strongly developed surface. This is the most probable scenario for the increased photoresponsivity within the energy range of the spectrum shown in Fig. 6.8 below the band gap of ZnO.

The ordinary photoeffect in both components of the two-layer structure under study allows us to distinguish the plasmonic strengthening in the two layers, mostly because of the different band gaps in ZnO and Si. When metallic nanoparticles were deposited on the top of the spacer, a significant increase (by a factor 2) of the photocurrent was observed in the frequency region relating to ZnO. The region relating to Si exhibited significantly reduced changes, demonstrating that the separation of the metallic nanocomponents by a distance of 1 micrometre strongly reduced the plasmonic effect in the bottom Si layer. This finding indicates that the plasmonic photovoltaic effect may be particularly beneficial in ultrathin-film solar cell technology.

6.5 Calculation of the Matrix Element for the Fermi Golden Rule Expression (6.37)

Let us calculate in detail the probability per time unit for interband transitions of electrons in a substrate semiconductor on which metallic nanospheres with radius a are deposited. These transitions are induced by photon-excited surface plasmons in the metallic component. The surface plasmons excited in the nano-sphere couple to the semiconductor band electrons in the near-field regime. The situation is similar to the ordinary photoeffect, though the perturbation of the electron system in the semiconductor is not of plane-wave form as in the case of the direct illumination of a semiconductor by incident photons, but attains the form of dipole-type near-field electric interaction [55, 23]. This causes a change in the matrix element in the relevant Fermi golden rule formula. The potential for the near-field interaction of the surface Mie-type plasmons with the band electrons can be written as [55]

$$w = e\psi(\mathbf{R}, t) = \frac{e}{\varepsilon_0 R^2} \hat{\mathbf{n}} \cdot \mathbf{D}_0 \sin(\omega t + \alpha) = w^+ e^{i\omega t} + w^- e^{-i\omega t},$$

$$w^+ = (w^-)^* = \frac{e}{\varepsilon_0 R^2} \frac{e^{i\alpha}}{2i} \hat{\mathbf{n}} \cdot \mathbf{D}_0.$$

(6.51)

The terms w^+, w^- correspond to emission and absorption, respectively ($\hat{\mathbf{n}} = \mathbf{R}/R$, $\mathbf{D}_0$ is the dipole surface plasmon amplitude). We choose the first term in order to consider the emission of energy from the plasmon oscillations and its transfer to the electron system in the semiconductor substrate. We model the semiconductor band system using the simplest single-band parabolic effective-mass approximation. The interband transition probability is given by the Fermi golden rule:

$$w(\mathbf{k}_1, \mathbf{k}_2) = \frac{2\pi}{\hbar} |\langle \mathbf{k}_1 | w^+ | \mathbf{k}_2 \rangle|^2 \delta(E(\mathbf{k}_1) - E(\mathbf{k}_2) + \hbar\omega), \qquad (6.52)$$

where the Bloch states in the conduction and valence bands are assumed to be plane waves for the sake of simplicity,

$$\Psi_{\mathbf{k}_1} = \frac{1}{(2\pi)^{3/2}} e^{i\mathbf{k}_1 \cdot \mathbf{R} - iE(\mathbf{k}_1)t/\hbar}, \qquad \Psi_{\mathbf{k}_2} = \frac{1}{(2\pi)^{3/2}} e^{i\mathbf{k}_2 \cdot \mathbf{R} - iE(\mathbf{k}_2)t/\hbar},$$

$$E(\mathbf{k}_1) = -\frac{\hbar^2 \mathbf{k}_1^2}{2m_p^*} - E_g, \qquad E(\mathbf{k}_2) = \frac{\hbar^2 \mathbf{k}_2^2}{2m_n^*},$$

(6.53)

the indices n, p refer to electrons from the conduction and valence bands, respectively, and E_g is the forbidden gap.

The electron wave functions are normalized to the Dirac delta function, which corresponds to an infinite movement and a continuous energy spectrum. The modulus squared of the wave functions do not have, in this case, a probabilistic interpretation (for which normalization to unity is required); therefore the expression,

$$w(\mathbf{k}_1, \mathbf{k}_2) = \frac{2\pi}{\hbar} |<\mathbf{k}_1 | w^+ | \mathbf{k}_2>|^2 \delta(E(\mathbf{k}_1) - E(\mathbf{k}_2) + \hbar\omega)$$

6.5 Calculation of Fermi Golden Rule Expression

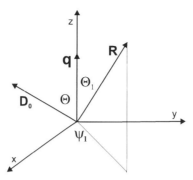

Figure 6.9 The reference frame is chosen such that the vector **q** is oriented along the z axis, while the vector $\mathbf{D}_0$ lies in the xz plane.

has to be divided by the delta Dirac function squared, i.e., by the factor $\left(V/(2\pi)^3\right)^2$. This factor corresponds to proper normalization of the probability, because

$$\frac{1}{(2\pi)^3}\int d^3 r e^{i\mathbf{k}\cdot\mathbf{r}} = \delta(\mathbf{k}) \simeq \text{(for } \mathbf{k}=0\text{)} \frac{V}{(2\pi)^3} \quad (V\to\infty). \tag{6.54}$$

Note that the same factor occurs also in the density of states when one integrates over all the initial and final states $\mathbf{k}_1, \mathbf{k}_2$. The two integrals both give the factor $\left(2V/(2\pi)^3\right)^2$ (the factor 2 is caused by spin degeneracy). Thus, the renormalization factors cancel (independently of $V\to\infty$).

Taking into account the renormalization described above, one finds that

$$\langle \mathbf{k}_1|w^+|\mathbf{k}_2\rangle = \frac{1}{(2\pi)^3}\int d^3 R \frac{e}{\varepsilon_0 2i} e^{i\alpha}\hat{\mathbf{n}}\cdot\mathbf{D}_0 \frac{1}{R^2} e^{-i(\mathbf{k}_1-\mathbf{k}_2)\cdot\mathbf{R}}. \tag{6.55}$$

Let us introduce the vector $\mathbf{q} = \mathbf{k}_2 - \mathbf{k}_1$. One can choose the coordinate system in such a way that the vector $\mathbf{q}$ is oriented along the z axis, and the vector $\mathbf{D}_0$ lies in the zx plane (as depicted in Fig. 6.9). Then, $\mathbf{q} = (0,0,q)$, $\hat{\mathbf{n}} = \mathbf{R}/R = (\sin\Theta_1\cos\psi_1, \sin\Theta_1\sin\psi_1, \cos\Theta_1)$, $\mathbf{R} = R(\sin\Theta_1\cos\psi_1, \sin\Theta_1\sin\psi_1, \cos\Theta_1)$, $\mathbf{D}_0 = D_0(\sin\Theta, 0, \cos\Theta)$ and

$$\mathbf{q}\cdot\mathbf{R} = qR\cos\Theta_1, \tag{6.56}$$

$$\hat{\mathbf{n}}\cdot\mathbf{D}_0 = D_0(\sin\Theta\sin\Theta_1\cos\psi_1 + \cos\Theta\cos\Theta_1). \tag{6.57}$$

Hence,

$$\langle \mathbf{k}_1|w^+|\mathbf{k}_2\rangle = \frac{1}{(2\pi)^3}\int d^3 R \frac{e}{\varepsilon_0 2i} e^{i\alpha}\hat{\mathbf{n}}\cdot\mathbf{D}_0 \frac{1}{R^2} e^{-i(\mathbf{k}_1-\mathbf{k}_2)\cdot\mathbf{R}}$$

$$= \frac{1}{(2\pi)^3}\frac{ee^{i\alpha}}{\varepsilon_0 2i} D_0 \int_a^\infty \frac{R^2}{R^2} dR \times \int_0^\pi \sin\Theta_1\, d\Theta_1$$

$$\times \int_0^{2\pi} d\psi_1\{\cos\Theta\cos\Theta_1 + \sin\Theta\sin\Theta_1\cos\psi_1\} e^{iqR\cos\Theta_1}$$

$$= \frac{1}{(2\pi)^3} \frac{ee^{i\alpha}}{\varepsilon_0 2i} D_0 \cos\Theta 2\pi \int_a^\infty dR \int_0^\pi \cos\Theta_1 \sin\Theta_1 d\Theta_1 e^{iqR\cos\Theta_1}.$$

(6.58)

In the integral over $d\psi_1$ the second term in the braces vanishes and only the first term contributes, with a factor 2π. Note that

$$\int_0^\pi \cos\Theta_1 \sin\Theta_1 d\Theta_1 e^{ix\cos\Theta_1} = -i\frac{d}{dx}\int_0^\pi \sin\Theta_1 d\Theta_1 e^{ix\cos\Theta_1} = -i\frac{d}{dx}\frac{2\sin x}{x}.$$

(6.59)

Thus

$$\langle \mathbf{k}_1|w^+|\mathbf{k}_2\rangle = \frac{-1}{(2\pi)^3} \frac{ee^{i\alpha}}{\varepsilon_0} D_0 \cos\Theta (2\pi) \int_a^\infty dR \frac{1}{q}\frac{d}{dR}\frac{\sin qR}{qR}$$

$$= \frac{1}{(2\pi)^2} \frac{ee^{i\alpha}}{\varepsilon_0} D_0 \cos\Theta \frac{1}{q}\frac{\sin qa}{qa} \xrightarrow{a\to 0} \frac{1}{(2\pi)^2} \frac{ee^{i\alpha}}{\varepsilon_0} D_0 \cos\Theta \frac{1}{q}.$$

(6.60)

The lower limit of the integral with respect to R is taken as a (the nanosphere radius). In this manner one can constrain the accessible space for planar waves in the semiconductor (if one assumes, for a model, a nanosphere completely embedded in the surrounding material). This is, however, a crude approximation as the tunnelling effect would certainly allow access to the inner nanoparticle space, to some extent at least. Moreover, if the nanosphere is deposited on the semiconductor surface one can confine the integration with respect to $d\Theta_1$ to the segment $[0, \pi/2]$, instead of $[0, \pi]$. In this case we obtain the following matrix element:

$$\langle \mathbf{k}_1|w^+|\mathbf{k}_2\rangle = \frac{1}{(2\pi)^3} \frac{ee^{i\alpha}}{2i\varepsilon_0} D_0 \cos\Theta (2\pi) \int_a^\infty dR \frac{1}{q}\frac{d}{dR}\frac{1-e^{iqR}}{qR}$$

$$= \frac{1}{(2\pi)^2} \frac{ee^{i\alpha}}{2i\varepsilon_0} D_0 \cos\Theta \frac{1}{q}\frac{1-e^{iqa}}{qa} \xrightarrow{a\to 0} \frac{1}{(2\pi)^2} \frac{ee^{i\alpha}}{2i\varepsilon_0} D_0 \cos\Theta \frac{1}{q},$$

(6.61)

because

$$\int_0^{\pi/2} \cos\Theta_1 \sin\Theta_1 d\Theta_1 e^{ix\cos\Theta_1} = -i\frac{d}{dx}\int_0^{\pi/2} \sin\Theta_1 d\Theta_1 e^{ix\cos\Theta_1} = \frac{d}{dx}\frac{1-e^{ix}}{x}.$$

(6.62)

In the limit $a = 0$, the modulus of the matrix element will be twice as small as in the case when the integration over $d\Theta_1$ is taken over the whole space.

6.5 Calculation of Fermi Golden Rule Expression

Now we will integrate over all initial and final states of both bands. One can substitute the related integration over $\mathbf{k}_1, \mathbf{k}_2$ with integration over $\mathbf{q}, \mathbf{k}_2$. Scalar products are invariant under coordinate system rotation; therefore the result of integration will be the same whether $\mathbf{q}$ is along the z axis or $\mathbf{D}_0$ is the along the z axis. In the latter case Θ gives the deviation of $\mathbf{q}$ with respect to z direction – this choice of reference frame is convenient for integration with respect to $d\mathbf{q}$.

Thus we arrive at a formula for the transition probability in the following form:

$$\delta w = \int d^3k_1 \int d^3k_2 \left(f_1(1-f_2) w(\mathbf{k}_1, \mathbf{k}_2) - f_2(1-f_1) w(\mathbf{k}_2, \mathbf{k}_1) \right), \quad (6.63)$$

where f_1, f_2 are the temperature-dependent Fermi–Dirac distribution functions for the initial and final states, respectively. Emission and absorption are included in the overall transition probability, but for room temperatures one can assume $f_2 \simeq 0$ and $f_1 \simeq 1$, which leads to

$$\delta w = \int d^3k_1 \int d^3k_2 \, w(\mathbf{k}_1, \mathbf{k}_2). \quad (6.64)$$

The above formula does not include the density state factors since they are cancelled by the probability renormalization, as mentioned above.

We thus have to calculate the following integral:

$$\delta w = \int d^3k_2 \int d^3q \, \frac{e^2}{(2\pi)^3 \hbar \varepsilon_0} \frac{D_0^2 \cos^2\Theta}{q^2} \frac{\sin^2(qa)}{(qa)^2} \delta \left(\frac{\hbar^2 k_1^2}{2m_n^*} + \frac{\hbar^2 k_2}{2m_p^*} - (\hbar\omega - E_g) \right), \quad (6.65)$$

with

$$\delta \left(\frac{\hbar^2 k_1^2}{2m_n^*} + \frac{\hbar^2 k_2}{2m_p^*} - (\hbar\omega - E_g) \right) = \frac{1}{\alpha+\beta} \frac{1}{2\beta' q k_2} \delta \left(\cos\Theta_2 - \frac{k_2^2 + \beta' q^2 - \gamma'}{2\beta' q k_2} \right),$$

where $\alpha = \hbar^2/2m_n^*$, $\beta = \hbar^2/2m_p^*$, $\gamma = \hbar\omega - E_g$, $\alpha' = \alpha/(\alpha+\beta)$, $\beta' = \beta/(\alpha+\beta)$, $\gamma' = \gamma/(\alpha+\beta)$. For each integral with respect to $\mathbf{k}_2$ and $\mathbf{q}$ the coordinate system can be rotated independently, and for the integration over $d\mathbf{k}_2$ we choose again the orientation of $\mathbf{q}$ along the z axis, which leads to the spherical angle Θ_2 in the product $\mathbf{k}_2 \cdot \mathbf{q} = k_2 q \cos\Theta_2$.

The expression for δw thus attains the form

$$\delta w = \frac{e^2 D_0^2}{(2\pi)^3 \hbar \varepsilon_0^2} \int d^3q \, \frac{\sin^2(qa) \cos\Theta}{q^2 a^2 \, q^2} \times \int_0^\infty dk_2 k_2^2 \int_0^\pi d\Theta_2 \sin\Theta_2$$

$$\times \int_0^{2\pi} d\psi_2 \frac{1}{\alpha+\beta} \frac{1}{2\beta' q k_2} \delta \left(\cos\Theta_2 - \frac{k_2^2 + \beta' q^2 - \gamma'}{2\beta' q k_2} \right). \quad (6.66)$$

Integration over $d\Theta_2$ employs the Dirac delta. The relevant nonzero contribution (due to integration over $d\cos\Theta_2$) is conditioned by the inequality

$$-1 < \frac{k_2^2 + \beta' q^2 - \gamma'}{2\beta' q k_2} < 1, \qquad (6.67)$$

which resolves itself into the constraint

$$|\beta' q - \sqrt{\gamma' - (1-\beta')\beta' q^2}| < k_2 < \beta' q + \sqrt{\gamma' - (1-\beta')\beta' q^2}. \qquad (6.68)$$

Thus

$$\delta w = \frac{1}{(\alpha+\beta)2\beta'} \frac{e^2 D_0^2}{2\pi \hbar \varepsilon_0^2} \int_0^{\sqrt{\gamma'/(\beta'(1-\beta'))}} dq \, q^2 \frac{\sin^2 qa}{(qa)^2}$$

$$\times \int_0^\pi d\Theta \sin\Theta \cos^2\Theta \frac{1}{q^3} \int_{|\beta' q - \sqrt{\gamma' - (1-\beta')\beta' q^2}|}^{\beta' q + \sqrt{\gamma' - (1-\beta')\beta' q^2}} dk_2 k_2 \qquad (6.69)$$

and hence

$$\delta w = \frac{2}{3} \frac{\sqrt{\gamma'}}{\alpha+\beta} \frac{e^2 D_0^2}{2\pi \hbar \varepsilon_0^2} \frac{1}{\zeta} \int_0^1 dx \frac{\sin^2(xa/\zeta)}{x^2(a/\zeta)^2} \sqrt{1-x^2}, \qquad (6.70)$$

where $\zeta = \sqrt{(1-\beta')\beta'/\gamma'}$. Therefore

$$\delta w = \frac{4}{3} \frac{\mu^2(m_n^* + m_p^*) 2(\hbar\omega - E_g) e^2 D_0^2}{\sqrt{m_n^* m_p^*} 2\pi \hbar^5 \varepsilon_0^2} \int_0^1 dx \frac{\sin^2(xa/\zeta)}{(xa/\zeta)^2} \sqrt{1-x^2}$$

$$= \frac{4}{3} \frac{\mu^2}{\sqrt{m_n^* m_p^*}} \frac{e^2 D_0^2}{2\pi \hbar^3 \varepsilon_0^2} \xi^2 \int_0^1 dx \frac{\sin^2(xa\xi)}{(xa\xi)^2} \sqrt{1-x^2}, \qquad (6.71)$$

where $\xi = 1/\zeta = \sqrt{2(\hbar\omega - E_g)(m_n^* + m_p^*)/\hbar}$.

In the limiting cases we finally obtain

$$\delta w = \begin{cases} \dfrac{1}{3} \dfrac{\mu^2(m_n^* + m_p^*) 2(\hbar\omega - E_g) e^2 D_0^2}{\sqrt{m_n^* m_p^*} 2\hbar^5 \varepsilon_0^2} & \text{for } a\xi \ll 1, \\[2ex] \dfrac{4}{3} \dfrac{\mu^2(m_n^* + m_p^*) 2(\hbar\omega - E_g) e^2 D_0^2}{\sqrt{m_n^* m_p^*} 4a\xi \hbar^5 \varepsilon_0^2} & \text{for } a\xi \gg 1. \end{cases} \qquad (6.72)$$

In the latter case the following approximation has been applied:

$$\int_0^1 \frac{dx \sin^2(xa\xi)}{(xa\xi)^2} \approx \text{(for } a\xi \gg 1\text{)} \frac{1}{a\xi} \int_0^\infty d(xa\xi) \frac{\sin^2(xa\xi)}{(xa\xi)^2} = \frac{\pi}{2a\xi},$$

while in the former case, $\int_0^1 dx \sqrt{1-x^2} = \pi/4$.

6.5 Calculation of Fermi Golden Rule Expression

The result must be multiplied by 4 (owing to spin degenerary). In the considered limiting cases we thus obtain

$$\delta w = \begin{cases} \dfrac{4\mu\sqrt{m_n^* m_p^*}(\hbar\omega - E_g)e^2 D_0^2}{3\hbar^5 \varepsilon_0^2} & \text{for } a\xi \ll 1, \\ \dfrac{4\mu^{3/2}\sqrt{2}\sqrt{\hbar\omega - E_g}e^2 D_0^2}{3 a\hbar^4 \varepsilon_0^2} & \text{for } a\xi \gg 1. \end{cases} \quad (6.73)$$

the formula applies to the case of a completely embedded nanosphere.

Calculation of the matrix element that appears in the Fermi golden rule (6.37) can be also performed using Fourier transforms. Equation (6.32) can be rewritten in the following manner:

$$\langle \mathbf{k}_1 | w^+ | \mathbf{k}_2 \rangle = \frac{1}{(2\pi)^{3/2}} \frac{ee^{i\alpha}}{\varepsilon 2i} \mathbf{D}_0 \cdot \left(\frac{1}{(2\pi)^{3/2}} \int d^3 R \, e^{-i\mathbf{q}\cdot\mathbf{R}} \frac{\mathbf{R}}{R^3} \chi(\mathbf{R}) \right),$$

where $\mathbf{q} = \mathbf{k}_1 - \mathbf{k}_2$ and χ defines the integration range for a specific type of deposition of the nanosphere. In the case where the nanoparticle is completely embedded in the semiconductor medium, the integration runs over the space outside the nanosphere, i.e., $\chi(\mathbf{R}) = \theta(R - a)$, where θ is the Heaviside step function. This is an idealization that does not account for the possible penetration of the nanosphere interior caused by tunnelling of the electron wave functions; however, this effect can be neglected for the present model. The factor in parentheses is the unitary three-dimensional Fourier transform of

$$\frac{1}{(2\pi)^{3/2}} \int d^3 R \, e^{-i\mathbf{q}\cdot\mathbf{R}} \frac{\mathbf{R}}{R^3} \chi(\mathbf{R}) = F\left[\frac{\mathbf{R}}{R^3} \chi(\mathbf{R}) \right].$$

Note that the property $F[\nabla f(\mathbf{R})] = i\mathbf{q} F[f(\mathbf{R})]$ yields the following transformation:

$$F\left[\frac{\mathbf{R}}{R^3} \right] = -F\left[\nabla \frac{1}{R} \right] = \frac{-i\mathbf{q}}{(2\pi)^{3/2}} \frac{4\pi}{q^2}.$$

To find $F\left[\frac{\mathbf{R}}{R^3}\chi(\mathbf{R})\right]$, we will decompose $\mathbf{R}/R^3 \chi(\mathbf{R})$ into a convolution of two functions: we can show that

$$\frac{\mathbf{R}}{R^3} \chi(\mathbf{R}) = \frac{\mathbf{R}}{R^3} * \frac{\delta(R-a)}{4\pi a R} \quad (6.74)$$

through direct calculation, i.e.,

$$\frac{\mathbf{R}}{R^3} \chi(\mathbf{R}) = \int d^3 t \, \frac{\mathbf{R}-\mathbf{t}}{|\mathbf{R}-\mathbf{t}|^3} \frac{\delta(t-a)}{4\pi a t}.$$

One can perform this integration by transforming into spherical coordinates with the zenith direction along **R**,

$$\frac{\mathbf{R}}{R^3} * \frac{\delta(R-a)}{4\pi a R} = \frac{1}{4\pi a} \int_0^\pi d\theta \sin\theta \int_0^{2\pi} d\varphi \int_0^\infty dt\, \delta(t-a) t \frac{\mathbf{R}-\mathbf{t}}{|\mathbf{R}-\mathbf{t}|^3}$$

$$= \frac{1}{2}\hat{\mathbf{R}} \int_{-1}^{1} du \frac{R-au}{(a^2+R^2-2aRu)^{3/2}},$$

where $u = \cos\theta$. Noting that

$$\frac{R-au}{(a^2+R^2-2aRu)^{3/2}} = -\frac{d}{dR}\frac{1}{(a^2+R^2-2aRu)^{1/2}}$$

and substituting $v = a^2 + R^2 - 2aRu$ yields the following integral:

$$\frac{\mathbf{R}}{R^3} * \frac{\delta(R-a)}{4\pi a R} = \frac{1}{2}\hat{\mathbf{R}} \frac{d}{dR} \frac{1}{2aR} \int_{(a+R)^2}^{(a-R)^2} dv \frac{1}{v^{1/2}} = \frac{\mathbf{R}}{R^3}\theta(R-a).$$

Using the convolution formula $F[f(\mathbf{r}) * g(\mathbf{r})] = (2\pi)^{3/2} F[f(\mathbf{r})] F[g(\mathbf{r})]$, one obtains

$$F\left[\frac{\mathbf{R}}{R^3}\chi(\mathbf{R})\right] = (2\pi)^{3/2} F\left[\frac{\mathbf{R}}{R^3}\right] F\left[\frac{\delta(R-a)}{2aR}\right],$$

where

$$F\left[\frac{\hat{\mathbf{n}}}{R^2}\right] = -F\left[\nabla\frac{1}{R}\right] = \frac{-i\mathbf{q}}{(2\pi)^{3/2}} \frac{4\pi}{q^2}$$

and

$$F\left[\frac{\delta(R-a)}{4\pi a R}\right] = \frac{1}{(2\pi)^{3/2}} \int d^3R\, e^{-i\mathbf{q}\mathbf{R}} \frac{\delta(R-a)}{4\pi a R}.$$

We note that

$$F\left[\frac{\delta(R-a)}{4\pi a R}\right] = \frac{1}{2}\int_{-1}^{1} du\, e^{-iqau} = \frac{\sin qa}{qa},$$

and finally arrive at

$$\langle \mathbf{k}_1 | w^+ | \mathbf{k}_2 \rangle = -\frac{1}{\sqrt{2\pi}} \frac{ee^{i\alpha}}{\varepsilon} \frac{\mathbf{D}_0 \cdot \mathbf{q}}{q^2} \frac{\sin qa}{qa}. \qquad (6.75)$$

6.5 Calculation of Fermi Golden Rule Expression

To estimate the absorption probability per unit time, a summation over all initial and final states in both bands, including their filling probabilities, can be performed in the same manner, starting with formula (7.7).

The expressions (6.73) are distinct from the corresponding formula for the ordinary photoeffect. In the latter case the perturbation is given by the vector potential in the kinematic momentum (for the gauge with div $\mathbf{A} = 0$):

$$\hat{H} = \frac{(-i\hbar\nabla - (e/c)\mathbf{A}(\mathbf{R},t))^2}{2m^*} \simeq \frac{(-i\hbar\nabla)^2}{2m^*} + \frac{i\hbar e}{m^*c}\mathbf{A}(\mathbf{R},t) \cdot \nabla. \quad (6.76)$$

For a monochromatic plane wave this perturbation has the form

$$w_0(\mathbf{R},t) = \frac{ie\hbar}{cm^*}\cos(\omega t - \mathbf{k}\cdot\mathbf{R} + \alpha)(\mathbf{A}_0 \cdot \nabla)$$

$$= \frac{ie\hbar}{2cm^*}(e^{i(\omega t - \mathbf{k}\cdot\mathbf{R}+\alpha)} + e^{-i(\omega t - \mathbf{k}\cdot\mathbf{R}+\alpha)})(\mathbf{A}_0 \cdot \nabla). \quad (6.77)$$

Because in this case states of both the band electrons and the photon have the form of plane waves, the matrix element in the Fermi golden rule is proportional to the Dirac delta with respect to the momentum sum. This proportionality expresses momentum conservation in a translationally invariant system, which is the case for a photon interacting with a semiconductor:

$$\langle \mathbf{k}_1 | w_0(\mathbf{R},t) | \mathbf{k}_2 \rangle = \frac{i\hbar e}{2cm^*(2\pi)^3}\int d^3R\, e^{-i(\mathbf{k}_1+\mathbf{k})\cdot\mathbf{R}}(\mathbf{A}_0 \cdot \nabla)e^{i\mathbf{k}_2\cdot\mathbf{R}}$$

$$= -\frac{e\hbar}{2cm^*}\mathbf{A}_0 \cdot \mathbf{k}_2 \delta(\mathbf{k}_1 + \mathbf{k} - \mathbf{k}_2). \quad (6.78)$$

Because of the high value of the photon velocity c, the photon momentum $\mathbf{k}$ is negligibly small and thus only vertical transitions, with $\mathbf{k}_1 = \mathbf{k}_2$, are permitted in the ordinary photoeffect.

However, this is not the case for nanosphere plasmons interacting in the near-field regime with the semiconductor substrate. This system is not translationally invariant and excitations other than vertical interband electron excitations are permitted, which results in enhancement of the total transition probability.

For comparison with the ordinary photoeffect, let us recall the appropriate calculus based on the Fermi golden rule scheme for the ordinary photoeffect:

$$w_0(\mathbf{k}_1, \mathbf{k}_2) = \frac{\pi e^2 \hbar}{2c^2 m^{*2}} k_2^2 A_0^2 \cos^2\Theta\, \delta^2(\mathbf{k}_1 - \mathbf{k}_2)\delta(E_p - E_n + \hbar\omega), \quad (6.79)$$

where Θ is the angle between $\mathbf{k}_2$ and $\mathbf{A}_0$. One can use the following approximation to eliminate the Dirac delta squared:

$$\delta^2(\mathbf{k}_1 - \mathbf{k}_2) = \delta(0)\delta(\mathbf{k}_1 - \mathbf{k}_2) \simeq \frac{V}{(2\pi)^3}\delta(\mathbf{k}_1 - \mathbf{k}_2). \quad (6.80)$$

As in the previous discussion, probability normalization must be performed. Thus, we obtain

$$w_0(\mathbf{k}_1,\mathbf{k}_2) = \frac{\pi e^2 \hbar}{2c^2 m^{*2}} \frac{V}{(2\pi)^3} k_2^2 A_0^2 \cos^2\Theta\, \delta(\mathbf{k}_1 - \mathbf{k}_2)\delta(E_p - E_n + \hbar\omega). \quad (6.81)$$

Integration over all states in both electron bands results in

$$\delta w_0 = \int d^3 k_1 \int d^3 k_2 \frac{\pi e^2 \hbar}{2c^2 m^{*2}} \frac{V}{(2\pi)^3} k_2^2 A_0^2 \cos^2\Theta\, \delta(\mathbf{k}_1 - \mathbf{k}_2)\delta(E_p - E_n + \hbar\omega) \quad (6.82)$$

or

$$\delta w_0 = \int d^3 k_2 \frac{\pi e^2 \hbar}{2c^2 m_p^{*2}} \frac{V}{(2\pi)^3} k_2^2 A_0^2 \cos^2\Theta\, \delta(E_p(\mathbf{k}_2) - E_n(\mathbf{k}_2) + \hbar\omega), \quad (6.83)$$

where m_p^* is the effective mass in the initial valence state (moreover, $A_0 = cE_0/\omega$). We thus obtain

$$\delta w_0 = \frac{1}{12}\frac{e^2\hbar}{m_p^{*2}}\frac{V}{\pi}\frac{E_0^2}{\omega^2}\frac{2^{3/2}\mu^{5/2}}{\hbar^5}(\hbar\omega - E_g)^{3/2}. \quad (6.84)$$

This expression can be rewritten in the following form, which is convenient for the derivation of Eq. (6.39):

$$\delta w_0 = \frac{4\sqrt{2}}{3}\frac{\mu^{5/2}e^2}{m_p^{*2}\omega\varepsilon\hbar^3}\left(\frac{\varepsilon E_0^2 V}{8\pi\hbar\omega}\right)(\hbar\omega - E_g)^{3/2}; \quad (6.85)$$

the expression should be multiplied by 4 (because of spin degeneracy, although for circular polarization of the photons this factor is only equal to 2 because of the angular momentum selection rules).

7

Plasmon-Induced Efficiency Enhancement of Solar Cells Modified by Metallic Nanoparticles: Material Dependence

The plasmon-mediated photoeffect is presented in examples of commercial multi-crystalline Si and CIGS cells. The material dependence of the plasmon effect with respect to variation in the metallic components and semiconductor substrate is discussed and illustrated.

Sunlight-energy harvesting mediated by surface plasmons in metal-nano-modified solar cells is caused by three effects: (1) the strong concentration of the plasmon oscillation field close to the metallic components, with their large local curvature, (2) the large amplitude of plasmon oscillations in metallic nanoparticles, and (3) the enhancement of the probability of interband transitions in the semiconductor substrate caused by the breaking of the translational symmetry for a nanoparticle and the consequent dipole near-field coupling of surface plasmons with the semiconductor band electrons [32, 110, 2, 111, 112, 23]. The transition probability for the excitation of electrons from the valence band to the conduction band in a semiconductor, being essential for the efficiency of the photovoltaic effect, grows due to the electric field amplitude enhancement and due to the admission of oblique transitions not prohibited here by momentum conservation [23]. For the ordinary photoeffect [59] the interband transitions are confined to those that are vertical, between states with almost the same momentum owing to momentum conservation and the very small momentum of sunlight photons (due to the large velocity of light, c.) Therefore in the ordinary photoeffect these is almost no change in electron momentum at scattering: for an excitation energy $\hbar\omega$ beyond the forbidden gap E_g in the substrate semiconductor, the photon dispersion $\hbar\omega = cq$ gives $q \ll p$, where $p \sim \pi\hbar/l$ is the semiconductor-band quasimomentum scale in the Brillouin zone (l denotes here the linear size of the elementary cell). Thus the change in the band-electron momentum $\mathbf{q} = \mathbf{p}_2 - \mathbf{p}_1$ is negligible on the scale of the Brillouin zone and $\mathbf{p}_1 \simeq \mathbf{p}_2$ (because $c = 10^8$ m/s) and only the vertical, momentum conserving, interband transitions contribute to the ordinary photoeffect, i.e., the transition is caused by free photons with the momentum $\mathbf{q}$ and energy $\hbar\omega = cq$.

However, for the interaction of band electrons with surface plasmons from metallic nanoparticles deposited on the semiconductor surface, the situation changes significantly. In the near-field regime [55] the potential of the plasmon dipole on the nanosphere is proportional to $1/R^2$ (R is the distance from the nanosphere centre), which has an infinite decomposition in the Fourier picture and thus overlaps with all quasimomenta in the substrate semiconductor Brillouin zone. This is in contrast with the potential of the free photon, which contributes only via a single $e^{i(\mathbf{q}\cdot\mathbf{r}-\hbar\omega t)/\hbar}$ plane-wave Fourier component in the electromagnetic field vector potential entering the kinematic momentum [59].

The effect of oblique interband transitions can be accounted for via the Fermi golden rule (FGR). According to the FGR scheme [75] the probability of an interband transition is proportional to the matrix element of the perturbation potential between the initial and the final state; this is summed over all initial states in the valence band and over all final states in the conduction band, with energy conservation imposed, $E_p(\mathbf{p}_1) + \hbar\omega = E_n(\mathbf{p}_2)$. Where $E_{p(n)}(\mathbf{p})$ are the valence-band and conduction-band dispersions and $\hbar\omega$ is the excitation energy related to damped and forced surface plasmon oscillations with the bare self-energy, $\hbar\omega_1 = \hbar\omega_p/\sqrt{3}$ (i.e., the Mie energy [18, 17], $\hbar\omega_p = \hbar\sqrt{n_e e^2/(m^*\varepsilon_0)}$ is the bulk plasmon energy in a metal [16], where n_e is the density of the free electrons in a metal, m^* is the effective mass of an electron in a metal, e is the electron charge, and ε_0 is the dielectric constant); the momentum is undefined, however. The initial momentum $\mathbf{p}_1$ and the final momentum, $\mathbf{p}_2$ can be arbitrary because momentum conservation is ruled out by the matrix element of the local dipole interaction.

7.1 Plasmon-Mediated Photoeffect: Probability of Electron Interband Excitation Due to Plasmons

The perturbation of the electron band system in the substrate semiconductor due to the presence of dipole surface plasmon oscillations in a metallic nanosphere (with radius a) deposited on the semiconductor surface has the form of the potential of the electromagnetic field of an oscillating dipole, including retardation. The Fourier components of the electric field $\mathbf{E}_\omega$ and magnetic field $\mathbf{B}_\omega$ produced at a distance $\mathbf{R}$ from the centre of the nanosphere by a surface plasmon dipole D_0 with frequency ω have the form [55]

$$\mathbf{E}_\omega = \frac{1}{\varepsilon}\left\{\mathbf{D}_0\left(\frac{k^2}{R} + \frac{ik}{R^2} - \frac{1}{R^3}\right) + \hat{\mathbf{n}}(\hat{\mathbf{n}}\cdot\mathbf{D}_0)\left(-\frac{k^2}{R} - \frac{3ik}{R^2} + \frac{3}{R^3}\right)\right\}e^{ikR} \quad (7.1)$$

and

$$\mathbf{B}_\omega = \frac{ik}{\sqrt{\varepsilon}}[\mathbf{D}_0 \times \hat{\mathbf{n}}]\left(\frac{ik}{R} - \frac{1}{R^2}\right)e^{ikR}, \quad (7.2)$$

7.1 Probability of Electron Interband Excitation Due to Plasmons

where ε is the dielectric permittivity of the surroundings. In the case of spherical symmetry, the plasmon dipole may be considered as pinned to the centre of the nanosphere (the origin of the reference frame system): $\mathbf{D} = \mathbf{D}_0 e^{-i\omega t}$. In Eqs. (7.1) and (7.2) we used the notation for the retarded argument, $i\omega(t - R/c) = i\omega t - ikR$, $\hat{\mathbf{n}} = \mathbf{R}/R$, $\omega = ck$, and the momentum $\mathbf{p} = \hbar\mathbf{k}$. The terms with denominators R^3, R^2, and R refer to the near-, medium-, and far-field zones of the dipole radiation, respectively. Because we are considering the interaction with an adjacent layer of the substrate semiconductor, all terms with denominators R^2 and R can be neglected as small in comparison to the term with denominator R^3 – this is the near-field-zone approximation (in which the magnetic field disappears and the electric field is of static dipole form [55]). Therefore the related perturbation potential to be added to the Hamiltonian of the system has the form

$$w = e\psi(\mathbf{R}, t) = \frac{e}{\varepsilon R^2}\hat{\mathbf{n}} \cdot \mathbf{D}_0 \sin(\omega t + \alpha) = w^+ e^{i\omega t} + w^- e^{-i\omega t}. \tag{7.3}$$

The term

$$w^+ = (w^-)^* = \frac{e}{\varepsilon R^2} \frac{e^{i\alpha}}{2i} \hat{\mathbf{n}} \cdot \mathbf{D}_0$$

describes emission, i.e., the case of interest.

According to the Fermi golden rule [75], the interband transition probability is proportional to

$$w(\mathbf{k}_1, \mathbf{k}_2) = \frac{2\pi}{\hbar} \left|\langle \mathbf{k}_1 | w^+ | \mathbf{k}_2 \rangle\right|^2 \delta(E_p(\mathbf{k}_1) - E_n(\mathbf{k}_2) + \hbar\omega), \tag{7.4}$$

where the Bloch states in the conduction and valence bands are assumed to be planar waves for simplicity, $\Psi_{\mathbf{k}} = (2\pi)^{-3/2} e^{i\mathbf{k}\cdot\mathbf{R} - iE_{n(p)}(\mathbf{k})t/\hbar}$, $E_p(\mathbf{k}) = -\hbar^2\mathbf{k}^2/(2m_p^*) - E_g$, and $E_n(\mathbf{k}) = \hbar^2\mathbf{k}^2/(2m_n^*)$ (the indices n, p refer to electrons from the conduction and valence bands, respectively, and E_g is the forbidden gap).

The matrix element

$$\langle \mathbf{k}_1 | w^+ | \mathbf{k}_2 \rangle = \frac{1}{(2\pi)^3} \int d^3 R \frac{e}{\varepsilon 2i} e^{i\alpha} \hat{\mathbf{n}} \cdot \mathbf{D}_0 \frac{1}{R^2} e^{-i(\mathbf{k}_1 - \mathbf{k}_2)\cdot\mathbf{R}} \tag{7.5}$$

can be found analytically by direct integration, which gives the formula ($\mathbf{q} = \mathbf{k}_1 - \mathbf{k}_2$),

$$\begin{aligned}\langle \mathbf{k}_1 | w^+ | \mathbf{k}_2 \rangle &= \frac{-1}{(2\pi)^3} \frac{ee^{i\alpha}}{\varepsilon} D_0 \cos\Theta (2\pi) \int_a^\infty dR \frac{1}{q} \frac{d}{dR} \frac{\sin(qR)}{qR} \\ &= \frac{1}{(2\pi)^2} \frac{ee^{i\alpha}}{\varepsilon} \frac{\mathbf{D}_0 \cdot \mathbf{q}}{q^2} \frac{\sin(qa)}{qa}.\end{aligned} \tag{7.6}$$

Next, we must sum over all initial and final states in both bands. Thus, for the total interband transition probability we obtain

$$\delta w = \int d^3k_1 \int d^3k_2 \left[f_1(1-f_2)w(\mathbf{k}_1,\mathbf{k}_2) - f_2(1-f_1)w(\mathbf{k}_2,\mathbf{k}_1) \right], \quad (7.7)$$

where f_1, f_2 are temperature-dependent Fermi–Dirac distribution functions for the initial and final states, respectively. At room temperature $f_2 \simeq 0$ and $f_1 \simeq 1$, which leads to

$$\delta w = \int d^3k_1 \int d^3k_2 w(\mathbf{k}_1,\mathbf{k}_2). \quad (7.8)$$

After analytical integration of the above formula (as presented in the previous chapter), we arrive at the expression

$$\delta w = \frac{4}{3} \frac{\mu^2 (m_n^* + m_p^*) 2(\hbar\omega - E_g)e^2 D_0^2}{\sqrt{m_n^* m_p^*} 2\pi \hbar^5 \varepsilon^2} \int_0^1 dx \frac{\sin^2(xa\xi)}{(xa\xi)^2} \sqrt{1-x^2}$$

$$= \frac{4}{3} \frac{\mu^2}{\sqrt{m_n^* m_p^*}} \frac{e^2 D_0^2}{2\pi \hbar^3 \varepsilon^2} \xi^2 \int_0^1 dx \frac{\sin^2(xa\xi)}{(xa\xi)^2} \sqrt{1-x^2}, \quad (7.9)$$

according to the assumed band dispersions; m_n^* and m_p^* denote the effective masses of electrons and holes, $\mu = m_n^* m_p^*/(m_n^* + m_p^*)$ is the reduced mass, and the parameter $\xi = \sqrt{2(\hbar\omega - E_g)(m_n^* + m_p^*)/\hbar}$. In the limiting cases, for a nanoparticle of radius a we finally obtain

$$\delta w = \begin{cases} \dfrac{4}{3} \dfrac{\mu \sqrt{m_n^* m_p^*}(\hbar\omega - E_g)e^2 D_0^2}{\hbar^5 \varepsilon^2} & \text{for } a\xi \ll 1, \\[1em] \dfrac{4}{3} \dfrac{\mu^{3/2} \sqrt{2} \sqrt{\hbar\omega - E_g} e^2 D_0^2}{a\hbar^4 \varepsilon^2} & \text{for } a\xi \gg 1. \end{cases} \quad (7.10)$$

In the latter case the following approximation was applied,

$$\int_0^1 dx \frac{\sin^2(xa\xi)}{(xa\xi)^2} \sqrt{1-x^2} \approx (\text{for } a\xi \gg 1) \frac{1}{a\xi} \int_0^\infty d(xa\xi) \frac{\sin^2(xa\xi)}{(xa\xi)^2} = \frac{\pi}{2a\xi},$$

whereas in the former case we have $\int_0^1 dx \sqrt{1-x^2} = \pi/4$.

With regard to two limiting cases $a\xi \ll 1$ and $a\xi \gg 1$, for $\xi = \sqrt{2(\hbar\omega - E_g)(m_n^* + m_p^*)/\hbar}$, we see that

$$a \simeq \frac{1}{\xi} \simeq \begin{cases} > 2 \times 10^{-9} \text{ (m)} & \text{for } \dfrac{\hbar\omega - E_g}{E_g} < 0.02, \\[1em] < 2 \times 10^{-9} \text{ (m)} & \text{for } \dfrac{\hbar\omega - E_g}{E_g} > 0.02, \end{cases}$$

and the value of this range weakly depends on the effective masses and E_g. Thus for nanoparticles with radii $a > 2$ nm the first regime holds only close to E_g (less

7.1 Probability of Electron Interband Excitation Due to Plasmons

than 2% of the distance of $\hbar\omega$ to the limiting E_g), whereas the second regime holds in the rest of the ω domain. For comparison,

$$a \simeq \frac{1}{\xi} \simeq \begin{cases} > 0.5 \times 10^{-9} \text{ (m)} & \text{for } \dfrac{\hbar\omega - E_g}{E_g} < 0.5, \\ < 0.5 \times 10^{-9} \text{ (m)} & \text{for } \dfrac{\hbar\omega - E_g}{E_g} > 0.5; \end{cases}$$

the first region widens considerably, to about 50% of the relative distance to E_g, but this holds only for ultrasmall nanoparticles ($a < 0.5$ nm). For larger nanoparticles, e.g., with $a > 10$ nm, the second regime thus dominates.

Note that the above formula, Eq. (7.9), and its explicit form in limiting situations given by Eq. (7.10), is a generalization of the ordinary photoeffect, for which the transition probability is different [59]:

$$\delta w_0 = \frac{4\sqrt{2}}{3} \frac{\mu^{5/2} e^2}{m_p^{*2} \omega \varepsilon \hbar^3} \left(\frac{\varepsilon E_0^2 V}{8\pi \hbar \omega} \right) (\hbar\omega - E_g)^{3/2}. \tag{7.11}$$

The number of photons of a frequency-ω electromagnetic wave with electric field component amplitude E_0 contained in the volume V equals $\varepsilon E_0^2 V / (8\pi \hbar\omega)$; hence, the probability of single-photon absorption by the semiconductor per time unit attains the following form for the ordinary photoeffect [59]:

$$q_0 = \delta w_0 \left(\frac{\varepsilon E_0^2 V}{8\pi \hbar \omega} \right)^{-1} = \frac{4(4)\sqrt{2}}{3} \frac{\mu^{5/2} e^2}{m_p^{*2} \omega \varepsilon \hbar^3} (\hbar\omega - E_g)^{3/2}; \tag{7.12}$$

as before, the factor 4 corresponds to the spin degeneracy of the band electrons.

In the case of plasmon mediation, all oblique interband transitions contribute, not only vertical transitions (as in the ordinary photoeffect). This results in an enhancement of the transition probability for near-field coupling in comparison with the plane-wave absorption rate in a semiconductor in the ordinary photoeffect. This enhancement is, however, gradually quenched with the growth in the radius a, as expressed by Eq. (7.10).

The probability q_m of energy absorption in a semiconductor, via the mediation of surface plasmons, per single photon incident on the metallic nanospheres equals the product of δw, given by Eq. (7.10), and the number N_m of metallic nanoparticles divided by the photon density and multiplied by the additional phenomenological factor β responsible for all effects not directly accounted for (such as deposition separation and surface properties reducing the coupling strength, as well as energy losses due to electron scattering and radiation from the far-field zone, i.e., Lorentz friction [55], into the upper hemisphere if the metallic nanoparticle is not completely embedded in the substrate semiconductor medium):

$$q_m = \beta N_m \delta w \left(\frac{\varepsilon E_0^2 V}{8\pi \hbar \omega} \right)^{-1}. \tag{7.13}$$

7.2 Damping Rate for Plasmons in a Metallic Nanoparticle Deposited on a Semiconductor

Assuming that the energy acquired by the semiconductor band system, $\mathcal{A}$, is equal to the output of the plasmon oscillation energy, which results in plasmon damping, one can estimate the corresponding damping rate of the plasmon oscillations. Namely, at the damped plasmon amplitude $D_0(t) = D_0 e^{-t/\tau'}$, one finds for the total transmitted energy

$$\mathcal{A} = \beta \int_0^\infty \delta w \hbar \omega \, dt = \beta \hbar \omega \delta w \tau'/2 = \begin{cases} \dfrac{2}{3} \dfrac{\beta \omega \tau' \mu \sqrt{m_n^* m_p^*}(\hbar\omega - E_g) e^2 D_0^2}{\hbar^4 \varepsilon^2} & \text{for } a\xi \ll 1, \\[1em] \dfrac{2}{3} \dfrac{\beta \omega \tau' \mu^{3/2} \sqrt{2}\sqrt{\hbar\omega - E_g}\, e^2 D_0^2}{a \hbar^3 \varepsilon^2} & \text{for } a\xi \gg 1, \end{cases} \tag{7.14}$$

where τ' is the damping time rate and β accounts for losses not included in the model. Comparing the value of $\mathcal{A}$ given by the formula (7.14) with the energy loss by plasmon damping estimated in [23] (i.e., the initial energy of the plasmon oscillations, which is transferred step by step to the semiconductor, $\mathcal{A} = D_0^2/(2\varepsilon a^3)$), one finds

$$\frac{1}{\tau'} = \begin{cases} \dfrac{4\beta \omega \mu \sqrt{m_n^* m_p^*}(\hbar\omega - E_g) e^2 a^3}{3\hbar^4 \varepsilon} & \text{for } a\xi \ll 1, \\[1em] \dfrac{4\beta \omega \mu^{3/2} \sqrt{2}\sqrt{\hbar\omega - E_g}\, e^2 a^2}{3\hbar^3 \varepsilon} & \text{for } a\xi \gg 1. \end{cases} \tag{7.15}$$

By τ' we denote here the large damping rate of plasmons due to energy transfer to the semiconductor substrate, greatly exceeding the internal damping, characterized by τ, the latter being due to the scattering of electrons inside the metallic nanoparticle [23] ($1/\tau \ll 1/\tau'$). We have neglected the radiation of plasmon energy to the far-field upper-hemisphere zone, i.e., the Lorentz friction effect, which is also smaller than the near-field-zone energy transfer to the substrate [23].

For example, for nanospheres of Au deposited on an Si layer we obtain, for $\omega = \omega_1$ (the Mie self-frequency), with a in nm,

$$\frac{1}{\tau' \omega_1} = \begin{cases} 44.092 \beta a^3 \dfrac{\mu}{m} \dfrac{\sqrt{m_n^* m_p^*}}{m} & \text{for } a\xi \ll 1, \\[1em] 13.648 \beta a^2 \left(\dfrac{\mu}{m}\right)^{3/2} & \text{for } a\xi \gg 1, \end{cases} \tag{7.16}$$

for light (heavy) carriers in Si, $m_n^* = 0.19m(0.98m)$, $m_p^* = 0.16m(0.52m)$; m is the bare electron mass, $\mu = m_n^* m_p^*/(m_n^* + m_p^*)$, $E_g = 1.14$ eV, and $\hbar \omega_1 = 2.72$ eV. For these parameters and for nanospheres with radius a in the range 5–50 nm, the lower case of Eq. (7.16) applies (at $\omega = \omega_1$). The parameter β fitted from the experimental data [33, 23] equals to about 0.001.

7.2 Plasmons in a Metallic Nanoparticle Deposited on a Semiconductor

In another scenario, when the output of the plasmon energy is recovered by continuous income from sunlight, one can consider an energy-balanced regime. In an idealized case, the whole incoming energy of the monochromatic electromagnetic wave of frequency ω is transferred to the semiconductor via plasmons, and we are dealing with the stationary state of a driven and damped plasmon oscillator. Even though a free undamped plasmon has the Mie self-resonance frequency $\omega_1 = \omega_p/\sqrt{3}$, where ω_p is the bulk frequency, the frequency of plasma oscillations equals the driven electric field frequency, ω, of the incident electromagnetic wave. Because of instantaneous leakage of the plasmon energy in the near field to the semiconductor substrate, this large damping of plasmons causes a redshift and a widening of the resonance, as is the case for every damped and driven oscillator. The widened resonance enables the energy transfer from plasmons to electrons to embrace also frequencies lower or larger than the Mie frequency but it is limited from below by the semiconductor forbidden gap $E_g/\hbar$.

The incident sunlight frequency dispersion covers the visible spectrum and also some UV and infrared tails. The total efficiency of the plasmon channel corresponds to a sum (integration) over all Fourier components $\omega > E_g/\hbar$ of the light including its interference with the intensity distribution in the spectrum of sunlight. To model this behaviour it is necessary to consider separately each single monochromatic electromagnetic mode, i.e., the Fourier component ω. The electric field of such a mode excites plasmons with the same frequency, and this plasmon is damped at a rate $1/\tau'$ (Eq. (7.15)). This damping causes a redshift of the resonance and reduces the resonance amplitude, which in turn accommodates itself to balance the energy transfer to the semiconductor from the incident sunlight at frequency ω. Within this damped and driven oscillator model, the amplitude of plasmon oscillations $D_0(\omega)$ is constant in time and shaped by

$$f(\omega) = \frac{1}{\sqrt{(\omega_1^2 - \omega^2)^2 + 4\omega^2/\tau'^2}}.$$

The extreme of the redshifted resonance is attained at $\omega_m = \omega_1\sqrt{1 - 2(\omega_1\tau')^{-2}}$, with corresponding amplitude $\sim \tau'/\left(2\sqrt{\omega_1^2 - \tau'^{-2}}\right)$. The redshift is proportional to $1/(\omega_1\tau'^2)$. In the case of the energy transfer balance described above, one obtains, according to Eq. (7.10),

$$q_m = \begin{cases} \beta C_0 \dfrac{128}{9}\pi^2 a^3 \dfrac{\mu\sqrt{m_n^* m_p^*}}{m^2}(\hbar\omega - E_g)\dfrac{e^6 n_e^2 \omega}{\hbar^4 \varepsilon^3} f^2(\omega) & \text{for } a\xi \ll 1, \\[2ex] \beta C_0 \dfrac{128}{9}\sqrt{2\pi}^2 a^2 \dfrac{\mu^{3/2}}{m^2}\sqrt{\hbar\omega - E_g}\dfrac{e^6 n_e^2 \omega}{\hbar^3 \varepsilon^3} f^2(\omega) & \text{for } a\xi \gg 1, \end{cases} \quad (7.17)$$

where $f(\omega)$ is the amplitude factor given above for the driven damped oscillator and $D_0 = (e^2 n_e E_0 4\pi a^3/(3m)) f(\omega)$ in Eq. (7.10); the amplitude of the electric field, E_0, in the incident electromagnetic wave cancels from Eq. (7.17) due to normalization to a single photon as in Eq. (7.13); $C_0 = N_m 4/3\pi a^3/V$ where V is the volume of the semiconductor and N_m is the number of metallic nanospheres.

The ratio q_m/q_0 revealing the improvement of the plasmon-mediated photoeffect over the ordinary photoeffect can be expressed as follows:

$$\frac{q_m}{q_0} = \begin{cases} \dfrac{4\sqrt{2}\pi^2 a^3 \beta C_0 \sqrt{m_n^* m_p^*}(m_p^*)^2 e^4 n_e^2 \omega^2 f^2(\omega)}{3\mu^{3/2} m^2 \sqrt{\hbar\omega - E_g}\hbar\varepsilon^2} & \text{for } a\xi \ll 1, \\ \dfrac{8\pi^2 a^2 \beta C_0 (m_p^*)^2 e^4 n_e^2 \omega^2 f^2(\omega)}{3\mu m^2 (\hbar\omega - E_g)\varepsilon^2} & \text{for } a\xi \gg 1. \end{cases} \quad (7.18)$$

This ratio turns out to be of the order of $10^4 \beta 40/H(\text{nm})$ for a surface density $n_s \sim 10^8/\text{cm}^2$ of nanoparticles (as in an experiment in [33]); note that $C_0 = n_s 4\pi a^3/(3H)$, where H is the thickness of the semiconductor layer. On inclusion of the phenomenological factor β and the thickness H (it was confirmed experimentally that the range of the near-field zone exceeds the Mie wavelength, i.e., it is not shorter than 1 µm), the above ratio is sufficient to explain the scale of the experimentally observed strong enhancement of the absorption rate in semiconductors due to plasmons. This strong enhancement of the transition probability is linked with the admission of momentum-non-conserving transitions, which are, however, reduced with the growth of the radius a. The strengthening of the near-field-induced interband transitions, in the case of large nanospheres, is, however, still significant as the quenching of oblique interband transitions is partly compensated by the $\sim a^3$ growth of the amplitude of dipole plasmon oscillations. The trade-off between these two competing size-dependent factors is responsible for the experimentally observed enhancement of light absorption and emission in diode systems mediated by surface plasmons in nanoparticle surface coverings [32–34, 117–119].

7.3 Efficiency of the Light Absorption Channel via Plasmons for Various Materials

The metals most often used in plasmon photovoltaics are nanoparticles of Au and Ag (sometimes also of Cu) because their surface plasmon resonances are located within the visible light spectrum. These nanoparticles can be deposited on various semiconductor substrates with different material parameters. We will list here the appropriate parameters for comparison with experiment, for various configurations of plasmon solar-cell systems. In order to compare with experiment one can estimate the photocurrent in the case of a semiconductor photodiode with a metallically modified photoactive surface. This photocurrent is given by $I' = |e|N(q_0 + q_m)A$, where N is the number of incident photons and q_0 and q_m are respectively the probabilities of single-photon absorption in the ordinary photoeffect [59] and of single-photon absorption mediated by the presence of metallic nanospheres, as

7.3 Efficiency of the Light Absorption Channel for Various Materials

Table 7.1 *Plasmon energies measured in metals*

Metal	Bulk plasmon energy (eV)	Surface plasmon energy (eV)
Li	6.6	3.4
Na	5.4	3.3
K	3.8	2.4
Mg	10.7	6.7
Al	15.1	8.8
Fe	10.3	5.0
Cu	6	3.5
Ag	3.8	3.5
Au	4.67	2.7

Table 7.2 *Mie frequency ω_1 in Eq. (7.18)*

Metal	Au	Ag	Cu
Mie frequency	4.11×10^{15}/s	5.2×10^{15}/s	5.7×10^{15}/s

Table 7.3 *Substrate material parameters for Eq. (7.18) ($m = 9.1 \times 10^{-31}$ kg, the mass of a bare electron; lh, light holes; hh, heavy holes; L, longitudinal; T, transverse)*

Semiconductor	m_n^*	m_p^*	E_g
Si	0.9m L[101], 0.19m T[110]	0.16m lh, 0.49m hh	1.12 eV
GaAs	0.067m	0.08m lh, 0.45m hh	1.35 eV
CIGS	0.09m–0.13m	0.72m	1–1.7 eV

derived in the previous section: $A = \tau_f^n/t_n + \tau_f^p/t_p$ is the amplification factor, where $\tau_f^{n(p)}$ is the annihilation time for both sign carriers and $t_{n(p)}$ is the drive time for carriers, i.e., the time from traversing the distance between the electrodes. From the above formulae, it follows that (here $I = I'(q_m = 0)$, i.e., I is the photocurrent without metallic modifications),

$$\frac{I'}{I} = 1 + \frac{q_m}{q_0}, \quad (7.19)$$

where the ratio q_m/q_0 is given by Eq. (7.18).

In Tables 7.1 and 7.3 we list parameters for metallic nanoparticles and for semiconductor substrates, which allows comparison of the ratio $q_m/q_0(\omega)$ for various material configurations using Eq. (7.18). Besides the explicit entering of the effective masses m_n^* and m_p^* and of E_g into Eq. (7.18), these parameters contribute also to the damping rate $1/\tau'$ via the function $f(\omega)$ (defined below Eq. (7.17)), utilizing Eq. (7.15), and finally again Eq. (7.18).

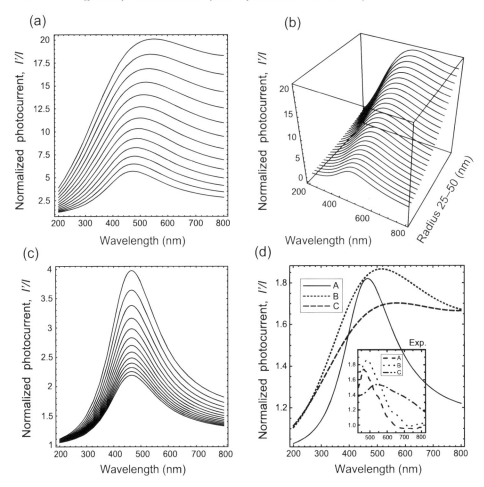

Figure 7.1 (a) Dependence of the normalized photocurrent $(I'/I)(\lambda)$ on wavelength for radius a increasing from 25 nm (bottom) to 50 nm (top); the step is 2 nm ($H = 4$ μm, $n_s = 1.6 \times 10^8/\text{cm}^2$). (b) The same as in (a) with three-dimensional visualization of the redshift of the maximum with radius growth. (c) Dependence of the normalized photocurrent $(I'/I)(\lambda)$ on wavelength for increasing layer depth H, from 4 μm (top) to 10 μm (bottom); the step is 0.5 μm ($a = 20$ nm, $n_s = 1.6 \times 10^8$ 1/cm^2). (d) Comparison with the experimental data, which are given in the inset [33]: A, $a = 25$ nm (fitting is better for 19 nm), $n_s = 6.6 \times 10^8$ 1/cm^2; B, $a = 40$ nm, $n_s = 1.6 \times 10^8$ 1/cm^2; C, $a = 50$ nm, $n_s = 0.8 \times 10^8$ 1/cm^2 ($H = 3$ μm).

All the material parameters as well as the geometry and size parameters modify both Eqs. (7.18) and (7.17), which give the eventual relatively complicated dependence of the plasmon-mediated photoeffect efficiency $(q_m/q_0)(\omega)$ with respect to material, size, and deposition type. We illustrate this dependence in the series of Figs. 6.5 and 7.1–7.6.

7.3 Efficiency of the Light Absorption Channel for Various Materials 143

The formula (7.18) is exemplified in Fig. 6.5 for Au nanoparticles deposited on an Si semiconductor (continuous line) which reproduces well the experimental behaviour (dashed line) [33]. Both channels of photon absorption resulting in photocurrent in the semiconductor sample are included, the direct ordinary photoeffect absorption with transition probability given by q_0 and the plasmon-mediated absorption with probability q_m, respectively. Note also that some additional effects such as the reflection of incident photons or destructive interference on the metallic net will contribute and are accounted for phenomenologically in the plasmon-mediated channel by an experimentally fitted factor β. The collective-interference-type corrections are rather weak for the considered low densities of metallic coverings, of the order of $10^8/cm^2$, and nanosphere sizes much lower than the resonance wavelength, though for larger concentrations and larger nanosphere sizes the reflection of photons would play a stronger reducing role [111]. The resonance threshold is accounted for by the damped resonance envelope function in Eq. (7.19), which includes also the semiconductor band-gap limit. As indicated in Fig. 6.5, the relatively high value of $q_m/q_0 \sim 10^4 40\beta/H(nm)$ enables a significant growth in efficiency of photoenergy transfer mediated by surface plasmons in nanoparticles deposited on an active layer by an increase in β or a reduction in H (at constant n_s). However, since an enhancement of β easily induces the overdamped regime of plasmon oscillations, the more suitable would be the lowering of H, especially convenient in thin-film solar cells. The overall behaviour of $(I'/I)(\omega) = 1 + q_m/q_0$ calculated according to the relation (7.19), and depicted in Fig. 6.5, agrees quite well with the experimental observations [33] in the position, height, and shape of the photocurrent curves for distinct samples (the strongest enhancement is achieved for $a = 40$ nm, for Au nanoparticles and an Si substrate). This corresponds to the rather surprising dependence, nonmonotonic and with a local maximum, of the efficiency measure, $(q_m/q_0)(\omega)$, with respect to metallic nanoparticle size a as expressed by Eq. (7.18). Such a size dependence has been confirmed experimentally [33].

In Fig. 7.2 we present the spectral dependence of the plasmonic efficiency enhancement q_m/q_0 for different substrates (Si, CIGS, GaAs) for Au nanoparticles with radius $a = 50$ nm and nanoparticle concentration $n_s = 10^8/cm^2$. Note that for the CIGS substrate (copper indium gallium diselenide) the spectral characteristic is narrower and blueshifted in comparison to those for Si and GaAs. Figure 7.3 reveals the increase in efficiency of the plasmon effect with growth of the forbidden gap E_g with other parameters unchanged. An especially significant material parameter is the hole mass, see Fig. 7.4; this can be seen from Eq. (7.18). The hole mass m_p^* enters the denominator in Eq. (7.12) for the ordinary photoeffect and also the numerator in Eq. (7.18). The higher is the mass m_p^* the lower is the efficiency q_0 of the ordinary photoeffect, and, in turn, the higher is the ratio q_m/q_0.

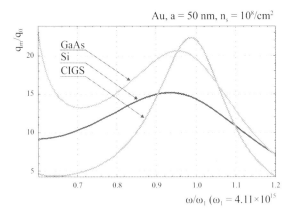

Figure 7.2 For comparison, the effectiveness of the plasmon channel expressed by q_m/q_0 (see Eq. (7.18)) for Si, GaAs, and CIGS substrates, for Au nanoparticles with radius 50 nm and surface density $10^8/\text{cm}^2$.

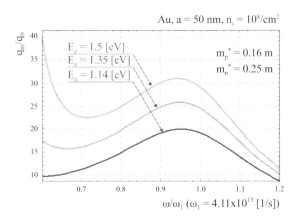

Figure 7.3 For comparison, the effectiveness of the plasmon channel expressed by q_m/q_0 (see Eq. (7.18)) for varying E_g but with the same effective masses of electrons and holes of the substrate semiconductors, covered with the same Au nanoparticles with radius 50 nm and surface density $10^8/\text{cm}^2$.

In Fig. 7.5 a comparison for Au, Ag, and Cu nanoparticles is presented for two sizes ($a = 50, 25$ nm). The blueshift of the spectral characteristics for Ag and Cu in comparison to Au is noticeable (see also Fig. 7.6) and is even clearer for lower-radius nanoparticles, owing to the narrowing of the spectral curves (see Fig. 7.6, lower panels). From a comparison in Fig. 7.6 of the upper and lower panels, for the Si and CIGS substrates with Au, Ag, Cu nanoparticles of size $a = 50, 25$ nm (at nanoparticle concentration $n_s = 10^8/\text{cm}^2$) it can be seen that the Au nanoparticles utilize the visible spectrum in a better manner than the Ag or Cu nanoparticles. The advantage presented by Au nanoparticles is greater in the case of an Si substrate but is reduced for a CIGS substrate because of blueshift of E_g

7.3 Efficiency of the Light Absorption Channel for Various Materials

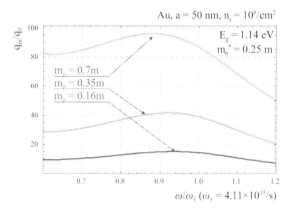

Figure 7.4 For comparison, the effectiveness of the plasmon channel expressed by q_m/q_0 (see Eq. (7.18)) for varying hole mass m_p^* but the same electron mass m_n^* and the same value of E_g, for a substrate semiconductor covered with the same Au nanoparticles with radius 50 nm and surface density $10^8/\text{cm}^2$.

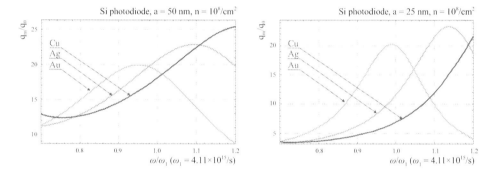

Figure 7.5 For comparison, the effectiveness of the plasmon channel expressed by q_m/q_0 (see Eq. (7.18)) for the same substrate semiconductor, Si, with Au, Ag, and Cu nanoparticles of the same radius, 50 nm (left) and 25 nm (right), and the same surface density, $10^8/\text{cm}^2$.

in CIGS with respect to Si. In the case of CIGS (especially for large nanoparticles, $a = 50$ nm), the advantage of Au over Ag in the overall utilization of the sunlight spectrum disappears, whereas it is pronounced in the case of an Si substrate. This is so because of the cut-off of the near-infrared part of sunlight in spectral absorption of CIGS, in contrast with Si, and which favours Ag more than Au nanoparticles; see Fig. 7.6. Below we describe an experimental confirmation of this behaviour of Si and CIGS substrates, under laboratory sunlight-type illumination, by Yamashita DensoYSS-50 Aunder AM1.5 [31].

For nanoparticles of Au and Ag of size $a = 50$ nm, optimized using Eq. (7.18), deposited on multicrystalline Si (mc-Si) and on copper indium gallium diselenide (CIGS) solar cells, the measured [31] overall increase in cell efficiency is as much

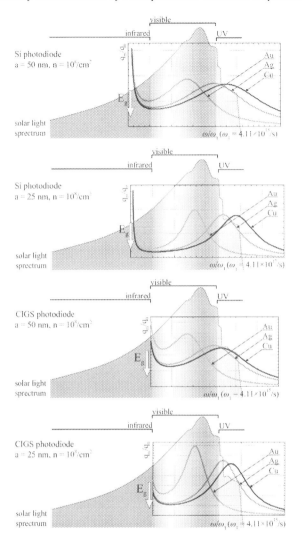

Figure 7.6 For comparison, the effectiveness of the plasmon channel expressed by $(q_m/q_0)(\omega)$ (see Eq. (7.18)) for the same substrate semiconductor Si (upper) and CIGS (lower) with Au, Ag, and Cu nanoparticles of the same radius, 50 nm (left) and 25 nm (right), and the same surface density, $10^8/cm^2$. Plots are made versus the sunlight spectrum as background. The figures illustrate the accommodation of the spectral characteristics of the plasmon-mediated photoeffect, $(q_m/q_0)(\omega)$, to the sunlight spectrum for different materials and covering parameters. The white arrows indicate the positions of the forbidden gap for Si and CIGS, respectively, on the background of the solar light spectrum. These positions accommodate themselves to the frequency ratio ω/ω_1 on the horizontal axis. The vertical axis shows the efficiency growth on the same scale for Au, Ag, and Cu and is not related to the vertical height of the solar spectrum. By comparison of the top panel with Fig. 7.5 (left) the influence of the deposition-type parameter β, taken here five times larger than in Fig. 7.5, can be seen.

7.3 Efficiency of the Light Absorption Channel for Various Materials 147

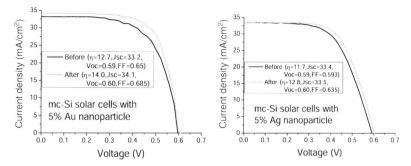

Figure 7.7 For comparison, the solar-cell efficiency due to plasmon modification for a multi-crystal Si solar cell, expressed as an increase in the area beneath the grey curve with respect to the area below the black curve. The photoactive surface is modified by (left) Au nanoparticles and (right) Ag nanoparticles [31].

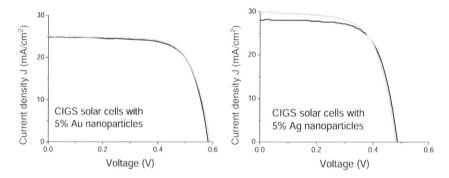

Figure 7.8 For comparison, the solar-cell efficiency due to plasmon modification for the CIGS cell expressed as an increase in the area beneath the grey curve with respect to the area below the black curve. The photoactive surface is modified by (left) Au nanoparticles and (right) Ag nanoparticles [31].

as 5%. The application of suitable concentrations of Au and Ag nanoparticles onto mc-Si solar cells increases their efficiency by 5.6% and 4.8%, respectively [31]. The application of Au and Ag nanoparticles onto the surface of CIGS solar cells improves their efficiency by 1.2% and 1.4%, respectively [31]. This can be seen in Figs. 7.7 and 7.8 giving the increase in solar-cell overall efficiency (the ratio of the area beneath the I–V curve for a metallically improved solar cell and for a clean solar cell; the same size (radius 50 nm) and the same concentration from the 5% colloidal solution sputtering over the surface has been applied to different samples). For more details see [31].

Worth noting is the agreement of the experimentally observed difference in the efficiency increase due to the plasmon effect in both cases, that of mc-Si cells and that of CIGS cells, if one compares the results of applying Au and Ag

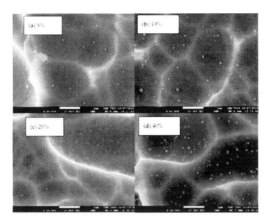

Figure 7.9 Visualization by electron microscopy (EM) photography of the distribution of metallic nanoparticles of Au on the surface of the multi-crystalline Si solar cell under study for the plasmonic effect, as shown in Fig. 7.7; the very low surface concentration of metallic components is visible.

particles (at the same nanoparticle size and the same surface concentration). This behaviour agrees with theoretical studies of the dependence on material of the plasmon effect, as shown above. From Figs. 7.2–7.5 we see that, for an Si substrate, Au nanoparticles with radii 50 nm utilize the solar light spectrum better than Ag or Cu particles (see Fig. 7.5), and indeed in the experiment for Au nanoparticles (see Fig. 7.7) the efficiency growth is about 10% larger than for Ag nanoparticles of the same size and concentration on the substrate mc-Si solar cell. Interestingly, for the substrate CIGS cell the effect is weaker and inverted; see Fig. 7.8. This is also noticeable from the theoretical modelling, owing to the different E_g and effective masses of carriers for CIGS with respect to those for mc-Si. The maxima of the efficiency enhancement curves for Au and Ag mutually shift in such a way that, for CIGS, Ag nanoparticles are somewhat better suited to the solar light spectrum than Au nanoparticles. However, to analyse these effects in more detail, measurements of the spectral characteristics of all the structures considered, with varying but monochromatic illumination should be performed.

Summarizing this section, it has been demonstrated using the Fermi golden rule that the efficiency of energy transfer between surface plasmon oscillations in a metallic nanoparticle and the substrate semiconductor depends on the parameters describing the nanoparticles (their radius and material) and on the semiconductor parameters (the energy gap and the effective masses of electrons and holes). The formula which generalizes the ordinary photoeffect to the plasmon-mediated photoeffect agrees well with experimental measurements on a laboratory photodiode configuration. The measured ratio of the photocurrents in setups with and

7.3 Efficiency of the Light Absorption Channel for Various Materials

without metallic nanocomponents was compared with the theoretically predicted scenario. Quantitative consistency is achieved both for the shape of the spectral characteristics and for the particle-size dependence (as illustrated for an Si diode with deposited Au nanoparticles having radii 25, 40, and 50 nm). Qualitative agreement has been demonstrated also for complete solar cells, where the plasmon effect is obscured by other elements of the extensive series of effects, in addition to the efficiency of the absorption of photons only, resulting in the overall solar-cell efficiency. Experimental data have been compared for a multi-crystalline Si solar cell and a CIGS (copper indium gallium diselenide) solar cell covered, or not covered, with Au or Ag nanoparticles with radii of the order of 50 nm. The increase in the overall photovoltaic efficiency for metallically modified cells varies between 1.5% (CIGS) and 6%(Si), depending on the nanoparticle concentration (for concentrations that are too dense the efficiency drops, owing to destructive interference and reflection). There is a better increase in efficiency (by about 10%) for Au nanoparticles in comparison to Ag nanoparticles, for an Si cell, whereas for a CIGS cell the difference between the effects of Ag nanoparticles and Au nanoparticles is inverted and strongly reduced. This also agrees qualitatively with the theoretical predictions, taking into account the differences in Mie frequency in Au and Ag and also the different semiconductor parameters for Si and CIGS.

As has been shown, the mediation of plasmons in metallic nanoparticles on top of the photodiode can enhance the efficiency of the photoeffect, even by a factor 2 (i.e., 100% increase), which has been confirmed experimentally for an Si photodiode covered with Au nanoparticles with radius 40 nm and surface concentration $\sim 10^8/cm^2$. Such a large increase is, however, reduced in commercial Si solar cells to about a 5% overall efficiency increase (and to about 2% in CIGS cells). Such an increase is not large but still of interest as in the competitive energy market the green renewable energy of solar cells is at present too costly, and any cost reduction, including via an increase in efficiency, is strongly desirable. The application of metallic low-density and low-cost nanocoverings with Au particles might allow an increase of 5% of the typical efficiency of a commercial Si cell [31], which captures about 14%–17% of the energy of illumination [136]. For thin-film cells and flexible plastic cells the efficiency is much lower. Thus, increasing it by even a few per cent by easily applied metallic coverings would seem to be significant. There are, however, problems, e.g., with the durability of such coverings deposited by sputtering onto the surface of a cell. In this regard a very interesting idea is to embed metallic nanoparticles in a thin film of a solar cell of perovskite type, which might be three-dimensionally printed together with nanoparticles inside, taking advantage of the low-temperature and chemical-type manufacturing of such cells (perovskite cells are especially in prospect now, because their efficiency is still growing and exceeds even 23%).

8
Numerical Simulation of Plasmon Photoeffect

We demonstrate in this chapter that the direct application of numerical packages such as COMSOL to the plasmonic effect in solar cells metallically modified at the nanoscale may be significantly inaccurate if quantum corrections are neglected. The near-field coupling of surface plasmons in metallic nanoparticles deposited on the top of a solar cell with the band electrons in the semiconductor substrate strongly enhances the damping of the plasmons in the metallic components. This effect is not accounted for in standard numerical packages using a Drude-type dielectric function for the metal (taken from measurements in bulk or for thin layers) as the prerequisite for the numerical electromagnetic field calculus. Inclusion of the proper corrections for plasmon damping causes enhancement of the plasmon-induced photoeffect efficiency of a metallized photodiode by several dozen per cent, at least, in comparison to the effect induced by the electric field concentration near the metallic nanoparticles. This is consistent with the experimental observations, which cannot be explained only by local the increases in the electrical field near the curvature of metallic nanoparticles, as determined by the finite element solution of the Maxwell–Fresnel boundary problem given by a numerical system like COMSOL. The proper damping rate for plasmons can be identified by application of the Fermi golden rule to the plasmon coupling with the band electrons. We demonstrate this effect by comparison of the COMSOL simulation of the plasmonic photoeffect with and without quantum corrections verified by the related experimental data.

The strengthening of sunlight energy harvesting in solar cells mediated by surface plasmons in metal nanoparticles deposited on a photoactive solar-cell surface is caused by three effects: (1) the strong concentration of the electric field of the incident electromagnetic wave close to regions of the metallic components with a large local curvature, as is the case for small nanoparticles, (2) the large amplitude of plasmon oscillations in metallic nanoparticles (particularly in larger nanoparticles, with a larger number of electrons), and (3) the enhancement of the probability of interband transitions in a semiconductor substrate caused by the breaking of translational symmetry for a nanoparticle coupled, in the near-field regime of the surface plasmons, with the substrate semiconductor-band electrons

[32, 2, 110–112, 23]. The transition probability for electrons excited from the valence band to the conduction band in a semiconductor, essential for the efficiency of the photovoltaic effect, grows owing to the electric field amplitude enhancement and to the admission of oblique transitions not prohibited here by momentum conservation [23]. For the ordinary photoeffect [59], when photons directly induce interband transitions in a semiconductor, these interband transitions are confined to only vertical jumps between states with almost the same momentum, owing to the momentum conservation rule and the very small momentum of sunlight photons (due to the large velocity of light, $c = 3 \times 10^8$ m/s), which hardly changes the electron momentum during scattering. For an excitation energy $\hbar\omega$ beyond the forbidden gap E_g of the substrate semiconductor, the photon dispersion $\hbar\omega = cq$ gives $q \ll p$, where $p \sim \pi\hbar/l$ is the semiconductor-band quasimomentum scale in the Brillouin zone (l denotes the semiconductor's elementary cell linear size). Thus, the change in the band electron momentum $\mathbf{q} = \mathbf{p}_2 - \mathbf{p}_1$ is negligible on the scale of the Brillouin zone and so $\mathbf{p}_1 \simeq \mathbf{p}_2$ (because of the large value of c); thus only the vertical, momentum-conserving, interband transitions contribute to the ordinary photoeffect when the transition is caused by free photons with momentum $\mathbf{q}$ and energy $\hbar\omega = cq$ (described by a plane wave $\sim e^{i(\mathbf{q}\cdot\mathbf{r}-\hbar\omega t)/\hbar}$).

The interaction of band electrons with surface plasmons from the metallic nanoparticles deposited on the semiconductor surface opens another channel for energy transfer of the solar light to the semiconductor substrate. Plasmons in metallical nanoparticles of several dozen nanometres size are strongly damped by Lorentz friction; thus, they also heavily absorb electromagnetic radiation near the plasmonic resonance. Hence, even sparse metallic coverings with such particles (Au, Ag, or Cu nanoparticles of 10–50 nm radius are to be preferred, because of the location of their surface plasmon resonances in the visible spectrum of light) can capture a large fraction of the incident solar light and quickly transfer this energy to the substrate semiconductor owing to the strong near-field-zone coupling of plasmons with band electrons in the semiconductor. In the near-field regime [55, 76] the potential of a plasmon dipole in a nanosphere is proportional to $1/R^2$ (where R is the distance from the sphere's centre), which in the Fourier picture overlaps with all quasimomenta in the substrate semiconductor's Brillouin zone. This is unlike the direct interaction of incident photons with the semiconductor, which contributes only via a single plane-wave Fourier component [59]. The plasmon-mediated channel of energy transfer dominates the absorption of light and changes the efficiency of the photoeffect. This channel is, however, not accounted for in the conventional solution of the Maxwell–Fresnel problem by, e.g., the numerical COMSOL system, which utilizes predefined dielectric functions of the metal and semiconductor; they are modelled

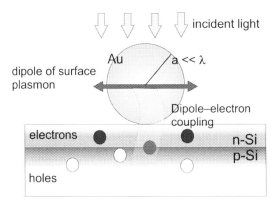

Figure 8.1 A spherical metallic nanoparticle (Au, Ag, or Cu) with radius a is deposited on a semiconductor substrate (for example, Si) with an active p–n junction embedded close (at a depth < 1 μm) to the upper surface bearing the nanoparticles. The high plasmon-absorption rate of a nanoparticle with size $a \in (10, 70)$ nm (i.e., Au) [26] guarantees a high level of excitation of the dipole plasmon mode by the incident light. A very effective channel for energy transfer from the dipole surface plasmon to the band electrons is opened via the near-field coupling (which is at subphoton distances, much less than λ, the plasmon resonance wavelength) of the dipole plasmon oscillations with the band electrons in the semiconductor substrate especially in the p–n junction; this results in a photocurrent for a photodiode configuration.

using experimental data from bulk and separate measurements without inclusion of the above-mentioned plasmon–band-electron coupling (schematically depicted in Fig. 8.1).

The considerable increase in light energy transfer via plasmons can be quantitatively accounted for via the Fermi golden rule determining the probability of interband transitions induced by the plasmon coupling. This transition probability is proportional to the matrix element of the time-dependent perturbation potential [75]. Here the perturbation potential is the electric field coupling of the dipole surface plasmon to charge carriers in the substrate semiconductor band system. The matrix element is calculated between electron initial and final states and is summed over all initial states in the valence band and over all final states in the conduction band; energy conservation is imposed, i.e., $E_p(\mathbf{p}_1) + \hbar\omega = E_n(\mathbf{p}_2)$, where $E_{p(n)}(\mathbf{p})$ are the valence-band and conduction-band dispersions. The energy $\hbar\omega$ refers to the energy of the surface plasmon oscillations, which have been induced in the metallic nanoparticle by the incident sunlight photons with electromagnetic wave $\sim e^{i\omega t}$, were ω is close to the surface plasmon resonance expressed as $\omega_1 = \omega_p/\sqrt{3}$ (ω_1, is the Mie frequency [18, 17], $\omega_p = \hbar\sqrt{n_e e^2/(m^*\varepsilon_0)}$ is the bulk-plasmon frequency in the metal [16], n_e is the density of the collective (nearly free) electrons in the metal, m^* is the effective mass of an electron in the metal,

e is the electronic charge, and ε_0 is the dielectric constant). The relation between ω and ω_1 can be expressed by a damped and forced oscillation scheme giving the amplitude of plasmon oscillations as

$$D_0(\omega) = \text{const.} \times E_0 \frac{1}{\sqrt{(\omega_1^2 - \omega^2)^2 + 4\omega^2/\tau^2}},$$

where E_0 is the amplitude of the electric component of the incident electromagnetic wave. Notice that the amplitude $D_0(\omega)$, is accommodated to the damping rate of the surface plasmons, $1/\tau$. This damping rate comprises all energy losses of plasmon oscillations, as follows.

1. The damping rate $1/\tau_0$ for the losses caused by the electrons in the metallic nanoparticle scattering on phonons, crystal imperfections, other electrons, and the nanoparticle boundary.
2. The damping rate $1/\tau_L$ due to the Lorentz friction losses corresponding to far-field-zone radiation (mostly in directions in the upper hemisphere for a nanoparticle deposited on the top of the semiconductor substrate).
3. The damping rate $1/\tau'$ for the energy losses of plasmons corresponding to the transfer of their energy to band electrons in the substrate semiconductor.

Thus,

$$\frac{1}{\tau} = \frac{1}{\tau_0} + \frac{1}{\tau_L} + \frac{1}{\tau'}.$$

We will show that the last channel is the most effective: practically all the plasmon energy is quickly, on the shortest time scale τ', transferred to the semiconductor electrons; the Lorentz friction is also important, however, for larger metallic nanoparticles with $a > 15$ nm [27, 26]. The energy transfer is strongest via the near-field coupling of plasmons to band electrons in the semiconductor substrate; this causes a strong damping of plasmons, which is what lies behind the experimentally observed giant plasmon enhancement of the photoeffect [137, 118, 2, 110–112, 23, 32–36] but is not accounted for in a conventional COMSOL simulation.

Note that this strong damping of plasmons (due to radiation) gives rise to a related strong absorption rate of the metallic components, which act as effective collectors of the incident sunlight even if their surface density in the solar-cell covering is low.

The surface plasmon oscillations have un-defined momentum, as localized oscillations. The band electrons are assigned, however, with quasimomenta in the crystal, an initial one, $\mathbf{p}_1$, and a final one, $\mathbf{p}_2$. These can be arbitrarily chosen because momentum conservation is ruled out by the matrix element of the local dipole interaction for the plasmon-induced transition of electrons, in contrast with the direct coupling with plane-wave photons in the ordinary photoeffect (the condition $\mathbf{p}_1 = \mathbf{p}_2$ strongly limits the photoeffect efficiency).

In the present chapter we briefly summarize the theoretical scheme of the plasmon-aided photoeffect and compare the analytical model with numerical simulations using COMSOL. In what follows we provide a description of the Lorentz friction and the plasmon damping caused by these radiation losses, in order to emphasize the large absorption ability of metallic nanoparticles of appropriately selected size. Then we describe the energy transfer from plasmons to band electrons in metallically improved solar cells, which is crucial for the efficiency enhancement due to the plasmon photoeffect. Next, we provide an analysis of the numerical modelling of the plasmon photoeffect and draw a conclusion about how to correct the COMSOL software, widely utilized in plasmon photovoltaic system simulation, via inclusion of an appropriately lifted dielectric function for the metallic nanoparticles. The latter corrections make the COMSOL simulation more realistic (they change the result for the plasmon-induced efficiency of the photoeffect by at least one order of magnitude), so improving the ability of COMSOL simulations to be compared with experiment [31–33, 111, 137–144].

Typical metals suitable for the plasmonic photoeffect are Au, Ag, and also Cu, whose surface plasmon resonances in nanoparticles overlap with the visible light spectrum. The typical size of the nanoparticles varies between 10 nm and 100 nm in diameter and the surface density of the metallic coverages is usually $10^{8-10}/\text{cm}^2$. It is remarkable that such sparse coverings with very small particles on the surface of a photocell can significantly enhance its photoefficiency (e.g., by a factor 2 in the photodiode setup reported in [33]). This phenomenon corresponds to the exceptionally high radiative abilities of metallic nanoparticles of such sizes. Neglect of it is the source of the failure of conventional COMSOL simulations. Note also that strengthening the nanoparticle surface concentration does not lead to further growth in the efficiency; conversely, too dense a coverage appears to be counter-productive because reflection and destructive interference reduces the plasmon photovoltaic effect.

The possibility of improving the efficiency of solar cells via some inexpensive and technologically feasible methods is now of considerable significance because the efficiency of commercial solar cells is still not high enough. To increase the efficiency, various strategies have been considered, including, quantum dot admixtures in multi-layer cells to better adjust the absorption spectrum to the dispersion of the solar light on the earth's surface or a metallic nanoscale improvement of the light absorption efficiency. We will demonstrate that the latter effect has a quantum character and, in order for the theoretical modelling to agree with the experimental observations, one must abandon the conventional classical methods of plasmonics using the finite element method (as, e.g., in COMSOL); the discrepancy of the classical simulation with the exact quantum approach, and also with the experimental data, reaches several dozen per cent. So, a large error is caused by the neglect, in the classical methods, of the very effective quantum channel for energy transfer between the plasmons in metallic nanoparticles and the band electrons in the semiconductor substrate. This channel can be accounted for in terms of the Fermi golden rule as follows. The golden rule is applied to the interband transition of electrons in the semiconductor substrate. These transitions are induced by the coupling of these electrons to plasmons excited by incident solar-light photons in the metallic nanoparticles. This channel for energy transfer appears to overwhelm the solar-light absorption in solar cells thus metallically modified.

8.1 Lorentz Friction Channel for Energy Losses of Surface Plasmons in a Metallic Nanoparticle

Plasmon oscillations in metallic nanoparticles have been widely analysed using various methods: the numerical Kohn–Sham-type approach [19, 20] but restricted to ultrasmall nanoparticles only (because of numerical constraints); the random phase approximation approach (RPA) [14, 16, 21, 23]; and classical solution of the Fresnel–Maxwell equations [17]. The latter approach is usable for arbitrary-sized particles but is limited by its neglect of quantum effects important at the nanoscale for metallic particles. The Mie approach [17], like the numerical solution of the Fresnel–Maxwell boundary problem by the finite element method utilized by, e.g, the COMSOL system, suffers from the need to use a phenomenologically predefined dielectric function for the metallic components as a prerequisite for the calculation algorithm. This dielectric function (in Drude form [17]) should comprise all quantum effects related to plasmons, but this is not so because the assumed frequency and damping of plasmons are taken from bulk metal values (at best from measurements on thin films), not from metallic nanoparticle measurement. It has been demonstrated that the size effect for metallic nanoparticles is predominant for the range of nanoparticle radius $a \in (15, 100)$ nm (for Au in vacuum) [27, 26], resulting in a different Mie response and COMSOL estimations, in comparison with models with the bulk dielectric function taken as the prerequisite for the calculus.

The reason for this discrepancy is the strongly growing radiation losses of plasmons (i.e., the Lorentz friction losses) in the case of nanoparticles with

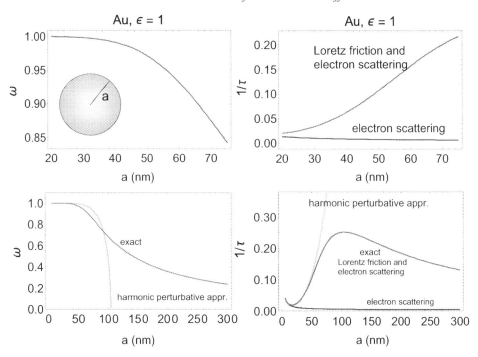

Figure 8.2 Size dependence of the resonance surface plasmon oscillations (dipole mode) including the Lorentz friction in a metallic (Au) nanoparticle in vacuum, for radius $a \in (20, 70)$ nm (upper) and $a \in (5, 300)$ nm (lower). The anharmonic oscillation regime caused by the third-order time derivative for the Lorentz friction force [55, 76] (black line) is apparently (see Eq. (8.1)) quite different from the harmonic approximation for the Lorentz friction (grey line) [27, 26]. The overdamped regime does not occur for the exact inharmonic oscillations (black curve), whereas the approximate harmonic oscillations (grey curve) stop when the damping rate $1/\tau$ reaches 1 (the resonance frequency ω and the damping rate $1/\tau$ are expressed in dimensionless units, i.e., they are divided by ω_1). The additional curve at lower right shows the electron scattering damping, which is much less than the Lorentz friction losses for $a > 20$ nm and diminishes with growth in a.

$a \in (15, 100)$ nm (for Au in vacuum but similarly also for Ag and Cu and other dielectric surroundings); these losses are not present either in smaller nanoparticles or in the bulk metal (as is noticeable in Fig. 8.2).

This figure presents the resonance frequency and related damping obtained by solution of the RPA dynamic equation for surface plasmons in metallic nanoparticles [23, 27, 26] with inclusion of the Lorentz friction force, which is proportional to the third-order time derivative of the plasmon dipole [55, 76]. The related anharmonic behaviour of the plasmon oscillations caused by this third-order time derivative is a remarkable effect. The plasmon frequency and

8.1 Lorentz Friction Channel for Energy Losses

damping are no longer linked by the ordinary relation, $\omega = \omega_1\sqrt{1 - (1/\tau_0\omega_1)^2}$, which conventionally results in an overdamped regime of quenched oscillations when $1/\tau_0\omega_1 > 1$. The exact solution including the Lorentz friction is always of an oscillatory type, with complex frequency (without any quenching of the oscillations) as follows:

$$\omega + i\frac{1}{\tau}$$
$$= -\frac{i}{3l} + \frac{i(1 + i\sqrt{3})(1 + 6lq)}{3 \times 2^{2/3}l\left(2 + 27l^2 + 18lq + \sqrt{4(-1-6lq)^3 + (2+27l^2+18lq)^2}\right)^{1/3}}$$
$$+ \frac{i(1 - i\sqrt{3})\left(2 + 27l^2 + 18lq + \sqrt{4(-1-6lq)^3 + (2+27l^2+18lq)^2}\right)^{1/3}}{6 \times 2^{1/3}l},$$
(8.1)

where $q = 1/(\tau_0\omega_1)$ and $l = (2/3)(a\omega_1/c)^3$. The functions ω and $1/\tau$ (in dimensionless units, i.e., divided by ω_1, the bare frequency, not shifted by damping) are plotted in Fig. 8.2 versus the nanosphere radius a. A strong deviation from harmonic behaviour is apparent for $a > 40$ nm. The absence of an overdamped regime is very clear. The electron scattering losses

$$\frac{1}{\tau_0} = \frac{v_F}{2l_B} + \frac{Cv_F}{2a}$$

(where v_F is the Fermi velocity in the metal, l_B is the mean free path in the metal, and $C \simeq 1$ is a factor depending on the reflection type of the electron scattering at the nanoparticle boundary) [24, 23] are displayed in Fig. 8.2, and it is noticeable that this loss channel is negligible for $a > 25$ nm (for Au in vacuum) in comparison with the Lorentz friction losses. It must be emphasized that in the conventional Mie and COMSOL approaches [17] only $1/\tau_0$ is accounted for, and this is a source of the above-mentioned discrepancies with the experimental observations.

It is clear that such strong plasmon damping as that via the Lorentz friction channel must be included for a realistic description of plasmon phenomena at the scale of metallic nanoparticles $a \in (15, 100)$ nm. If such nanoparticles are deposited on a semiconductor surface to mediate the photoeffect, the situation changes again in a significant manner. The coupling of plasmons with the substrate band electrons opens an especially efficient channel for energy transfer, and this channel dominates the plasmon damping – it is stronger than the Lorentz friction damping. To describe this channel the quantum approach must be applied in the framework of the Fermi golden rule.

8.2 Fermi Golden Rule for Probability of Electron Interband Excitation due to Plasmons in a Metallic Nanoparticle

Let us consider now a metallic nanoparticle (of Au, Ag, or Cu) of spherical shape with radius $a \in (5, 70)$ nm deposited on a semiconductor substrate with an embedded p–n junction for the photodiode setup schematically presented in Fig. 8.1.

We assume that in the metallic nanoparticle incident sunlight excites a surface plasmon of dipole type. Such a plasmon is excited by incident photons with frequency close to the dipole surface plasmon resonance, which in Au, Ag, or Cu nanoparticles with radii of several tens of nanometres falls at a wavelength ~ 500 nm, i.e., it is much greater than the nanoparticle dimension. Thus, the electrical component of the electromagnetic wave is almost homogeneous along the whole nanoparticle and the dipole regime is fulfilled, i.e., only the surface plasmon dipole mode can be excited by this electromagnetic wave. Repeating the derivation presented in Section 7.1, 7.2, and 7.3 of the previous chapter one can estimate the probability of the plasmon-induced transitions of band electrons between the valence and conduction bands in the semiconductor substrate via application of the Fermi golden rule [75].

As we have already explained in previous chapters, in the case of metallic nanoparticles deposited on a photoactive semiconductor layer, plasmons induce in the substrate oblique interband transitions of the band electrons, not only vertical transitions as in the ordinary photoeffect. This results in the enhancement of the transition probability for the near-field coupling of plasmons with band electrons in comparison with the photon (plane-wave) absorption rate in a semiconductor in the conventional photoeffect. This enhancement of the transition probability due to the admission of non-momentum-conserving interband hopping is, however, gradually quenched with growth in the radius a, which leads to a specific trade-off for the efficiency increase because larger nanoparticles are preferred owing to their stronger dipole amplitude, proportional to a^3.

From the algebra detailed in Sections 7.1–7.3 we obtained an analytical formula for the efficiency ratio of the plasmon-mediated photoeffect transition probability per single incident photon and the ordinary photoeffect transition probability per single incident photon including the above-mentioned trade-off:

$$\frac{q_m}{q_0} = \begin{cases} \dfrac{4\sqrt{2}\pi^2 a^3 \beta C_0 \sqrt{m_n^* m_p^*}(m_p^*)^2 e^4 n_e^2 \omega^2 f^2(\omega)}{3\mu^{3/2} m^2 \sqrt{\hbar\omega - E_g} \hbar \varepsilon^2} & \text{for } a\xi \ll 1, \\[2ex] \dfrac{8\pi^2 a^2 \beta C_0 (m_p^*)^2 e^4 n_e^2 \omega^2 f^2(\omega)}{3\mu m^2 (\hbar\omega - E_g) \varepsilon^2} & \text{for } a\xi \gg 1, \end{cases} \quad (8.2)$$

where q_m is the probability of interband transitions in the substrate semiconductor induced by the dipole surface plasmon in the nanoparticle (per single incident photon exciting the plasmon),

8.2 Fermi Golden Rule for Probability of Electron Interband Excitation 159

$$q_m = \begin{cases} \beta C_0 \dfrac{128}{9} \pi^2 a^3 \dfrac{\mu \sqrt{m_n^* m_p^*}}{m^2} (\hbar\omega - E_g) \dfrac{e^6 n_e^2 \omega}{\hbar^4 \varepsilon^3} f^2(\omega) & \text{for } a\xi \ll 1, \\ \beta C_0 \dfrac{128}{9} \sqrt{2\pi}^2 a^2 \dfrac{\mu^{3/2}}{m^2} \sqrt{\hbar\omega - E_g} \dfrac{e^6 n_e^2 \omega}{\hbar^3 \varepsilon^3} f^2(\omega) & \text{for } a\xi \gg 1, \end{cases} \quad (8.3)$$

where

$$f(\omega) = \dfrac{1}{\sqrt{(\omega_1^2 - \omega^2)^2 + 4\omega^2/\tau'^2}}$$

is the amplitude factor for the driven and damped oscillator and $D_0 = (e^2 n_e E_0 4\pi a^3/(3m)) f(\omega)$ in Eq. (7.10); the amplitude of the electric field, E_0, in the incident electromagnetic wave is obtained from Eq. (7.17) by the normalization per single photon as in Eq. (7.13); $C_0 = N_m 4/3\pi a^3/V$ where V is the volume of the semiconductor and N_m is the number of metallic nanospheres. The denominator q_0 is the probability of an interband transition induced by a single incident photon in the ordinary photoeffect,

$$q_0 = \dfrac{4(4)\sqrt{2}}{3} \dfrac{\mu^{5/2} e^2}{m_p^{*2} \omega \varepsilon \hbar^3} (\hbar\omega - E_g)^{3/2}. \quad (8.4)$$

The plasmon damping rate is estimated as

$$\dfrac{1}{\tau'} = \begin{cases} \dfrac{4\beta\omega\mu \sqrt{m_n^* m_p^*}(\hbar\omega - E_g) e^2 a^3}{3\hbar^4 \varepsilon} & \text{for } a\xi \ll 1, \\ \dfrac{4\beta\omega\mu^{3/2} \sqrt{2}\sqrt{\hbar\omega - E_g} e^2 a^2}{3\hbar^3 \varepsilon} & \text{for } a\xi \gg 1. \end{cases} \quad (8.5)$$

By τ' we denoted here the short time scale for the large damping rate, $1/\tau'$, of plasmons due to the energy transfer to the semiconductor substrate, which greatly exceeds the internal (within metal) damping characterized by $1/\tau_0$, the latter being due to the scattering of electrons inside the metallic nanoparticle [23] ($1/\tau_0 \ll 1/\tau'$). The radiation from the far-field zone towards the upper hemisphere (i.e., the Lorentz friction for the plasmon) is also smaller than the near-field-zone energy transfer to the substrate [23]. Let us repeat here our comment on the two limiting cases, $a\xi \ll 1$ and $a\xi \gg 1$, given in the above formulae, with $\xi = \sqrt{2(\hbar\omega - E_g)(m_n^* + m_p^*)}/\hbar$. Notice that

$$a \simeq \dfrac{1}{\xi} \simeq \begin{cases} > 2 \times 10^{-9} \text{ m} & \text{for } \dfrac{\hbar\omega - E_g}{E_g} < 0.02, \\ < 2 \times 10^{-9} \text{ m} & \text{for } \dfrac{\hbar\omega - E_g}{E_g} > 0.02, \end{cases}$$

and this range depends weakly on the effective masses and E_g. Thus for nanoparticles with radii $a > 2$ nm the first regime holds only close to E_g (for less than 2%

of the distance to the limiting E_g), whereas the second regime holds for the rest of the ω domain. For comparison,

$$a \simeq \frac{1}{\xi} \simeq \begin{cases} > 0.5 \times 10^{-9} \text{ m} & \text{for } \dfrac{\hbar\omega - E_g}{E_g} < 0.5, \\ < 0.5 \times 10^{-9} \text{ m} & \text{for } \dfrac{\hbar\omega - E_g}{E_g} > 0.5. \end{cases}$$

The first region widens considerably, to about 50% of the relative distance to E_g, but holds only for ultra-small size of nanoparticles ($a < 0.5$ nm. For larger nanospheres, e.g., for $a > 10$ nm, the second regime is dominates.

The ratio q_m/q_0 turns out to be of order of $10^4 \times 40\beta/H$ (nm) for a surface density of nanoparticles (as in the experiment reported in [33]) $n_s \sim 10^8/\text{cm}^2$; note that $C_0 = n_s 4\pi a^3/(3H)$ in Eq. (8.2) and H is the thickness of the semiconductor layer. Including the phenomenological factor β and the thickness H (it has been confirmed experimentally that the range of the near-field zone exceeds the Mie wavelength, i.e., it is not shorter than 1 μm), formulae (8.2) and (8.5) are sufficient to explain the scale of the experimentally observed strong enhancement of the absorption rate in semiconductors due to plasmons. This strong enhancement is linked with the admission of momentum-non-conserving transitions but is, however, reduced with growth in the radius a. The strengthening of the near-field-induced interband transitions in the case of large nanospheres is, however, still significant as the quenching of the oblique interband transitions is partly compensated by the $\sim a^3$ growth in the amplitude of the dipole plasmon oscillations. The optimized trade-off between these two competing size-dependent factors is responsible for the experimentally observed enhancement of light absorption and emission in diode systems mediated by surface plasmons in nanoparticle surface coverings [32–34, 118, 119].

The photocurrent in the photodiode can be expressed as $I' = |e|N(q_0 + q_m)A$, where N is the number of incident photons and q_0 and q_m are the probabilities of single-photon absorption in the ordinary photoeffect [59] and of single-photon absorption mediated by the presence of metallic nanospheres, respectively, as derived above; $A = \tau_f^n/t_n + \tau_f^p/t_p$ is the amplification factor, where $\tau_f^{n(p)}$ are the annihilation times for the carriers, and $t_{n(p)}$ are the drive times for the carriers (the times for traversing the distance between the electrodes). From the above definitions, it follows that the efficiency measure for the plasmon photovoltaic effect attains the form (here $I = I'(q_m = 0)$, i.e., I is the photocurrent without metallic modifications)

$$\frac{I'}{I} = 1 + \frac{q_m}{q_0}, \tag{8.6}$$

where the ratio q_m/q_0 is given by Eq. (8.2).

8.2 Fermi Golden Rule for Probability of Electron Interband Excitation 161

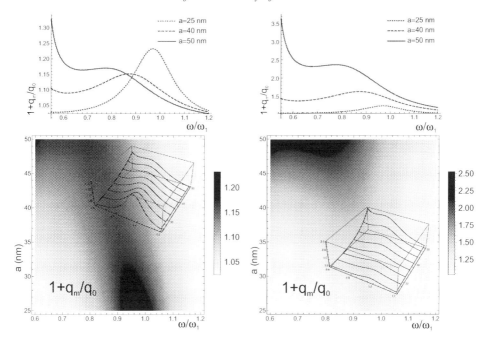

Figure 8.3 Size dependence of the photoeffect efficiency enhancement due to plasmon mediation in light absorption, (left) for constant total mass of all metallic nanoparticles deposited on the Si substrate and (right) for constant surface concentration of particles (surface layer with thickness $H = 3$ μm, proximity factor $\beta = 0.0005$, density of Au nanoparticles (left) $n_s = 50 \times 10^8 (25/a)^3/\text{cm}^2$ and (right) $50 \times 10^8/\text{cm}^2$ with radius $a \in (25, 50)$ nm). The redshift with growth in nanoparticle size is noticeable. The efficiency increase $I/I_0 = 1 + q_m/q_0$ (see Eqs. (8.6) and (8.2)) depends on the nanoparticle size and material, on the semiconductor parameters, and on n_s (ω is the frequency of the incident light in dimensionless units, i.e., divided by ω_1).

The material parameters as well as the geometry and size parameters modify both Eqs. (8.2) and (8.3), which gives an eventual relatively complicated dependence of the plasmon-mediated photoeffect efficiency $(q_m/q_0)(\omega)$ with respect to the material, size, and deposition type. We illustrate this dependence in Fig. 8.3. The formula (8.2) is exemplified in this figure for Au nanoparticles deposited on an Si semiconductor, which reproduces well the experimental behaviour [33] (the required material data are listed in Tables 7.2 and 7.3). Both channels of photon absorption resulting in photocurrent in the semiconductor sample are included, direct photoeffect absorption for which the probability of transitions is given by q_0 and plasmon-mediated absorption with probability q_m, respectively. Note also that some additional effects such as the reflection of incident photons or destructive interference on the metallic net would contribute, and these are accounted for phenomenologically in the plasmon-mediated channel by an experimentally fitted factor β. The collective-interference-type corrections are rather weak for the low densities of metallic coverings considered, of the order of $10^{8-9}/\text{cm}^2$, and for nanosphere sizes much lower than the resonance wavelength (~ 500 nm); however, for larger concentrations and larger nanosphere sizes, they would play a stronger reducing role (by reflecting photons) [111, 31]. The resonance threshold is accounted for by the damped resonance envelope function in Eq. (8.6), including the semiconductor band-gap limit. As indicated in Fig. 8.3, the relatively high value of $q_m/q_0 \sim 10^4 \times 40\beta/H(\text{nm})$ enables significant growth in efficiency of photoenergy transfer to a semiconductor mediated by surface plasmons by an increase in β or a reduction in H (at constant n_s). Because an enhancement of β could easily induce the overdamped regime of plasmon oscillations the more effective prospect would be the lowering of H, especially convenient in thin-film solar cells. This reflects the fact that the damping of plasmons due to their coupling with band electrons in the substrate semiconductor can greatly exceed the Lorentz friction. Electron scattering losses do not play a significant role in comparison with the dominant channel of energy flow. The overall behaviour of $I'/I = 1 + q_m/q_0$ calculated according to the relation (8.6), and depicted in Fig. 8.3 agrees well with the experimental observations [33], in the position, height, and shape of the photocurrent curves for distinct samples.

In the previous chapter we compared the spectral dependence of the plasmonic efficiency enhancement on substrate (Si, CIGS – copper indium gallium diselenide), for the same Au, Ag, and Cu nanoparticles with radius $a = 50, 25$ nm and the same nanoparticle concentration $n_s = 10^8/\text{cm}^2$. From this comparison, in Fig. 7.6, for Si and CIGS substrates with Au, Ag, or Cu nanoparticles of size $a = 50, 25$ nm (at nanoparticle concentration $n_s = 10^8/\text{cm}^2$) one notices that the Au nanoparticles utilize the visible sunlight spectrum better than Ag or Cu nanoparticles. The advantage of Au nanoparticles is greater in the case of an Si substrate and is reduced for a CIGS substrate, because of the blueshift of E_g in CIGS with respect to Si. In the case of CIGS (especially for larger nanoparticles, $a = 50$ nm), the advantage of Au over Ag in the overall utilization of the sunlight spectrum disappears, whereas it is pronounced in the case of an Si substrate. This is so because of the cut-off of the near infrared part of sunlight in the spectral absorption of CIGS, in contrast with that of Si, which favours Ag more than Au nanoparticles; see Fig. 7.6. This behaviour agrees with experimental observations [31].

For nanoparticles of Au and Ag of size $a = 50$ nm, optimized due to Eq. (8.2), that are deposited on multi-crystalline silicon (mc-Si) and on CIGS solar cells, the measured [31] overall increase in cell efficiency is as much as 5%. The application of Au and Ag nanoparticles to mc-Si solar cells increases their efficiency by 5.6% and 4.8%, respectively [31]. The application of Au and Ag nanoparticles to the surfaces of CIGS solar cells improves their efficiency by 1.2% and 1.4%, respectively [31].

8.3 Numerical Modelling of the Plasmon Photoeffect by COMSOL

The commercial numerical package COMSOL is a convenient tool for solving, by the finite element method, differential equations with imposed arbitrary boundary conditions. It can be utilized to solve the Maxwell–Fresnel problem for arbitrary geometry and material composition of the analysed systems (see COMSOL Multiphysics 5.0, Wave Optics module, http://www.comsol.com). In particular, solar cells can be modelled in this way, including the plasmonic effect induced by the metallic nanoparticles deposited on a cell surface. The great advantage of COMSOL is its flexibility with regard to nanoparticle size and shape, deposition type, and the material of the metallic components and of the semiconductor substrate.

In order to simulate a solar cell using the COMSOL system, an appropriate geometric arrangement is required for the setup and also the predefinition of the material parameters, i.e., the dielectric functions for all components of the investigated device.

The dielectric function defines the electric susceptibility of a material, linking the displacement electric field **D** with the external electric field **E**,

$$\mathbf{D}(\omega, \mathbf{k}) = \varepsilon(\omega, \mathbf{k})\mathbf{E}(\omega, \mathbf{k}),$$

where ω and **k** are Fourier variables in the time and space domains, respectively. The real part of the dielectric function defines the refractive index of a material, whereas the imaginary part describes light absorption. All macroscopic processes of the light–matter interaction are included in an effective manner in this function, which also includes quantum effects. Such compressed material microscopic information allows for a completely classical consideration of the Maxwell–Fresnel problem of light scattering and absorption at interfaces in complex multi-material systems with various configurations.

The photon absorption probability (per single photon and time unit) is given for the case of the ordinary photoeffect by Eq. (7.11), i.e.,

$$\delta\omega_0 = \frac{4\sqrt{2}}{3} \frac{\mu^{5/2} e^2}{m_p^2 \omega \varepsilon \hbar^3} \left(\frac{4\pi \varepsilon_0 \varepsilon E_0^2 V}{8\pi \hbar \omega} \right) (\hbar\omega - E_g)^{3/2}, \tag{8.7}$$

and for the case when the photoeffect is mediated by metallic nanoparticles it is given by Eq. (7.10), i.e.,

$$\delta\omega = \frac{4}{3} \frac{\mu^{3/2} \sqrt{2} \sqrt{\hbar\omega - E_g} e^2 D_0^2}{a 16\pi^2 \varepsilon_0^2 \varepsilon^2 \hbar^4}. \tag{8.8}$$

These functions can be used to assess the imaginary part of the dielectric function of materials in the photovoltaic problem including the plasmon effect. The absorption coefficient, $\alpha(\omega) = \hbar\omega\delta\omega$, links with the imaginary part of the dielectric function as follows: $\varepsilon''(\omega) = (nc/\omega)\alpha(\omega)$, where n is the refractive index, $\varepsilon = \varepsilon' + i\varepsilon''$. The plasmon dipole amplitude, D_0 in Eq. (8.8), can be found from the irradiated total power:

$$D_0^2 = \frac{4\pi\varepsilon_0\lambda^4}{(2\pi)^4 c} P = \frac{4\pi\varepsilon_0\lambda^4}{(2\pi)^4 c} \int_\Sigma \mathbf{S} \cdot d\sigma, \qquad (8.9)$$

where P is the total power radiated by the dipole, $\mathbf{S}$ is the Poynting vector, and Σ is the nanoparticle surface.

For COMSOL simulation, the total energy of the light absorbed in the semiconductor substrate can be calculated by integrating the square of the electric field over the semiconductor volume:

$$Q_{with(without)}(\omega) = \frac{\omega}{4\pi} \int_V n_s \varepsilon''_{with(without)}(\omega) E^2 dV, \qquad (8.10)$$

where ε'' is the imaginary part of the dielectric function of the semiconductor substrate either with or without metallic nanoparticles deposited, respectively ('without' will be noted as '0'). The enhancement of the light absorption due to plasmonic elements (called the efficiency enhancement of the plasmon photoeffect) can be defined as the ratio of the light energy absorbed by the semiconductor substrate covered with metallic nanoparticles and that without any metallic coverage:

$$\mathcal{A}(\omega) = \frac{Q_{with}(\omega)}{Q_0(\omega)}. \qquad (8.11)$$

One can next estimate the enhancement of the short-circuit current, I_{sc}, in the cell due to the presence of metallic nanoparticles, averaging over the solar spectrum,

$$I_{sc} = \frac{\int Q_{with}(\omega)\mathcal{F}(\omega)d\omega}{\int Q_0(\omega)\mathcal{F}(\omega)d\omega}, \qquad (8.12)$$

where $\mathcal{F}(\omega)$ is the standard solar global spectrum, conventionally named as AM1.5G.

For COMSOL calculations a unit cell must be defined, e.g., as shown in Fig. 8.4. Simulations with such a unit cell consist of three domains: (1) the air surroundings, (2) the semiconductor substrate, and (3) the metallic nanoparticle. One can consider two models: a model consisting of a single metallic nanoparticle on a substrate (which can then be multiplied by the nanoparticle concentration, neglecting, however, all interparticle electromagnetic interaction), and a model consisting of a periodic metallic nanoparticle array deposited on a semiconductor substrate. The latter model allows inclusion of the interparticle electromagnetic interference effect, which is sensitive to the particle separation.

Despite which model is considered, the simulation goes via two steps. In the first step the background electric field distribution must be calculated, i.e., the

8.3 Numerical Modelling of the Plasmon Photoeffect by COMSOL

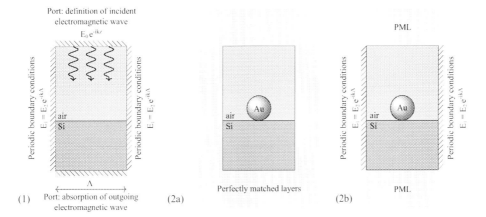

Figure 8.4 The scheme of the unit cell of the numerical model for COMSOL. (1) The first calculation step – evaluation of the background electric field as the reference field distribution for the second step. (2a) The second step for the single-particle model (no periodic conditions are imposed). (2b) The second step for the metallic nanoparticle (MNP) array model with the periodic Floquet conditions imposed. (Figs. 8.4–8.10 after [145]).

electric field distribution in the case of a plane electromagnetic wave incident on a semiconductor substrate without plasmonic nanoparticles. This referential field distribution is used for comparison in the second step of simulation, the evaluation of the scattered electric field caused by the the presence of metallic nanoparticles deposited on the semiconductor surface.

In both the models of system arrangements mentioned above, the first step is identical. The incident light is assumed to be a plane wave propagating vertically along the z axis using two ports. At the top boundary the incoming plane wave parameters are defined and, at the bottom boundary, the parameters of the wave transmitted through the substrate are defined. On the lateral boundaries the Floquet periodic boundary conditions are conventionally assumed, which allows one to simulate a large system via the effective multiplication of the calculation unit cell.

In the single-particle model, for a simulation of one independent particle in the calculation cell, in the second step one defines an additional domain surrounding the calculation cell and absorbing all outgoing light (the so-called perfectly matched layers, PMLs), instead of the periodic conditions. The PML assumption allows us to reduce the simulation area, which is important for temporal optimization of the numerical algorithm. The nanoparticles in this model are completely independent and occur as a multiplication of the single nanoparticle. The neglect of interparticle effects corresponds to low concentrations of the metallic coverage. The width of the computation unit cell was assumed to be 350 nm, the height of the air surroundings 300 nm, the Si substrate thickness is set to 200 nm, and the PML thickness to 150 nm.

In the model with a periodic array of metallic nanoparticles, in the second step of the simulation we use again the Floquet periodic boundary conditions on the lateral boundaries and set an additional PML at the top and at the bottom of the computational cell. The width of the computational cell was assumed to vary in the range $\Lambda \in (90, 360)$ nm, the height of the air surroundings is set to 300 nm, the Si substrate thickness to 400 nm, and the PML thickness to 150 nm.

Outside the PML domains, tetrahedral mesh elements were used with size equal to $a/5$ inside the metallic nanoparticle and of radius a inside the Si substrate.

The simulations were performed assuming the dielectric function for Si with modified absorption is given by the formula (7.10) and the results were compared with a similar simulation using a dielectric function for Si without any metallic modification taken from [146]. The dielectric function of the Au was modelled within the Drude–Lorentz approximation with damping rate given by Eq. (6.43).

The results of the COMSOL simulations are presented in Figs. 8.5–8.10 and in Tables 8.1 and 8.2. Notice that the neglect of plasmon damping causes a

Table 8.1 *The photocurrent enhancement for the model of an MNP array on an Si substrate (see Eq. (8.12)), for MNP radii $a = 20, 30, 40, 50, 60$ nm and array periods $\Lambda = 3a, 4a$. The results were obtained for the model using dielectric functions correctly modified in their imaginary part. We used solar spectrum AM1.5G to calculate the photocurrent enhancement from Eq. (8.12).*

Radius (nm)	$\Lambda = 3a$	$\Lambda = 4a$
20	1.66	1.44
30	1.87	1.71
40	1.97	1.91
50	1.98	1.97

Table 8.2 *The photocurrent enhancement for the model of an MNP array on an Si substrate. The MNP radii were taken as $a = 20, 30, 40, 50, 60$ nm and the array periods as $\Lambda = 3a, 4a$. The results were obtained for the model using the bulk semiconductor dielectric function without metallic components [146] and the dielectric function of the metal without the proper plasmon damping. The solar spectrum AM1.5G was used.*

Radius [nm]	$\Lambda = 3a$	$\Lambda = 4a$
20	0.996	0.999
30	1.006	1.010
40	1.011	1.014
50	1.006	1.014

8.3 Numerical Modelling of the Plasmon Photoeffect by COMSOL

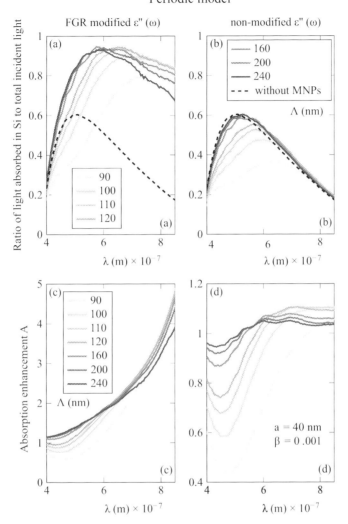

Figure 8.5 The periodic model. (a, b) The enhancement of light absorption in the Si substrate as a function of the incident wavelength for various sizes Λ of the unit cell. (c, d) The efficiency enhancement of the system given in Eq. (8.11) as a function of the incident wavelength. (a, c) Results obtained for the model using the dielectric function of the semiconductor $\varepsilon''(\omega)$ modified by the plasmon-damping contribution (the energy leaving the plasmons is absorbed by the semiconductor substrate). (b, d) Results obtained for the model using the non-modified dielectric function for the bulk semiconductor [146]. The calculations were made for MNPs of radii $a = 40$ nm and for array periods $\Lambda = 90, 100, 110, 120, 160, 200, 240$ nm describing the lowering of the nanoparticle concentration.

Periodic model

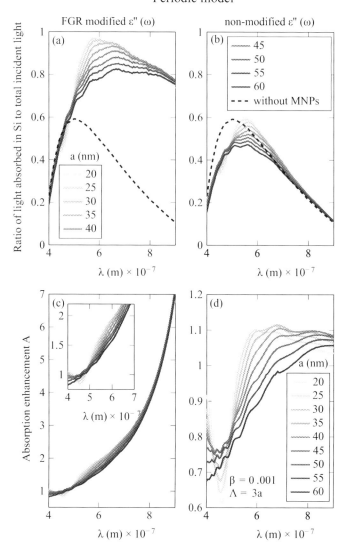

Figure 8.6 The periodic model. (a, b) Absorption enhancement in the Si substrate as a function of the incident wavelength for varying radii of the MNPs. (c, d) The efficiency enhancement from Eq. (8.11) as a function of the incident wavelength for varying radii of the nanoparticles. (a, c) Results obtained in the model using a modified $\varepsilon''(\omega)$. (b, d) Results obtained in the model using an inappropriate (non-modified) dielectric function taken from bulk measurements [146]. The calculation was made for MNP arrays with the period $\Lambda = 3a$ and radius $a = 20, 25, 30, 35, 40, 45, 50, 55, 60$ nm.

8.3 Numerical Modelling of the Plasmon Photoeffect by COMSOL 169

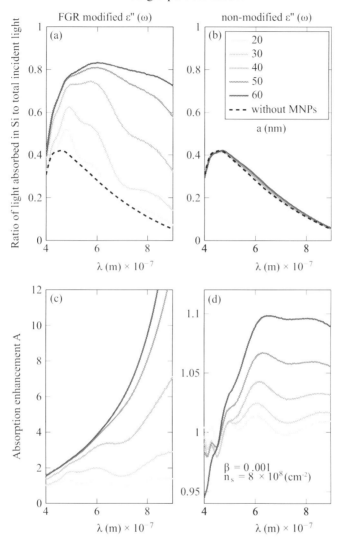

Figure 8.7 Single-particle model. (a, b): The absorption enhancement in the Si substrate as a function of the incident wavelength for varying radii of the MNPs (Λ chosen to conserve surface density n_s at $8 \times 10^8/\text{cm}^2$). (c, d) The efficiency enhancement factor from Eq. (8.11) as a function of the incident wavelength for varying sizes of nanoparticles and constant concentration. (a, c) Results obtained for the model using modified $\varepsilon''(\omega)$. (b, d) Results obtained for the model using the irrelevant non-modified bulk dielectric function [146]. The calculations were made for a single MNP deposited on an Si substrate with radius $a = 20, 30, 40, 50, 60$ nm, $\beta = 0.001$ and for an effective MNP concentration (accommodated by changes in Λ) $n_s = 8 \times 10^8/\text{cm}^2$.

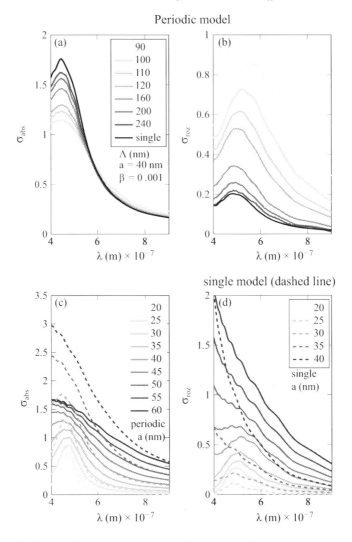

Figure 8.8 The periodic model. (a, c) The absorption cross section of an Au nanoparticle deposited on an Si substrate as a function of the incident wavelength for varying single-cell size (a) and varying nanoparticle radius (c). (b, d) The scattering cross section for an Au nanoparticle deposited on an Si substrate; the corresponding cross sections obtained in the case of the single-particle model are indicated by dashed lines. (a, b) Results obtained from the periodic MNP array model with constant MNP radii $a = 40$ nm and array period taken from the set $\Lambda = 90, 100, 110, 120, 160, 200, 240$ nm. (c, d) Results obtained from the periodic MNP array model with varying MNP radii $a = 20, 25, 30, 35, 40, 45, 50, 55, 60$ nm and array period $\Lambda = 3a$. All cross sections are normalized to the particle surface area.

8.3 Numerical Modelling of the Plasmon Photoeffect by COMSOL

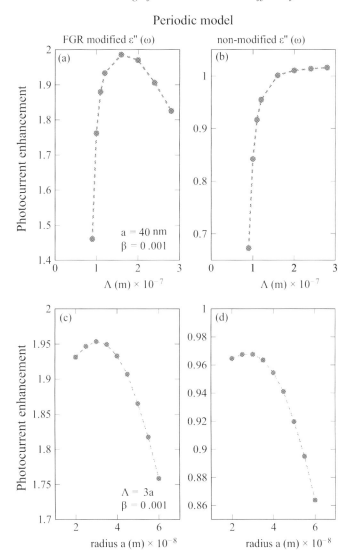

Figure 8.9 The periodic model. (a, b) The photocurrent enhancement from Eq. (8.12) as a function of MNP array period. The radii of the MNPs are assumed to be $a = 40$ nm. (c, d) The photocurrent enhancement as a function of MNP radius. (a, c) Results obtained in the model using modified $\varepsilon''(\omega)$. (b, d) Results obtained in the model using the non-modified bulk dielectric function [146]. We used the solar spectrum AM1.5G.

reduction in the efficiency of the photoeffect by at least one order of magnitude, both in the single-particle model (i.e., neglecting the electromagnetic interaction between particles), see Figs. 8.7 and 8.10, and in the periodic array model (including the electromagnetic interaction of the metallic nanoparticles); see

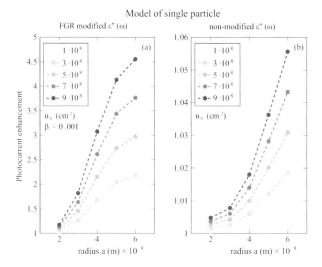

Figure 8.10 The single-particle model. (a, b) The photocurrent enhancement (Eq. (8.12)) as a function of MNP radius calculated for the model of the single MNP on Si substrate. The photocurrent enhancement is showed for various MNPs concentrations n_s. We used solar spectrum AM1.5G. (a) Results obtained in the model using the modified $\varepsilon''(\omega)$ by the plasmon damping. (b) Results obtained in the model using non-modified inappropriate dielectric function taken from the measurement in bulk [146].

Figs. 8.5, 8.6, 8.8, 8.9. The effect caused by plasmon damping is similarly strong in the case of absorption in the Si substrate including the mediation of nanoparticles (Figs. 8.5–8.7) and in the case of the absorption cross section, Fig. 8.9. In both cases a significant size effect is visible, manifesting itself in a strong dependence on the nanoparticle radius and on the calculation of the cell dimension (changing the latter is equivalent to varying the nanoparticle concentration). For the photocurrent, the strengthening achieves as much as two orders of magnitude in comparison to including the field concentration close to the nanoparticles only according to the Maxwell–Fresnel solution without quantum corrections (see, e.g., Fig. 8.10). A quantitative comparison is exemplified in Table 8.1 for the COMSOL simulation with inclusion of quantum corrections versus a similar simulation without quantum corrections; see Table 8.2. This comparison agrees with the presentation in Figs. 8.5–8.9.

An advantage of the COMSOL simulation is its ability to account for the electromagnetic interaction of the nanoparticles (simplified, however, to nearest neighbours only, via the periodic conditions imposed in the nanoparticle array model). The reducing role of a strong density of metallic coverage is noticeable; this is probably due to collective reflection, see Figs. 8.5, 8.6, 8.8, 8.9. This reducing effect of too large a concentration is, however, considerably weakened by the inclusion of quantum corrections related to the giant damping of plasmons described above.

The COMSOL simulation presented here is confined to spherical-shaped metallic nanoparticles. It makes little difference, however, to simulate arbitrary-shaped nanoparticles with arbitrary-type deposition on the semiconductor substrate (in particular, it is easy to include the case of a nanoparticle partly or completely embedded in a semiconductor medium). This flexibility of COMSOL simulation is its great advantage. It is reasonable to expect that a variation in nanoparticle shape would not cause significant changes with respect to the behaviour described for spherical nanoparticles, as the shape change would influence only the self-modes of the dipole surface plasmons; this would merely shift the resonance frequency slightly, practically without any modification of the plasmon damping due to coupling with the band electrons in the semiconductor substrate. The latter can be, however, more strongly affected by the type of deposition of the nanoparticles. The near-field coupling of plasmon dipole oscillations in the metallic nanoparticle with the band electrons in the semiconductor substrate can be considerably enhanced if the nanoparticle is embedded in a semiconductor medium. The practical realization of this stronger coupling may be arranged by using a sandwich structure with the metallic nanoparticles located in between the semiconductor layers. Though such an experiment has not been performed, one might expect a growth in the plasmon-mediated photoeffect owing to the strengthening of coupling of plasmons to electrons in all directions in the surrounding semiconductor.

8.4 Comparison with Experiment

Observation of the strong enhancement of the photoeffect induced by the mediation of light energy transfer to the semiconductor substrate by surface plasmons in metallic nanoparticles deposited on the photoactive surface has been reported for various experimental setups. Some of these observations are collected in Table 8.3. In various experiments also different methods of nanoparticle covering manufacturing are used. For instance, in Refs. [147, 148] gold nanoparticles on the semiconductor surface are produced by pulsed laser deposition. The size of particles is controlled by the number of ablation shots and the surrounding gas pressure. In [149] the method of magnetron sputtering production of $Au:TiO_2$ films was applied. Recently much attention has focused on perovskite solar cells due to the easy availability of low-temperature and cheap technology and simultaneously very high efficiency. The incorporation of Ag:Zn nanoparticles in the perovskite-based solar cell [150, 151] has led to improved device performance by 4.52% to 5.70%. Application of Au bar-shaped particles or multi-shell spherical Ag-based nanoparticles has led to a record increase of perovskite cell efficiency up to even 30%–40% [153]. The incorporation of bimetallic nanoparticles or core-shell nanoparticles in perovskite-based solar cells is a promising strategy as it allows for better

Table 8.3 Measured values of the photocurrent enhancement in Si solar cells and Si photodiodes with deposited metallic nanoparticles. For the various setups and nanoparticle deposition parameters, different increases in the photoeffect efficiency have been observed experimentally, as reported in the references indicated. The majority of the observed behaviour cannot be explained only by the local concentration of the electromagnetic field near the curvature of nanoparticles as accounted for by the conventional COMSOL modelling, and agreement of the experimental data with the theoretical simulation needs the inclusion of the plasmon damping contribution, as described earlier. However, some exceptionally low efficiency increases (or even decreases) evidence the complicated competition of various factors beyond the model considered, being apparently sensitive to the position of the p-n junction active layer in the substrate or to destructive interference effects or reflection from too dense coverings. (Table after [145].)

Metal	Size (nm)	Shape	Concentration (cm^{-2})	Enhancement	p-n Junction depth (nm)	Type	Ref.
Au	50	spherical	6.6×10^8	18% (max[1] 80%)			
Au	80	spherical	1.6×10^8	31% (max[1] 100%)	80	c-Si	[33]
Au	100	spherical	0.77×10^8	38% (max[1] 60%)			
Au	100	spherical	9.9×10^8	2.8%	500	c-Si[2]	[138]
Au	100	spherical	0.3×10^8	5.6%			
Au	100	spherical	1.5×10^8	2.0%			
Au	100	spherical	3.2×10^8	3.3%	> 500[3]	mc-Si[4]	[31]
Au	100	spherical	6.67×10^8	−2.8%			
Au	100	spherical	10×10^8	−7.5%			
Au	100	spherical	3.5×10^8	3.3%	500	c-Si[5]	[139]

[1] The maximal value of absorption enhancement per single wavelength.
[2] Commercialy available n/p Si solar cell produced by Silicon Solar Inc.
[3] Emitter thickness is 500 nm.
[4] Anti-reflection coating from SiN of thickness 70 nm.
[5] n-type Si wafer (001) of thickness 300 μm and donor concentration around 10^{15} cm^{-3}.

Au	diameter 20, height (2:3)	island	1.3×10^{11}	20% (max[1] 40%)	—	a-Si:H/c-Si[6]	[111]
Au	diameter 65	spherical	10×10^8	18%	300	c-Si[7]	[140]
Ag	40	island	124×10^8	127%	160	SOI[8]	[137]
Ag	66	island	67×10^8	283%	160	SOI[8]	
Ag	108	island	25×10^8	592%	160	SOI[8]	
Ag	thickness 12, 14, 16[9]	island	—	19%, 14%, 2%[10]	—	Si	[141]
Ag	diameter (120:140), height (45:60)	island	30×10^8	354%	95	SOI[8]	[32]
Ag	thickness 12 and 16[9]	island	—	33% and 16%[10]	1250	SOI[11]	[141]
Ag	25:91	spheroidal	38%[12]	17%	—	c-Si[13]	[144]
Al	22:81	spheroidal	40%	21%	—	c-Si[13]	
In	17:25	spheroidal	30%	23%	—	c-Si[13]	
Ag	∼ 60	island	31%	0%	$> 150^{14}$	c-Si	[142]
Al	diameter 190, height 70	cylindrical	19%[15]	49%		c-Si[16]	[143]

[6] Heterojunction a-Si:H/c-Si (p-type(100)): the thickness of the a-Si:H layer is 18 nm.
[7] Si wafer with transparent graphene electrode.
[8] Silicon-on-insulator (SOI).
[9] Nanoparticle coverage was fabricated by the annealing of silver film of a particular thickness (12, 14, 16 nm) at temperature 200°C for 50 minutes.
[10] Enhancement averaged over the solar spectrum AM1.5G.
[11] Silicon-on-insulator (SOI); top Si layer thickness 1250 nm.
[12] For dense coverings the percentage of metallized surface is shown.
[13] Solar cell covered by 20 nm layer of TiO$_2$ separating nanoparticles from silicon.
[14] Emitter thickness is 150 nm. [15] Periodic, period 380 nm. [16] Solar cell covered by 40 nm layer of SiO$_2$ separating nanoparticles from silicon.

accommodation to the absorption spectrum in the material. Caesium doping is also considered to improve perovskite solar cells [152]. Nano-modifications of dye solar cells have also investigated [154]; by a careful tuning of the amount of organic fluorophore in the hybrid coating material, a maximum increase in power conversion efficiency exceeding 4% is achieved in plastic organic cells [155]. Other metals have been also tested for use as nano-components including, e.g., aluminium and titanium besides gold and silver [156, 157], the most popular, as well as multi-shape particles (Ag) [158].

8.5 Conclusions

By COMSOL simulations it has been demonstrated that the neglect of quantum corrections to the dielectric functions of photocell materials causes a large error reaching more than one order of magnitude in the assessment of the photoeffect efficiency increase due to plasmon mediation in the light energy absorption. The dielectric functions of all materials in the photocell setup must be predefined as prerequisites for the COMSOL calculation. These dielectric functions are strongly affected by the quantum effect of near-field coupling between plasmons and band electrons, accounted for by the Fermi golden rule approach. At the nanoscale this coupling opens a highly effective channel of energy transfer, which modifies the dielectric functions of both the metallic nanoparticle and the coupled semiconductor substrate. They are different from the corresponding material characteristics taken from experiments in bulk and separately for the two coupled subsystems. This is a source of the above-mentioned large error in the simulation of the photovoltaic efficiency if the material dielectric functions are taken from conventional data tables. Thus it must be emphasized that the plasmon photovoltaic effect is caused not by the local concentration of electric field near the curvature of the metallic components but by the quantum effect of coupling between plasmons and electrons. The latter is beyond the ability of COMSOL simulation, which can properly include the local electric field strengthening near the metallic nanoparticles but cannot account for the coupling effect. This effect must be included by appropriate modification of the dielectric functions of both the metallic components and the semiconductor. As verified by COMSOL, such a modification of the dielectric functions gives realistic values of the plasmon-induced efficiency enhancement consistent with the experimental observations.

By application of the quantum Fermi golden rule scheme it has been demonstrated that the efficiency of the energy transfer channel between the surface plasmon oscillations in metallic nanoparticles and the substrate semiconductor is dependent on the material parameters of both components (the radius and material and surface concentrations of the metallic nanoparticles, the energy gap, the effective masses of the electrons and holes, and the dielectric permittivity of

8.5 Conclusions

the substrate). The analytical formulae for the plasmon–photoeffect efficiency and the time rate of the related plasmon damping allow optimization of the plasmon-mediated photoeffect in solar cells. For example, an Si photodiode covered with Au nanoparticles achieves maximal efficiency if covered with Au nanoparticles with radii of few tens of nm and surface density 10^8 ^{9}/cm^2. Other solar cells, such as CIGS (copper indium gallium diselenide) achieve a lower efficiency increase by metallization because of the worse accommodation of the semiconductor material parameters to the solar spectrum modified by plasmon mediation. The increase in the overall photovoltaic efficiency for the metallically modified cells considered by us varies between 1.5% (CIGS) and 6% (Si), depending on the nanoparticle concentration (note that concentrations that are too dense cause an efficiency decrease owing to destructive interference and reflection).

Thus, it has been demonstrated that metallic nanoparticles deposited on the surface of a photodiode can enhance the efficiency of the photoeffect by as much as a factor 2 (i.e., a 100% increase), which has been confirmed experimentally (for Au nanoparticles with a radius 40 nm deposited on Si at a concentration of 10^8/cm^2). Such a large increase is, however, reduced in commercial Si solar cells to $\sim 5\%$ of overall efficiency increase (and to $\sim 2\%$ in CIGS cells), because the photon absorption is only one in a long series of factors resulting in the final efficiency of a solar cell. The strong plasmon photovoltaic effect is caused mainly by quantum effects; the efficiency enhancement found by COMSOL without quantum corrections (i.e., when only the concentration of the electrical field close to the curvature of the metallic nanoparticles is accounted for) does not exceed 0.1, i.e., is lower than the quantum effect by at least one order of magnitude.

One can thus conclude that conventional numerical modelling (as in the COMSOL system with the material parameters supplied in packets attached to the system) is erroneous if carried out with the neglect of quantum-induced corrections to the dielectric functions of all components of metallized solar cells. The conventional COMSOL simulations without quantum corrections fail in comparison with experimental observations of the plasmon-mediated photoeffect. A realistic explanation of this effect is possible exclusively in the developed quantum approach.

Let us mention also the nonlocal plasmon effects in the metallic nanoparticles which also cause corrections to the conventional COMSOL (or Mie) approach. Nonlocal effects in metallic nanostructures are related to the spatial dispersion of the dielectric function, i.e., $\mathbf{D}(\mathbf{r}) = \int \varepsilon(\mathbf{r}, \mathbf{r}') \mathbf{E}(\mathbf{r}') d^3 r'$ (here, $\mathbf{D}$ and $\mathbf{E}$ denote the electric displacement field and the electric field, respectively, and $\varepsilon(\mathbf{r}, \mathbf{r}')$ denotes the nonlocal dielectric function). The nonlocal effects in the system can be accounted for via the hydrodynamic approach to electrodynamics, including the use of a quasiclassical kinetic Boltzmann-type equation (or Navier–Stokes equation) to describe the nonuniform electronic medium. As the latter involves the Fermi velocity of electrons and a microscopic scattering integral, the nonlocal approach is customarily called a quantum correction to the classical electromagnetic approaches to plasmons such as the Mie theory. This resembles the situation for the skin effect, in particular for its anomalous version [12] where a nonlocal electromagnetic response is essential. As in the case of discrimination between the normal and anomalous skin effect regimes, the role-play-time scales according to the electron scattering and the incident electromagnetic wave frequency, or the plasmon oscillation frequency in the case of plasmons. This gives an estimate for the applicability of nonlocal corrections to surface plasmons in nanoparticles: namely, the scattering time a/v_F, where a is the nanoparticle radius, should be lower than the plasmon time period $2\pi/\omega_1$ (where ω_1 is the Mie frequency), which gives approximately $a < 5$ nm for the nanoparticle radius for Au as the range of nonlocal plasmon effects. Indeed, the nonlocal corrections calculated using the hydrodynamic approach appear to be important for small metallic clusters with $a \sim 3$ nm. At this size the nonlocal corrections cause as much as a $\sim 10\%$ blueshift of the Mie plasmon frequency [159] but the effect quickly disappears at higher a: for $a > 5$ nm the nonlocal corrections are small and are completely negligible at the size $a \in (10, 100)$ nm regarded as of interest for applications due to the large Lorentz friction. Nonlocal plasmon effects in small clusters meet, however, with the spill-out effect [19] and Landau damping [123, 79], which are pronounced at the scale $a \sim 2$–3 nm. To precisely describe these phenomena, together time-dependent density operator theory [160] and a numerical *ab initio* calculation time-dependent local density approximation (TDLDA) are used, employing the Kohn–Sham equation approach [161]. This is the reason for addressing nonlocal effects in quantum regime, though they are treated as somewhat classical in the hydrodynamic approach [162–164]. Nonlocal effects manifest themselves also in the plasmon-mediated photovoltaic effect [165, 159] but only for small and ultrasmall nanoparticles (not larger than $a \sim 5$ nm) or for few-nanometre separations between particles. At larger size scales nonlocal corrections are small compared with the quantum effect of the strong coupling of plasmons with the substrate semiconductor presented above to simplify.

9
Plasmon–Polaritons in Metallic Nanoparticle Chains

In this chapter the collective plasmon modes in a metallic nano-chain are described in an accurate and far-reaching analytical formulation. The related problem of plasmon–polaritons in metallic chains is of considerable significance for applications in optoelectronics. Conventional optoelectroncs is limited by diffraction constraints to the micrometre size scale of optical fibres, greatly exceeding the nanoscale of the electronic elements. The exchange of photons by plasmon–polaritons with the same energy but with group velocity lower by orders of magnitude allows for the avoidance of diffraction constraints when an optical fibre is substituted by a metallic nano-chain. The latter appears to be an almost perfect plasmon–polariton wave guide without any radiation losses. The theoretical formulation of plasmon–polariton theory presented here supplements previous numerical simulations. The theory allows the explanation of the variety formerly observed in numerical models for plasmon–polariton properties, including some misleading artefacts resulting from approximate numerical solutions of the related singular nonlinear problem. In particular, superluminal plasmon–polariton propagation, observed in numerical experiments, has been excluded by an accurate solution using the theory presented in the chapter.

As discussed previously, the investigation of plasmon oscillations in metallic nanoparticles, besides its fundamental character, also has great significance for applications. Metallic surface modifications of solar cells at the nanoscale exhibit a growth in photovoltaic efficiency owing to plasmon mediation in sunlight energy harvesting [104, 32, 33, 105, 34, 35]. On the other hand, periodic linear structures of metallic nanoparticles – metallic nano-chains – serve as plasmon wave guides having low damping [3, 116, 4]. The wavelengths of plasmon–polaritons propagating in such structures are considerably shorter in comparison with light of the same frequency [166, 40, 38]. This allows for the avoidance of the diffraction limits for light circuits, when one transforms the light signal into a plasmon–polariton wave [1, 5, 6]. This is regarded as promising for the forthcoming construction of plasmon optoelectronic nanodevices that are not available in ordinary light wave guides because of diffraction constraints.

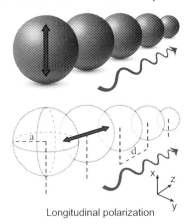

Figure 9.1 A chain of metallic nanospheres in which the plasmon–polariton can propagate with longitudinal polarization (plasmon dipole oscillations along the chain) or tranverse polarization (dipole oscillations perpendicular to the chain axis).

Metallic clusters with size 1–2 nm have been widely investigated both theoretically, within the so-called time-dependent local density approximation (TDLDA) [19, 71, 20, 79], and experimentally, especially for small clusters of alkali metals such as Na or K. Extremely small metal clusters contain a low number of electrons and have poor radiation properties; thus for energy transfer, by including plasmon–polaritons, larger particles are more suitable. Larger metallic nanoparticles of size several dozen nm manifest radiative properties that overwhelm the plasmon behaviour at this scale. For such nanoparticles, with a size of tens of nanometres, variants of the random phase approximation (RPA) have turned out to be appropriate and useful [19, 71, 21], as has been demonstrated in previous chapters. In particular, much attention has been focused on large nanoparticles of noble metals (Au, Ag, and Cu) because of the location of the surface plasmon resonances in particles of these metals within the visible light spectrum.

Owing to the intense radiation of surface plasmons in large nanoparticles, special interest has been paid to plasmon–polariton wave guides in the form of chains consisting of these nanoparticles [41, 167, 168]; see Fig. 9.1. Near-field microscopy applied to such systems has directly demonstrated the long-range propagation of a plasmon–polariton signal along these structures. For instance, in [169], the observation of 5 µm-range propagation in gold nanoparticles with average radius $a = 50$ nm aligned in an equidistant chain with the separation between neighbouring spheres $d = 200$ nm has been reported. In a series of papers [24, 170, 39, 171, 166], the practically non-damped propagation of collective plasmonic modes in Au and Ag nano-chains has been demonstrated over distances of ~ 0.5 µm, and energy transfer along the chain was directly confirmed. Such a long range of plasmon–polariton propagation, reaching visible light wavelengths and revealing a reduced group velocity, below $0.1c$ [170, 171, 166], fits well with the requirements of sub-diffraction nanophotonics. In [172], the weakly damped propagation of plasmon–polaritons along Ag nanowires with diameters of about 100 nm over distances of about 20 µm was reported. In [40], similarly long-range values for the practically undamped propagation of plasmon–polaritons in various metallic nanostructures were summarized. The theoretical estimation of plasmon–polariton attenuation in metallic nano-chains made within a dipole–dipole interaction model using finite element numerical methods leads to rather larger values for the attenuation take over ($\sim$ 3 dB/15 µm) than those observed experimentally [40, 169].

In the present treatise, we solve the problem of this discrepancy by means of a rigorous treatment of collective surface plasmon propagation along a nano-chain, including retardation effects for the near-, medium-, and far-field zones of plasmon dipoles and describing plasmon oscillations in individual nanospheres within the RPA approach via an analytical calculus. The radiative properties of collective surface plasmons in nanospheres aligned in the chain will be studied and described, taking into account the near-, medium-, and far-field contributions to the dipole coupling between elements of the chain and including retardation effects. The ideal vanishing of radiation losses in the chain for a wide spectrum of collective plasmonic wave-type modes will be demonstrated. The undamped plasmon–polariton modes induced by an external persistent oscillating electric forcing field can be identified. The frequency of these modes equals the frequency of the external field, and their amplitude is maximal for resonance with the plasmon–polariton self-mode frequency. The calculus of plasmon self-mode frequencies, their attenuation rates, and their group velocities will be detailed for longitudinal and transverse polarizations of collective excitations and for a variety of chain parameters. The problem of the logarithmic singular contribution caused by the far-field component of the dipole interaction will be solved for the self-energy of plasmon–polaritons with transverse polarization and the similar problem caused by the medium-field component for the group velocities of both polarizations, with a link to the numerical simulations of the long-range kinetics of plasmon–polaritons in metallic nano-chains will also be solved [173–176].

9.1 Plasmon Oscillations in Metallic Nanospheres

We will consider a chain of metallic (Au or Ag) nanospheres of radius a located in a dielectric environment, with dielectric constant ε, in the presence of a dynamic external electric field applied to some part of the chain. We will analyse a collective wave-type reaction of the electron liquid in the metallic components of the chain to this field. To describe this composite system, let us first briefly summarize the properties of plasmon oscillations for the single metallic nanosphere analysed in previous chapters, paying special attention to the surface plasmon damping caused by radiation of electromagnetic waves by the charged plasma oscillations.

Conventionally, a jellium model for each metallic nanosphere is assumed, in which the positive charge of the ion background is static and uniformly smeared over the whole nanosphere:

$$n_e(r) = n_e \Theta(a-r), \qquad n_e = N_e/V,$$

where N_e is the number of free electrons in the nanosphere, V is the volume of the nanosphere, Θ is the Heaviside step function, and $|e|n_e$ is the averaged density of the positive charge compensating the charge of the free electrons.

The fluctuations in the local electron density in the nanosphere can be divided into surface and volume parts:

$$\delta\rho(\mathbf{r},t) = \begin{cases} \delta\rho_1(\mathbf{r},t) & \text{for } r < a, \\ \delta\rho_2(\mathbf{r},t) & \text{for } r \geq a \ (r \to a+). \end{cases} \quad (9.1)$$

These two parts of the local electron density fluctuations satisfy equations derived in the framework of the random phase approximation (RPA),

$$\frac{\partial^2 \delta\rho_1(\mathbf{r},t)}{\partial t^2} = \frac{2}{3}\frac{\epsilon_F}{m}\nabla^2 \delta\rho_1(\mathbf{r},t) - \omega_p^2 \delta\rho_1(\mathbf{r},t) \quad (9.2)$$

and

$$\frac{\partial^2 \delta\rho_2(\mathbf{r},t)}{\partial t^2} = -\frac{2}{3m}\nabla\left\{\left[\frac{3}{5}\epsilon_F n_e + \epsilon_F \delta\rho_2(\mathbf{r},t)\right]\frac{\mathbf{r}}{r}\delta(r-a)\right\}$$
$$- \left[\frac{2}{3}\frac{\epsilon_F}{m}\frac{\mathbf{r}}{r}\nabla\delta\rho_2(\mathbf{r},t) + \frac{\omega_p^2}{4\pi}\frac{\mathbf{r}}{r}\nabla\int d^3r_1 \frac{1}{|\mathbf{r}-\mathbf{r}_1|}\right.$$
$$\left. \times (\delta\rho_1(\mathbf{r}_1,t)\Theta(a-r_1) + \delta\rho_2(\mathbf{r}_1,t)\Theta(r_1-a))\right]\delta(r-a), \quad (9.3)$$

where $\omega_p^2 = 4\pi n_e e^2/m$ is the bulk plasmon frequency; ϵ_F is the Fermi energy, n_e is the equilibrium density of electrons (equal to the jellium density), and m is the electron mass. The analysis and solution of the above equations has been given explicitly in previous chapters, resulting in the determination of the plasmon self-mode spectrum, for both the volume and surface modes.

Nevertheless, this RPA treatment does not account for plasmon damping. In order to include the damping of plasmons in a phenomenological manner, one adds an attenuation term to the plasmon dynamic oscillatory-type equations, i.e., the terms $(-2/\tau_0)\partial\delta\rho(\mathbf{r},t)/\partial t$ added to the right-hand sides of both Eqs. (9.2)

9.1 Plasmon Oscillations in Metallic Nanospheres

and (9.3). The damping rate $1/\tau_0$ (as already discussed in previous chapters) accounts for electron scattering losses [24],

$$\frac{1}{\tau_0} \simeq \frac{v_F}{2\lambda_b} + \frac{Cv_F}{2a}, \quad (9.4)$$

where C is a constant of order unity, a is the nanosphere radius, v_F is the Fermi velocity in the metal, λ_b is the electron mean free path in the bulk metal (including the scattering of electrons on other electrons, on impurities, and on phonons [24]). For example, for Au, $v_F = 1.396 \times 10^6$ m/s and $\lambda_b \simeq 53$ nm at room temperature; the latter term in Eq. (9.4) accounts for the scattering of electrons at the boundary of the nanoparticle, whereas the former term corresponds to scattering processes similar to those in the bulk. Other effects, such as Landau damping (especially important in small clusters [80, 79]), corresponding to the decay of a plasmon to a high-energy particle–hole pair, are of less significance for nanosphere radii larger than 2–3 nm [80] and are completely negligible for radii larger than 5 nm; here we consider large nanospheres with radii ≥ 10 nm. Note that for the growth in this radius range a lesser role is played also by the electron liquid spill-out effect [19, 20], though this is of primary importance for small clusters [19, 21].

Besides the homogeneous Eqs. (9.2) and (9.3) determining the self-frequencies of plasmon modes, dual inhomogeneous equations should be written with an explicit expression for the forcing factor. This factor would be the time-dependent electric field, e.g., the electric component of the incident electromagnetic wave. For the electromagnetic wave frequency that is in resonance with the plasmons in the metallic nanosphere, the wavelength (being of the order of 500 nm in this case) greatly exceeds the nanosphere size (with radius 10–50 nm); thus the dipole-regime conditions are fulfilled. For a forcing field $\mathbf{E}(t)$ that is almost uniform over the nanosphere, which corresponds to the dipole approximation, only the dipole surface mode can be excited. Then the electron-liquid response resolves itself into a single dipole-type mode described by the function $Q_{1m}(t)$ (i.e., with $l = 1$ and $m = -1, 0, 1$, the angular momentum numbers relating to spherical symmetry and describing the dipole mode). The function $Q_{1m}(t)$ satisfies the equation

$$\frac{\partial^2 Q_{1m}(t)}{\partial t^2} + \frac{2}{\tau_0}\frac{\partial Q_{1m}(t)}{\partial t} + \omega_1^2 Q_{1m}(t)$$
$$= \sqrt{\frac{4\pi}{3}\frac{en_e}{m}} \left[E_z(t)\delta_{m,0} + \sqrt{2}\left(E_x(t)\delta_{m,1} + E_y(t)\delta_{m,-1}\right) \right], \quad (9.5)$$

where $\omega_1 = \omega_p/\sqrt{3\varepsilon}$; this is a dipole surface plasmon Mie-type frequency [18] and ε is the dielectric constant of the nanosphere surroundings [23]. The function $Q_{1m}(t)$ is the only contribution to the plasmon response to a homogeneous uniform electric field. Thus, for such a uniform forcing field, the electron density fluctuations attain the form [23]

$$\delta\rho(\mathbf{r},t) = \begin{cases} 0 & \text{for } r < a \\ \sum_{m=-1}^{1} Q_{1m}(t) Y_{1m}(\Omega) & \text{for } r \geq a \ (r \to a+), \end{cases} \quad (9.6)$$

where $Y_{lm}(\Omega)$ is a spherical harmonic with $l = 1$.

For plasmon oscillations given by Eq. (9.6), one can calculate the corresponding dipole $\mathbf{D}(t)$, as demonstrated in previous chapters:

$$\begin{cases} D_x(t) = e \int d^3r\, x\, \delta\rho(\mathbf{r},t) = \sqrt{\frac{2\pi}{3}} e Q_{1,1}(t) a^3, \\ D_y(t) = e \int d^3r\, y\, \delta\rho(\mathbf{r},t) = \sqrt{\frac{2\pi}{3}} e Q_{1,-1}(t) a^3, \\ D_z(t) = e \int d^3r\, z\, \delta\rho(\mathbf{r},t) = \sqrt{\frac{4\pi}{3}} e Q_{1,0}(t) a^3, \end{cases} \quad (9.7)$$

and $\mathbf{D}(t)$ satisfies the equation (rewriting Eq. (9.5))

$$\left[\frac{\partial^2}{\partial t^2} + \frac{2}{\tau_0}\frac{\partial}{\partial t} + \omega_1^2\right]\mathbf{D}(t) = \frac{a^3 4\pi e^2 n_e}{3m}\mathbf{E}(t) = \varepsilon a^3 \omega_1^2 \mathbf{E}(t). \quad (9.8)$$

The phenomenological oscillatory-type damping term in the above equation includes all types of scattering phenomena, i.e., electron–electron, electron–phonon, electron–admixture interactions, as well as the contribution caused by the scattering of electrons at the nanoparticle boundary [24]. Electron scattering effects cause significant attenuation of plasmons, especially important in small metal clusters. These contributions to the damping time rate are proportional to $1/a$ and are of less significance with growth in radius. There is also, however, another important channel of plasmon damping caused by radiation losses, not included in the formula for τ_0. We will show that the contribution to the overall attenuation of plasmons caused by the radiation losses scales as a^3, and therefore for large nanospheres the radiative losses dominate plasmon damping.

The radiative losses of oscillating charges, related to the plasmon dipole variation in time, can be expressed by the Lorentz friction [55], i.e., by an equivalent fictitious electric field slowing down the motion of charges,

$$\mathbf{E}_L = \frac{2}{3c^3}\frac{\partial^3 \mathbf{D}(t)}{\partial t^3}. \quad (9.9)$$

Thus, one can rewrite Eq. (9.8) including the Lorentz friction term,

$$\left[\frac{\partial^2}{\partial t^2} + \frac{2}{\tau_0}\frac{\partial}{\partial t} + \omega_1^2\right]\mathbf{D}(t) = \varepsilon a^3 \omega_1^2 \mathbf{E}(t) + \varepsilon a^3 \omega_1^2 \mathbf{E}_L, \quad (9.10)$$

9.1 Plasmon Oscillations in Metallic Nanospheres

or more explicitly, for the case when $\mathbf{E} = 0$, since $\omega_1 = \omega_p/\sqrt{3\varepsilon}$,

$$\left[\frac{\partial^2}{\partial t^2} + \omega_1^2\right]\mathbf{D}(t) = \frac{\partial}{\partial t}\left[-\frac{2}{\tau_0}\mathbf{D}(t) + \frac{2}{3\omega_1\sqrt{\varepsilon}}\left(\frac{\omega_p a}{c\sqrt{3}}\right)^3\frac{\partial^2}{\partial t^2}\mathbf{D}(t)\right]. \quad (9.11)$$

Applying now a perturbation procedure for solving Eq. (9.11) and treating the right-hand side of this equation as the perturbation, one obtains in the zeroth step of the perturbation procedure $[\partial^2/\partial t^2 + \omega_1^2]\mathbf{D}(t) = 0$, from which $(\partial^2/\partial t^2)\mathbf{D}(t) = -\omega_1^2\mathbf{D}(t)$. Therefore, in the first step of the perturbation procedure, one substitutes the latter formula into the right-hand side of Eq. (9.11), obtaining

$$\left[\frac{\partial^2}{\partial t^2} + \frac{2}{\tau}\frac{\partial}{\partial t} + \omega_1^2\right]\mathbf{D}(t) = 0, \quad (9.12)$$

where

$$\frac{1}{\tau} = \frac{1}{\tau_0} + \frac{\omega_1}{3\sqrt{\varepsilon}}\left(\frac{\omega_p a}{c\sqrt{3}}\right)^3. \quad (9.13)$$

In this way, we have included the Lorentz friction in the total attenuation rate $1/\tau$, which is justified for nanospheres that are not too large, i.e., when the second term in Eq. (9.13), proportional to a^3, is sufficiently small to fulfil the perturbation procedure constraints. For nanospheres with radius 10–30 nm, this approximation is justified and has been verified experimentally for Au and Ag nanospheres [77]. It is worth emphasizing the ability of this perturbative approach to explain the experimentally observed redshift of the surface plasmon resonance frequency with growing a (visible in the extinction spectrum of light passing through water-colloidal solutions of metallic nanospheres with various radii [77]). Indeed, the solution of Eq. (9.12) is of the form $\mathbf{D}(t) = Ae^{-t/\tau}\cos(\omega_1' t + \phi)$, where $\omega_1' = \omega_1\sqrt{1 - 1/(\omega_1\tau)^2}$; the latter gives the experimentally observed redshift of the plasmon resonance due to $\sim a^3$ growth of the attenuation caused by the radiation losses. The Lorentz friction term in Eq. (9.13) dominates the plasmon damping for $a \geq 12$ nm (for Au and Ag) owing to this a^3 dependence. The plasmon damping grows rapidly with a and results in a pronounced redshift of the resonance frequency in good agreement with the experimental data for $10 < a < 30$ nm (Au and Ag) [77]. In this size range of metallic particles, we find a conspicuous cross-over of the attenuation rate in its dependence on a. The minimum of damping is achieved at

$$a^* = \left(\frac{9}{2}\varepsilon\frac{v_F c^3}{\omega_p^4}\right)^{1/4} \quad (9.14)$$

and, for $a < a^*$, the damping rate grows with decreasing a approximately as $1/a$, while for $a > a^*$ the rate grows as a^3 with increasing a. The value of a^* can be estimated for Au, Ag, and Cu; see Table 9.1.

Table 9.1 *Radius of the nanosphere corresponding to the minimal value of surface plasmon damping.*

$n = \sqrt{\varepsilon}$	Au	Ag	Cu
		a^*	
$n = 1$, vacuum	8.78	8.77	8.82
$n = 1.4$, water	10.4	10.42	10.5
$n = 2$	12.4	12.41	12.5

9.2 Radiative Properties of a Metallic Nanosphere in a Chain

When a metallic nanosphere is an element of a chain created by similar nanospheres equidistantly distributed along a line, one has to take into account that its radiation losses may be modified by energy income from other nanospheres when, in the total system, a collective plasmon excitation propagates. The periodicity of the chain makes the system similar to a one-dimensional crystal; see Fig. 9.2. The interaction between nanospheres can be considered as a dipole-type coupling. The minimal separation of nanospheres in the chain is $d = 2a$, where d is the distance between neighbouring sphere centres, but we will consider $d > 3a$, because the dipole approximation for plasmon interaction in the nanosphere chain is sufficiently accurate for $d > 3a$ since then the contribution of multipole interactions can be neglected. Various numerical large-scale calculations of the electromagnetic field distribution in such systems have been made, including dipole and also multipole interactions between the plasmonic oscillations in metallic components [41, 167, 168]. It is worth noting that a model of interacting dipoles [177, 178] was developed first for the investigation of stellar matter [179, 180] and only then was it adopted for metal particle systems [181, 182]. Numerical studies beyond the dipole model [41, 168] indicate that the dipole model is sufficiently accurate when the particle separation is not lower than the particle radius. Otherwise, multipole contributions to the interaction start to be important [183].

Assuming that in a sphere located at the point $\mathbf{r}$ we have an oscillating dipole $\mathbf{D}$, then, at another point at a position $\mathbf{r}_0$ with respect to $\mathbf{r}$, this dipole causes an electric field in the following form (it includes the relativistic retardation of electromagnetic signals) [55, 76],

$$\mathbf{E}(\mathbf{r}, \mathbf{r}_0, t) = \frac{1}{\varepsilon} \left(-\frac{\partial^2}{v^2 \partial t^2} \frac{1}{r_0} - \frac{\partial}{v \partial t} \frac{1}{r_0^2} - \frac{1}{r_0^3} \right) \mathbf{D}(\mathbf{r}, t - r_0/v)$$

$$+ \frac{1}{\varepsilon} \left(\frac{\partial^2}{v^2 \partial t^2} \frac{1}{r_0} + \frac{\partial}{v \partial t} \frac{3}{r_0^2} + \frac{3}{r_0^3} \right) \mathbf{n}_0 (\mathbf{n}_0 \cdot \mathbf{D}(\mathbf{r}, t - r_0/v)), \quad (9.15)$$

where $\mathbf{n}_0 = \mathbf{r}_0 / r_0$ and $v = c/\sqrt{\varepsilon}$. The above formula includes terms corresponding to the near-field zone (the denominator with r_0^3), the medium-field zone (the

9.2 Radiative Properties of a Metallic Nanosphere in a Chain

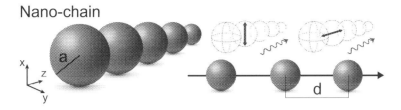

Figure 9.2 Metallic nanospheres with radius a are aligned along the chain axis with separation d between chain elements. At top right are shown the transverse and longitudinal plasmon polarizations.

denominator with r_0^2), and the far-field zone (the denominator with r_0) in the dipole field. This allows us to write out a dynamical equation for the plasmon oscillations in each nanosphere of the chain; they can be numbered by an integer l (d denotes the separation between nanospheres in the chain, $d > 3a$). The vectors $\mathbf{r}$ and $\mathbf{r}_0$ are collinear if the origin is associated with a nanosphere in the chain.

Therefore, the equation for surface plasmon oscillations on the lth sphere is as follows:

$$\left[\frac{\partial^2}{\partial t^2} + \frac{2}{\tau_0}\frac{\partial}{\partial t} + \omega_1^2\right] D_\alpha(ld,t) = \varepsilon\omega_1^2 a^3 \sum_{m=-\infty,\, m\neq l}^{m=\infty} E_\alpha\left(md, t - \frac{|l-m|d}{v}\right)$$
$$+ \varepsilon\omega_1^2 a^3 E_{L\alpha}(ld,t) + \varepsilon\omega_1^2 a^3 E_\alpha(ld,t). \quad (9.16)$$

The first term on the right-hand side of Eq. (9.16) describes the dipole-type coupling between nanospheres, and the other two terms correspond to the contribution to plasmon attenuation due to the Lorentz friction (as described in the previous section) and to the external forcing field. The index α enumerates the polarizations; for longitudinal polarization $\alpha = z$ and for transverse polarization $\alpha = x(y)$, with respect to the chain orientation (assumed to be in the z direction). According to Eq. (9.15), we have:

$$E_z(md,t) = \frac{2}{\varepsilon d^3}\left(\frac{1}{|m-l|^3} + \frac{d}{v|m-l|^2}\frac{\partial}{\partial t}\right) \times D_z(md, t - |m-l|d/v),$$

$$E_{x(y)}(md,t) = -\frac{1}{\varepsilon d^3}\left(\frac{1}{|m-l|^3} + \frac{d}{v|m-l|^2}\frac{\partial}{\partial t} + \frac{d^2}{v^2|l-d|}\frac{\partial^2}{\partial t^2}\right)$$
$$\times D_{x(y)}(md, t - |m-l|d/v). \quad (9.17)$$

Taking advantage of the chain periodicity (in analogy to the Bloch states in crystals with the reciprocal quasimomentum lattice), one can assume that

$$D_\alpha(ld,t) = D_\alpha(k,t)\, e^{-ikld} \qquad 0 \le k \le \frac{2\pi}{d}. \quad (9.18)$$

One can assert this in a more formal manner, taking the Fourier picture of Eq. (9.16) as follows. As the dipoles are localized at nanosphere centres, the system is discrete, similarly to the case of phonons in a one-dimensional crystal. One can thus apply a discrete Fourier transform (DFT) with respect to the positions, whereas an ordinary continuous Fourier transform (CFT) would be with respect to time. The DFT is defined for a finite set of numbers, so we can consider a chain with $2N+1$ nanospheres, i.e., of length $L = 2Nd$. Thus, for any discrete characteristics $f(l)$, $l = -N, \ldots, 0, \ldots, N$ of the chain, such as a selected polarization of the dipole distribution, one is dealing with the DFT picture:

$$f(k) = \sum_{l=-N}^{N} f(l) e^{ikld}, \qquad (9.19)$$

where $k = 2\pi/(2Nd)n$, $n = 0, \ldots, 2N$. This means that $kd \in [0, 2\pi]$ owing to the periodicity of the equidistant chain, with separation between nanosphere centres equal to d. The Born–Karman boundary condition is imposed on the whole system, resulting in the above form of k. In order to account for the infinite length of the chain, one can take the limit $N \to \infty$, which causes the variable k to be quasicontinuous, but we still have $kd \in [0, 2\pi]$.

As will be described in Section 9.3, one arrives at the Fourier picture of Eq. (9.16), with a DFT for positions and a CFT for time:

$$\left(-\omega^2 - i\frac{2}{\tau_0}\omega + \omega_1^2\right) D_\alpha(k,\omega) = \omega_1^2 \frac{a^3}{d^3} F_\alpha(k,\omega) D_\alpha(k,\omega) + \varepsilon a^3 \omega_1^2 E_{0\alpha}(k,\omega), \qquad (9.20)$$

with, setting $x = kd$ and $\xi = \omega d/v$,

$$F_z(k,\omega) = 4 \sum_{m=1}^{\infty} \left(\frac{\cos(mx)}{m^3} \cos(mx) + \xi \frac{\cos(m\xi)}{m^2} \sin(m) \right)$$

$$+ 2i \left[\frac{1}{3}\xi^3 + 2 \sum_{m=1}^{\infty} \left(\frac{\cos(mx)}{m^3} \sin(m\xi) - \xi \frac{\cos(mx)}{m^2} \cos(m\xi) \right) \right],$$

$$F_{x(y)}(k,\omega) = -2 \sum_{m=1}^{\infty} \left(\frac{\cos(mx)}{m^3} \cos(m\xi) + \xi \frac{\cos(mx)}{m^2} \sin(m\xi) \right.$$

$$\left. - \xi^2 \frac{\cos(mx)}{m} \cos(m\xi) \right)$$

$$- i \left[-\frac{2}{3}\xi^3 + 2 \sum_{m=1}^{\infty} \left(\frac{\cos(mx)}{m^3} \sin(m\xi) + \xi \frac{\cos(mx)}{m^2} \cos(m\xi) \right.\right.$$

$$\left.\left. - (\xi)^2 \frac{\cos(mx)}{m} \sin(m\xi) \right) \right]. \qquad (9.21)$$

9.2 Radiative Properties of a Metallic Nanosphere in a Chain

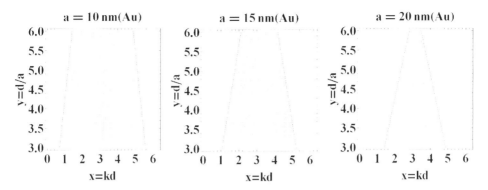

Figure 9.3 The shaded regions ($0 < kd \pm \omega_1 d/v < 2\pi$) in which the radiation losses vanish for infinite chains of Au nanospheres with radius $a = 10, 15, 20$ nm and chain separation $d/a \in [3, 6]$ ($\omega = \omega_1, v = c$).

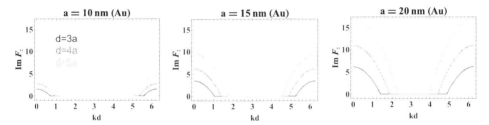

Figure 9.4 The function Im $F_z(k; \omega = \omega_1)$ for infinite chains of Au nanospheres with radius $a = 10, 15, 20$ nm and chain separation $d = 3a, 4a, 5a$ ([46]).

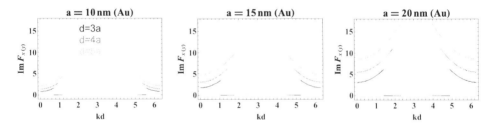

Figure 9.5 The function Im $F_{x(y)}(k; \omega = \omega_1)$ for infinite chains of Au nanospheres with radius $a = 10, 15, 20$ nm and chain separation $d = 3a, 4a, 5a$ ([46]).

A direct calculation of the functions Im $F_z(k, \omega)$ and Im $F_{x(y)}(k, \omega)$, which correspond to the radiative damping for the longitudinal and transverse plasmon–polariton polarizations, respectively, will be given in Section 9.3; see Eqs. (9.27) and (9.29). We show there that both these functions vanish when $0 < kd \pm \omega d/v < 2\pi$ (the corresponding region is indicated in Fig. 9.3). Outside this region, the radiative damping expressed by the functions Im $F_\alpha(k, \omega)$ is not zero, which for the longitudinal and transverse modes is illustrated in Figs. 9.4 and 9.5, correspondingly.

9.3 Calculation of the Radiative Damping of a Plasmon–Polariton in a Chain

Both sides of Eq. (9.16) can be multiplied by $e^{i(kld-\omega t)}/2\pi$ and then, one can perform a summation with respect to the nanosphere positions and an integration over t. Taking into account that

$$\frac{1}{2\pi}\int_{-\infty}^{\infty}\sum_{l=-N}^{N} D_\alpha\left(\pm md+ld; t-\frac{md}{v}\right) e^{-i(kld-\omega t)} dt = e^{i(\mp kmd+\omega md/v)} D_\alpha(k,\omega), \tag{9.22}$$

one thus obtains the following equation in Fourier representation (a DFT for the nanosphere positions and a CFT for time):

$$\left(-\omega^2 - i\frac{2}{\tau_0}\omega + \omega_1^2\right) D_\alpha(k,\omega) = \omega_1^2 \frac{a^3}{d^3} F_\alpha(k,\omega) D_\alpha(k,\omega) + \varepsilon a^3 \omega_1^2 E_{0\alpha}(k,\omega), \tag{9.23}$$

where $k = 2\pi n/(2Nd)$, $n = 0, 1, \ldots, 2N$, i.e., $kd \in [0, 2\pi]$, due to the periodicity of the chain with equidistant separations d of the nanospheres, and the form of k is due to the Born–Karman boundary condition with the period $L = 2Nd$. For $N \to \infty$ (the infinite-chain limit), k is a quasicontinuous variable. In Eq. (9.23) we have, again setting $x = kd$, $\xi = \omega d/v$,

$$F_z(k,\omega) = 4\sum_{m=1}^{\infty}\left(\frac{\cos(mx)}{m^3}\cos(m\xi) + \xi\frac{\cos(mx)}{m^2}\sin(m\xi)\right)$$

$$+ 2i\left[\frac{1}{3}\xi^3 + 2\sum_{m=1}^{\infty}\left(\frac{\cos(mx)}{m^3}\sin(m\xi) - \xi\frac{\cos(mx)}{m^2}\cos(m\xi)\right)\right],$$

$$F_{x(y)}(k,\omega) = -2\sum_{m=1}^{\infty}\left(\frac{\cos(mx)}{m^3}\cos(m\xi) + \xi\frac{\cos(mx)}{m^2}\sin(m\xi)\right.$$

$$\left. - \xi^2\frac{\cos(mx)}{m}\cos(m\xi)\right)$$

$$- i\left[-\frac{2}{3}\xi^3 + 2\sum_{m=1}^{\infty}\left(\frac{\cos(mx)}{m^3}\sin(m\xi) + \xi\frac{\cos(mx)}{m^2}\cos(m\xi)\right.\right.$$

$$\left.\left. - \xi^2\frac{\cos(mx)}{m}\sin(m\xi)\right)\right]. \tag{9.24}$$

9.3 Calculation of the Radiative Damping of a Plasmon–Polariton in a Chain

Some summations in the above equations can be done analytically [184]:

$$\begin{cases} \sum_{m=1}^{\infty} \frac{\sin(mz)}{m} = \frac{\pi - z}{2}, & \text{for } 0 < z < 2\pi, \\ \sum_{m=1}^{\infty} \frac{\cos(mz)}{m} = \frac{1}{2}\ln\left(\frac{1}{2 - 2\cos(z)}\right), \\ \sum_{m=1}^{\infty} \frac{\cos(mz)}{m^2} = \frac{\pi^2}{6} - \frac{\pi}{2}z + \frac{1}{4}z^2, & \text{for } 0 < z < 2\pi, \\ \sum_{m=1}^{\infty} \frac{\sin(mz)}{m^3} = \frac{\pi^2}{6}z - \frac{\pi}{4}z^2 + \frac{1}{12}z^3, & \text{for } 0 < z < 2\pi. \end{cases} \quad (9.25)$$

Using the above formulae, one can show that if $0 < kd \pm \omega d/v < 2\pi$ then

$$\operatorname{Im} F_z(k, \omega) = 2 \sum_{m=1}^{\infty} \left[\frac{\sin(m(x+\xi)) - \sin(m(x-\xi))}{m^3} \right.$$

$$\left. -\xi \frac{\cos(m(x+\xi)) + \cos(m(x-\xi))}{m^2} \right] + \frac{2}{3}\xi^3$$

$$= 2\left[\frac{\pi^2}{6}(x+\xi) - \frac{\pi}{4}(x+\xi)^2 + \frac{1}{12}(x+\xi)^3\right]$$

$$- 2\left[\frac{\pi^2}{6}(x-\xi) - \frac{\pi}{4}(x-\xi)^2 + \frac{1}{12}(x-\xi)^3\right]$$

$$- 2\xi\left[\frac{\pi^2}{6} - \frac{\pi}{2}(x+\xi) + \frac{1}{4}(x+\xi)^2\right]$$

$$- 2\xi\left[\frac{\pi^2}{6} - \frac{\pi}{2}(x-\xi) + \frac{1}{4}(x-\xi)^2\right] + \frac{2}{3}\xi^3 \equiv 0. \quad (9.26)$$

However, if $kd - \omega d/v < 0$ or $kd + \omega d/v > 2\pi$ for some values of the wave vector k then a more general formula is needed. By using the Heaviside step function, one can extend Eq. (9.25) to the second period of their left-hand sides; note that for $d/a \in [2, 10]$, $a < 25$ nm, the periods after the second are not reached for $kd \in [0, 2\pi)$. The extended form for $\operatorname{Im} F_z(k, \omega)$ is as follows, $x = kd$, $\xi = \omega d/v$,

$$\operatorname{Im} F_z(k, \omega) = \Theta(2\pi - x - \xi)2\left[\frac{\pi^2}{6}(x+\xi) - \frac{\pi}{4}(x+\xi)^2 + \frac{1}{12}(x+\xi)^3\right]$$

$$+ \Theta(-2\pi + x + \xi)2\left[\frac{\pi^2}{6}(x+\xi-2\pi) - \frac{\pi}{4}(x+\xi-2\pi)^2\right.$$

$$\left. + \frac{1}{12}(x+\xi-2\pi)^3\right]$$

$$-\Theta(x-\xi)2\left[\frac{\pi^2}{6}(x-\xi) - \frac{\pi}{4}(x-\xi)^2 + \frac{1}{12}(x-\xi)^3\right]$$

$$-\Theta(-x+\xi)2\left[\frac{\pi^2}{6}(x-\xi+2\pi) - \frac{\pi}{4}(x-\xi+2\pi)^2 \right.$$
$$\left. + \frac{1}{12}(x-\xi+2\pi)^3\right]$$

$$-\Theta(2\pi - x - \xi)2\xi\left[\frac{\pi^2}{6} - \frac{\pi}{2}(x+\xi) + \frac{1}{4}(x+\xi)^2\right]$$

$$-\Theta(-2\pi + x + \xi)2\xi\left[\frac{\pi^2}{6} - \frac{\pi}{2}(x+\xi-2\pi) + \frac{1}{4}(x+\xi-2\pi)^2\right]$$

$$-\Theta(-x+\xi)2\xi\left[\frac{\pi^2}{6} - \frac{\pi}{2}(x-\xi+2\pi) + \frac{1}{4}(x-\xi+2\pi)^2\right]$$

$$-\Theta(x-\xi)2\xi\left[\frac{\pi^2}{6} - \frac{\pi}{2}(x-\xi) + \frac{1}{4}(x-\xi)^2\right] + \frac{2}{3}\xi^3. \qquad (9.27)$$

The function given by Eq. (9.27) is depicted in Fig. 9.4. Equation (9.27) allows one to account for the inconsistency of the periodic functions given by the sums of the sines and cosines with the non-periodic Lorentz friction term and the inconsistency of the arguments $kd \pm \omega d/v$ of the trigonometric functions beyond the first period. Figure 9.3 gives the solution of the equation $(kd - \omega d/v)(kd + \omega d/v - 2\pi) = 0$, which determines the region for $x = kd$ versus d/a inside which exact cancellation of the Lorentz friction by the radiative energy income from other nanospheres takes place. In Fig. 9.4, this cancellation is presented for various nanosphere diameters, for longitudinally polarized plasmon collective excitations.

A similar analysis can be done for transversely polarized excitation, i.e., for Im $F_{x(y)}(k, \omega)$. This function is exactly zero only for the regions $0 < kd - \omega d/v < 2\pi$ and $0 < kd + \omega d/v < 2\pi$, where one can write

$$\text{Im } F_{x(y)}(k,\omega) = -\sum_{m=1}^{\infty}\left[\frac{\sin(m(x+\xi)) - \sin(m(x-\xi))}{m^3}\right.$$
$$-\xi\frac{\cos(m(x+\xi)) + \cos(m(x-\xi))}{m^2}$$
$$\left.-\xi^2\frac{\sin(m(x+\xi)) - \sin(m(x-\xi))}{m}\right] + \frac{2}{3}(\xi)^3$$
$$= -\left[\frac{\pi^2}{6}(x+\xi) - \frac{\pi}{4}(x+\xi)^2 + \frac{1}{12}(x+\xi)^3\right]$$
$$+ \left[\frac{\pi^2}{6}(x-\xi) - \frac{\pi}{4}(x-\xi)^2 + \frac{1}{12}(x-\xi)^3\right]$$

9.3 Calculation of the Radiative Damping of a Plasmon–Polariton in a Chain

$$+ \xi \left[\frac{\pi^2}{6} - \frac{\pi}{2}(x+\xi) + \frac{1}{4}(x+\xi)^2 \right]$$

$$+ \xi \left[\frac{\pi^2}{6} - \frac{\pi}{2}(x-\xi) + \frac{1}{4}(x-\xi)^2 \right]$$

$$+ \frac{1}{2}\xi^2[\pi - x - \xi] - \frac{1}{2}\xi^2[\pi - x + \xi] + \frac{2}{3}\xi^3$$

$$\equiv 0. \tag{9.28}$$

Nevertheless, outside the regions $0 < kd \pm \omega d/v < 2\pi$ the value of $\operatorname{Im} F_{x(v)}$ is nonzero, as is demonstrated in Fig. 9.5 and as can be accounted for by the following formula, again $x = kd$, $\xi = \omega d/v$):

$$\operatorname{Im} F_{x(v)}(k,\omega) = -\Theta(2\pi - x - \xi)\left[\frac{\pi^2}{6}(x+\xi) - \frac{\pi}{4}(x+\xi)^2 + \frac{1}{12}(x+\xi)^3\right]$$

$$- \Theta(-2\pi + x + \xi)\left[\frac{\pi^2}{6}(x+\xi - 2\pi) - \frac{\pi}{4}(x+\xi - 2\pi)^2 \right.$$

$$\left. + \frac{1}{12}(x+\xi - 2\pi)^3\right]$$

$$+ \Theta(x - \xi)\left[\frac{\pi^2}{6}(x-\xi) - \frac{\pi}{4}(x-\xi)^2 + \frac{1}{12}(x-\xi)^3\right]$$

$$+ \Theta(-x + \xi)\left[\frac{\pi^2}{6}(x-\xi + 2\pi) - \frac{\pi}{4}(x-\xi + 2\pi)^2 \right.$$

$$\left. + \frac{1}{12}(x-\xi + 2\pi)^3\right]$$

$$+ \Theta(2\pi - x - \xi)\xi\left[\frac{\pi^2}{6} - \frac{\pi}{2}(x+\xi) + \frac{1}{4}(x+\xi)^2\right]$$

$$+ \Theta(x - \xi)\xi\left[\frac{\pi^2}{6} - \frac{\pi}{2}(x-\xi) + \frac{1}{4}(x-\xi)^2\right]$$

$$+ \Theta(-2\pi + x + \xi)\xi\left[\frac{\pi^2}{6} - \frac{\pi}{2}(x+\xi - 2\pi) + \frac{1}{4}(x+\xi - 2\pi)^2\right]$$

$$+ \Theta(-x + \xi)\xi\left[\frac{\pi^2}{6} - \frac{\pi}{2}(x-\xi + 2\pi) + \frac{1}{4}(x-\xi + 2\pi)^2\right]$$

$$+ \Theta(2\pi - x - \xi)\frac{1}{2}\xi^2[\pi - x - \xi] + \Theta(-2\pi + x + \xi)$$

$$\times \frac{1}{2}\xi^2[3\pi - x - \xi] - \Theta(x - \xi)\frac{1}{2}(\xi)^2[\pi - x + \xi]$$

$$- \Theta(-x + \xi)\frac{1}{2}(\xi)^2[-\pi - x + \xi] + \frac{2}{3}\xi^3. \tag{9.29}$$

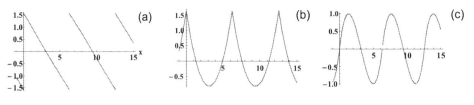

Figure 9.6 The sums (a) $\sum_{n=1}^{\infty} \frac{\sin(nx)}{n}$, (b) $\sum_{n=1}^{\infty} \frac{\cos(nx)}{n^2}$, (c) $\sum_{n=1}^{\infty} \frac{\sin(nx)}{n^3}$, for $x \in (-1, 15)$.

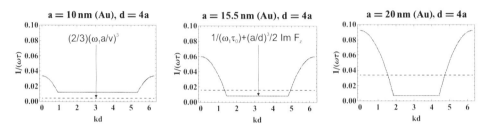

Figure 9.7 For comparison, the various contributions to collective plasmon damping in the chain, for longitudinal polarization. The solid line corresponds to the total damping rate (in dimensionless units) $1/\omega_1 \tau_z = 1/\omega_1 \tau_0 + (a^3/2d^3) \operatorname{Im} F_z(k)$, and the flat part of this line displays the remaining, ohmic, contribution at that part of the k-period where all the electromagnetic losses vanish; the dashed line indicates the Lorentz friction value (in dimensionless units), $\frac{1}{3}(\omega_1 a/v)^3$.

This function is plotted in Fig. 9.5. The discontinuities on the borders between the regions with vanishing radiative damping and with nonzero radiative attenuation are caused by the discontinuous function $\sum_{n=1}^{\infty} \frac{\sin(nz)}{n}$ (see Fig. 9.6), which enters $\operatorname{Im} F_{x(y)}$ but not $\operatorname{Im} F_z$ (see Eq. (9.24)).

In order to compare the magnitudes of the various contributions to the damping of the collective plasmons in the chain, let us plot the dimensionless values for the longitudinal polarization,

$$\frac{1}{\omega_1 \tau} = \frac{1}{\omega_1 \tau_0} + \frac{a^3}{2d^3} \operatorname{Im} F_z(k)$$

(in Fig. 9.7) in comparison to the Lorentz friction contribution $\frac{1}{3}\left(\frac{\omega_1 a}{v}\right)^3$ (broken line line in Fig. 9.7). In this figure note that for small nanospheres ($a = 10$ nm) the Lorentz term is lower than the ohmic attenuation (the flat part of the solid line, see also Eq. (9.4)), while for larger nanospheres ($a = 15, 20$ nm) the Lorentz friction dominates as it is proportional to a^3. For these larger nanospheres, the Ohmic damping is also lower owing to the a^{-1} term in Eq. (9.4). The same can be demonstrated for transverse polarization, as is shown in Fig. 9.8.

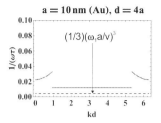 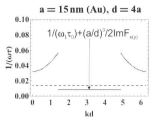 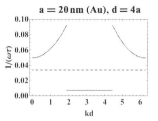

Figure 9.8 For comparison, the various contributions to collective plasmon damping in the chain, for transverse polarization: the solid line corresponds to the total damping rate (in dimensionless units) $1/\omega_1 \tau_{x(y)} = 1/\omega_1 \tau_0 + (a^3/2d^3)\mathrm{Im}\, F_{x(y)}(k)$, and flat part of this line displays the remaining, ohmic, contribution at that part of the k-period where all electromagnetic losses vanish; the dashed line indicates the Lorentz friction value (in dimensionless units) $\frac{1}{3}(\omega_1 a/v)^3$.

9.4 Plasmon–Polariton Self-Modes in a Chain Propagation

The real parts of the functions F_α renormalize the self-frequency of the plasmon–polaritons in the chain, while the imaginary parts renormalize the damping of these modes. $\mathrm{Re}\, F_\alpha(k,\omega)$ and $\mathrm{Im}\, F_\alpha(k,\omega)$ are functions of k and ω. For a first-order approximation in the perturbation approach, one can put $\omega = \omega_1$ in $\mathrm{Re}\, F_\alpha$ and also in the residual nonzero $\mathrm{Im}\, F_\alpha$, the latter being outside the region $0 < kd \pm \omega_1 d/v < 2\pi$. Let us emphasize, however, that the vanishing of $\mathrm{Im}\, F_\alpha(k,\omega)$ inside the region $0 < kd \pm \omega d/v < 2\pi$ holds for any value of ω [44] and thus also for the exact solution for the frequency, not only for $\omega = \omega_1$.

Therefore, one can rewrite the dynamic Eq. (9.20) for plasmon–polariton modes in the chain in the following form:

$$\left(-\omega^2 - i\frac{2}{\tau_\alpha(k)}\omega + \omega_\alpha(k)^2\right) D_\alpha(k,\omega) = \varepsilon a^3 \omega_1^2 E_{0\alpha}(k,\omega), \qquad (9.30)$$

where the renormalized attenuation rate (in the perturbation approach) is given by

$$\frac{1}{\tau_\alpha(k)} = \begin{cases} \frac{1}{\tau_0} & \text{for } 0 < kd \pm \omega_1 d/v < 2\pi, \\ \frac{1}{\tau_0} + \frac{a^3 \omega_1}{2d^3}\, \mathrm{Im}\, F_\alpha(k,\omega_1) & \text{for } kd - \omega_1 d/v < 0 \\ & \text{or } kd + \omega_1 d/a > 2\pi, \end{cases} \qquad (9.31)$$

and the renormalized self-frequency (in the perturbation approach) is given by

$$\omega_\alpha^2(k) = \omega_1^2\left(1 - \frac{a^3}{d^3}\mathrm{Re}\, F_\alpha(k,\omega_1)\right). \qquad (9.32)$$

Equation (9.30) can easily be solved for both the inhomogeneous and the homogeneous ($E_{0\alpha} = 0$) cases. The general solution of Eq. (9.30) has the form of a sum of the general solution for the homogeneous equation and a single particular solution of the inhomogeneous equation. The first of these includes initial conditions and describes damped self-oscillations with frequency

$$\omega'_\alpha = \sqrt{\omega_\alpha^2(k) - \frac{1}{\tau_\alpha^2(k)}}, \qquad (9.33)$$

i.e., for each k and α,

$$D'_\alpha(k,t) = A_{\alpha,k} e^{i(\omega'_\alpha t + \phi_{\alpha,k})} e^{-t/\tau_\alpha(k)}, \qquad (9.34)$$

where the constants $A_{\alpha,k}$ and $\phi_{\alpha,k}$ are adjusted to the initial conditions.

For the inhomogeneous case, the particular solution is as follows:

$$D''_\alpha(k,t) = \varepsilon a^3 \omega_1^2 E_{0\alpha}(k) e^{i(\gamma t + \eta_{\alpha,k})} \frac{1}{\sqrt{(\omega_\alpha^2(k) - \gamma^2)^2 + 4\gamma^2/\tau_\alpha^2(k)}}, \qquad (9.35)$$

for the assumed single Fourier time component $E_{0\alpha}(k,t) = E_{0\alpha}(k)e^{i\gamma t}$ and $tg(\eta_{\alpha,k}) = 2\gamma/\tau(\alpha,k)/(\omega(\alpha,k)^2 - \gamma^2)$, the usual expression for a forced oscillator. Let us emphasize that $E_{0\alpha}(k)$ is a real function since $E_{0\alpha}(ld)^* = E_{0\alpha}(ld) = E_{0\alpha}(-ld)$. An appropriate choice of the latter function, which in practice is a choice of the number of externally excited nanospheres in the chain, e.g., as made by a suitably focused laser beam, allows for the modelling of its Fourier picture $E_{0\alpha}(k)$. This gives the envelope of the wave packet if one inverts the Fourier transform in the solution Eq. (9.35) back to the position variable. For the case of the external excitation of a single nanosphere, the wave packet envelope includes homogeneously all the wave vectors $k \in [0, 2\pi)$. The larger the number of nanospheres that are simultaneously excited, the narrower is the envelope for the k wave packet. For $E_{0\alpha}(ld)^* = E_{0\alpha}(ld) = E_{0\alpha}(-ld)$, the Fourier transform has the same properties, i.e., $E_{0\alpha}(k)^* = E_{0\alpha}(k) = E_{0\alpha}(-k)$, and the latter equality can be rewritten, owing to the periodicity $2\pi/d$ of k, as $E_{0\alpha}(-k) = E_{0\alpha}(2\pi/d - k) = E_{0\alpha}(k)$. The inverse Fourier picture of Eq. (9.35) (its real part) has the form

$$D''_\alpha(ld,t) = \int_0^{2\pi/d} dk \cos(kld - \gamma t - \eta_{\alpha,k}) \varepsilon a^3 \omega_1^2 E_{0\alpha}(k)$$

$$\times \frac{1}{\sqrt{(\omega_\alpha^2(k) - \gamma^2)^2 + 4\gamma^2/\tau_\alpha^2(k)}}. \qquad (9.36)$$

This integral can be evalued by virtue of the mean value theorem as follows:

$$D''_\alpha(ld,t) = \frac{2\pi}{d} \cos(k^* ld - \gamma t - \eta_{\alpha,k^*}) \varepsilon a^3 \omega_1^2 E_{0\alpha}(k^*)$$

$$\times \frac{1}{\sqrt{(\omega_\alpha(k^*)^2 - \gamma^2)^2 + 4\gamma^2/\tau_\alpha^2(k^*)}}. \qquad (9.37)$$

9.4 Plasmon–Polariton Self-Modes in a Chain Propagation

The above expression describes undamped wave motion with frequency γ and velocity, amplitude, and phase shift determined by k^*. The amplitude attains its maximal value at the resonance, when

$$\gamma = \omega_\alpha(k^*)\sqrt{1 - \frac{2}{(\tau_\alpha(k^*)\omega_\alpha(k^*))^2}}. \tag{9.38}$$

In a chain subject to a persistent time-dependent electric field excitation applied to some number (even a small number) of nanospheres, one is dealing with an undamped wave packet propagating along the whole chain. Such modes depend on the particular shaping of the wave packet by the specific choice of chain excitation, which may be responsible for the experimentally observed long–range, practically undamped, plasmon–polariton propagation [169, 171, 39, 40].

The self-modes described by Eq. (9.34) are damped, and their propagation depends on appropriately prepared initial conditions admitting nonzero values of $A_{\alpha,k}$. These initial conditions might be prepared by switching off a time-dependent external electric field initially exciting some fragment of the chain. The resulting wave packet may embrace the wave numbers k from some specified region of $[0, 2\pi)$. If only the wave numbers k for which $0 < kd \pm \omega_1 d/v < 2\pi$ contribute to the wave packet, its damping is then only of ohmic type (as shown in Section 9.3). The corresponding value of $1/\tau_0$ decreases with growing a (see Eq. (9.4)); thus, a longer range of these excitations in the chain can be obtained for larger spheres. The limiting value of $1/\tau_0 = v_f/2\lambda_B \sim 10^{13}$/s gives the maximal range of propagation for these plasmon–polariton modes $\sim 0.1c\tau_0 \sim 10^{-6}$ m; for the group velocity of the wave packet, we have assumed $\sim 0.1c$ as its maximum value (though depending on the radius and separation of the nanospheres). The group velocity, calculated for both polarizations, is presented in Section 9.5.

Though the presented above analysis applies to chains consisting of ideal nanospheres, the conclusions hold also for other shaped particles and agree with the experimental observations, at least qualitatively. In [171], the propagation of plasmon–polaritons in a nano-chain of Ag rod-shaped particles (90 : 30 : 30 nm oriented with their longer axes perpendicular to the chain in order to enhance near-field coupling [39], with face-to-face separation of 70 nm) is evidenced by observing the luminescence of a dye particle, located in proximity to the transmitting electromagnetic signal but distant from the pointlike excitation source, over a range of 0.5 μm. The observed behaviour has been supported by finite difference time domain (FDTD) numerical simulations. Several samples of the chain were fabricated by electron beam lithography in the form of a two-dimensional matrix with sufficiently well-separated individual chains. The energy blueshift of the plasmon resonance for the nano-rods in the chain in comparison with that for a single particle was observed as about 0.1 eV (see Fig. 4 in [171]). This agrees with our estimation of the reduced radiated losses in a chain in comparison with the strong Lorentz friction for a single metallic nanoparticle and the related smaller redshift of damped oscillations. The position of maximum resonance in the chain is located at a higher energy for the transverse polarization mode than for the longitudinal mode [170], which also agrees with the theoretical predictions. In [39] it was indicated that FDTD simulations give lower values of the group velocities for both polarizations, and higher attenuation rates, in comparison with these quantities previously estimated [170, 24] with the simplified point-dipole model with near-field interactions only and neglecting retardation effects. Let us note, however, that the simplified approach, including only the near-field contribution to the electric fields of the interacting dipoles, leads to an artefact, i.e., for some values of the chain parameters d and a, an instability of the collective dynamics occurs [185]. This instability is completely removed by the inclusion of the medium- and far-field contributions to the electric field of the dipole and by rigorous inclusion of the relativistic retardation [44]. Nevertheless, the dipole interaction model, even if it includes, besides the near-field contribution, the medium- and far-field contributions and all retardation effects, still suffers from the absence of the magnetic field component needed for the complete description of far-field wave propagation, which might be of particular significance for ferromagnetic metallic nano-chains. Moreover, as has been demonstrated in [186–188], for large separations in a metallic nano-chain, the scattering of electromagnetic radiation dominates the signal behaviour in the chain, which then acts as a Bragg grating for plasmon–polaritons. For ellipsoidal Au nanoparticles (210 : 80 nm) deposited on the top of a silicon wave guide, the change of regime from collective plasmon–polariton guiding to the Bragg scattering scenario takes place for distances between nanoparticles exceeding about 1 µm [186, 188]. This proves that the dipole coupling model for a nano-chain works quite well for a wide range of chain parameters, in practice up to the order of a micron for the distance between the metallic elements in the chain, which supports the qualitative argument that the Bragg grating regime is not efficient for subwavelength distances and justifies the applicability of the present model. The SNOM measurements of near-field coupled plasmon modes in metallic nano-chains [166], interpreted within the classical field-susceptibility formalism [187], also support the sufficiency of the dipole approximation for nanosphere radii a of order of 10–30 nm and chain separations d not exceeding $\sim 10a$.

9.5 The Self-Frequencies and Group Velocities of Plasmon–Polaritons in a Nano-Chain

According to Eqs. (9.32) and (9.33), for the self-frequencies of plasmon–polaritons in a chain one can write out their explicit form using the expressions for $F_\alpha(k, \omega_1)$

9.5 Self-Frequencies and Group Velocities of Plasmon–Polaritons

given by Eq. (9.21). The real parts of the functions $F_\alpha(k, \omega_1)$ renormalize the corresponding frequencies for both polarizations, and they attain the following form (still with the perturbative approach, according to Eq. (9.32)), setting $x = kd$ and $\xi_1 = \omega_1 d/v$,

$$\omega_z^2(k) = \omega_1^2 \left[1 - \frac{a^3}{d^3} 4 \sum_{m=1}^{\infty} \left(\frac{\cos(mx)}{m^3} \cos(m\xi_1) + \xi_1 \frac{\cos(mx)}{m^2} \sin(m\xi_1)\right)\right]$$

$$= \omega_1^2 \left[1 - \frac{a^3}{d^3} 2 \sum_{m=1}^{\infty} \left(\frac{\cos(m(x+\xi_1)) + \cos(m(x-\xi_1))}{m^3}\right.\right.$$

$$\left.\left. + \xi_1 \frac{\sin(m(x+\xi_1)) - \sin(m(x-\xi_1))}{m^2}\right)\right], \quad (9.39)$$

$$\omega_{x(y)}^2(k) = \omega_1^2 \left[1 + \frac{a^3}{d^3} 2 \sum_{m=1}^{\infty} \left(\frac{\cos(mx)}{m^3} \cos(m\xi_1) + \xi_1 \frac{\cos(mx)}{m^2} \sin(m\xi_1)\right.\right.$$

$$\left.\left. - \xi_1^2 \frac{\cos(mx)}{m} \cos(m\xi_1)\right)\right]$$

$$= \omega_1^2 \left[1 + \frac{a^3}{d^3} \sum_{m=1}^{\infty} \left(\frac{\cos(m(x+\xi_1)) + \cos(m(x-\xi_1))}{m^3}\right.\right.$$

$$\left. + \xi_1 \frac{\sin(m(x+\xi_1)) - \sin(m(x-\xi_1))}{m^2}\right.$$

$$\left.\left. - \xi_1^2 \frac{\cos(m(x+\xi_1)) + \cos(m(x-\xi_1))}{m}\right)\right].$$

$$(9.40)$$

The shift in the self-frequencies of the plasmon–polaritons caused by their attenuation is accounted for by Eq. (9.33) where $\omega_\alpha(k)$ is given by Eqs. (9.39) and (9.40) and where $\tau_\alpha(k)$ has the form as in Eq. (9.31). Taking into account the explicit forms for Im $F_\alpha(k, \omega_1)$, i.e., Eqs. (9.27) and (9.29), one can easily calculate the self-frequencies $\omega'_\alpha(k)$; these functions are shown in Figs. 9.9 and 9.10 for longitudinal and transverse polarizations, respectively.

Note that, for the transverse polarization, in Eq. (9.40) the sum

$$\sum_{m=1}^{\infty} \frac{\cos(m(x+\xi_1)) + \cos(m(x-\xi_1))}{m} \quad (9.41)$$

can be performed analytically according to Eqs. (9.25), resulting in the contribution

$$-\frac{1}{2} \ln[(2 - 2\cos(x+\xi_1))(2 - 2\cos(x-\xi_1))] \quad (9.42)$$

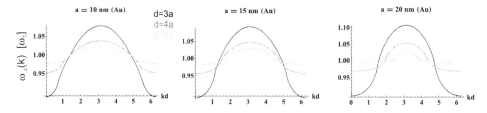

Figure 9.9 The self-frequency $\omega'_z(k)$ of a plasmon–polariton with longitudinal polarization in an infinite chain of Au nanospheres with radius $a = 10, 15, 20$ nm and chain separation $d = 3a, 4a, 5a$ calculated using the perturbation approach ([46]).

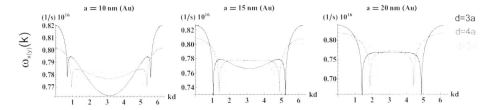

Figure 9.10 The self-frequency $\omega'_{x(y)}(k)$ of a plasmon–polariton with transverse polarization in an infinite chain of Au nanospheres with radius $a = 10, 15, 20$ nm and chain separation $d = 3a, 4a, 5a$ calculated using the perturbation approach. The logarithmic singularities are artefacts of the perturbation method of solution of Eq. (9.20); in the exact solution of this equation all singularities are quenched; see Section 9.6 ([46]).

(the other sums in Eqs. (9.39) and (9.40) have to be performed numerically). The logarithmic singularity in the self-frequencies for the transverse plasmon–polaritons on the rim of the region $0 < x - \xi_1 < 2\pi$ (inside which radiative damping vanishes) is indicated in Fig. 9.10. This singularity causes a hyperbolic discontinuity in the transverse-polarization-mode group velocity (see Fig. 9.12). Nevertheless, the logarithmic singularity in the self-energy and the related hyperbolic discontinuity in the group velocity for transverse polarization turn out to be artefacts of the perturbation solution of Eq. (9.20). The exact numerical solution of this equation demonstrates the effective quenching of the logarithmic singularity to a small local minimum, resulting in a finite group-velocity discontinuity, as will be shown in Section 9.6. This property of the transverse polarization mode of plasmon–polaritons in a chain, caused by interference of the far-field components of the dipole interactions between nanospheres, was analysed numerically in [43, 173] and commented on in [44]. Numerical studies of plasmon–polariton propagation in metallic nano-chains [43, 173–175] indicate a mode that is very narrow and weak but long range, in addition to the wide spectrum of quickly damped modes. This 'fainting' long-range mode has been associated

9.5 Self-Frequencies and Group Velocities of Plasmon–Polaritons 201

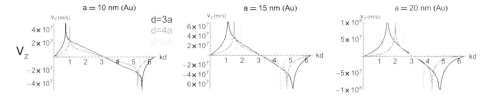

Figure 9.11 The group velocity $v_z(k)$ of a plasmon–polariton with longitudinal polarization in an infinite chain of Au nanospheres with radius $a = 10, 15, 20$ nm and chain separation $d = 3a, 4a, 5a$ calculated with the perturbation approach; the logarithmic singularity in the perturbation formula for v_z (though not in the formula for ω'_z) leads to c being exceeded locally; this artefact of the perturbation approach is removed by the exact solution (see Section 9.6).

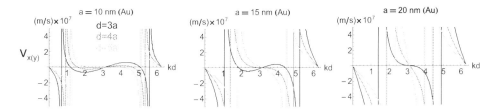

Figure 9.12 The group velocity $v_{x(y)}(k)$ of a plasmon–polariton with transverse polarization in an infinite chain of Au nanospheres with radius $a = 10, 15, 20$ nm and chain separation $d = 3a, 4a, 5a$ calculated with the perturbation approach. The hyperbolic singularity is an artefact of the perturbation method of solution of Eq. (9.20); the exact solution of this equation reduces this singularity to a narrow local jump of the group velocity not exceeding c (see Section 9.6). The small asymmetry of the approximated singularities is caused by the imposition of the hyperbolic singularity and the additional logarithmic singularity of $v_{x(y)}(k)$ at the same point.

with constructive interference of the far-field part of the dipole–dipole interaction of the nanospheres in a chain, resulting in local enhancement of the transverse-polarization-mode group velocity in the near vicinity of singularity points (on the light cone).

In order to find the group velocities of particular self-modes of plasmon–polaritons in a chain, the derivative of $\omega'_\alpha(k)$ with respect to k must be obtained, which, according to Eqs. (9.31), (9.33), (9.39), and (9.40), is straightforward though involving extended calculus. The sums in the formulae for $\omega_\alpha(k)$ still cannot be performed analytically, except for the sum with denominator m in Eq. (9.40). The resulting group velocity values calculated numerically for both polarizations and for $kd \in [0, 2\pi)$ and $d/a \in [3, 10]$ are presented in Figs. 9.11 and 9.12 (for $a = 10, 15, 20$ nm for both polarizations).

9.6 Exact Solution of Eq. (9.20) for the Plasmon–Polariton Self-Energy

The imaginary part of the complex variable ω, the solution of Eq. (9.20), defines the plasmon–polariton attenuation, while the real part of ω gives the self-frequency of these oscillations (in the case of the homogeneous equation, i.e., when $E_{0\alpha}(k,\omega) = 0$). The derivative of this self-frequency with respect to the wave vector k defines the group velocity of the particular mode. Because of the logarithmic singular term in the far-field transverse contribution to the dipole interaction in the chain ($x = kd$ and $\xi = \omega d/v$),

$$\sum_{m=1}^{\infty} \frac{1}{m} [\cos(m(x+\xi)) + \cos(m(x-\xi))]$$
$$= -\frac{1}{2} \ln[(2 - 2\cos(x+\xi))(2 - 2\cos(x-\xi))], \tag{9.43}$$

one cannot apply the perturbation method to solve the dynamical equation, at least in the region close to the singularity (on the light cone). Note that, with the perturbation approach, one substitutes ω by ω_1 (the Mie frequency) on the right-hand side of Eq. (9.20). This produces, however, a hyperbolic singularity in transverse group velocity by virtue of Eq. (9.43). Moreover, with the perturbation approach, the logarithmic singularity occurs for both polarizations, which is noticeable if one takes the derivative with respect to k of Eqs. (9.39) and (9.40). All these singularities occur at isolated points for which $kd \pm \omega_1 d/v = l\pi$, where l is an integer. Both hyperbolic and logarithmic divergences in the perturbation formula for group velocities at this point would result in the local exceeding of c by the corresponding group velocities. To resolve the problem of these unphysical divergences, the exact solution of Eq. (9.20) must be found; because of the divergence of Eq. (9.43) the corresponding contribution cannot be treated as a perturbation any longer. An exact solution of the highly nonlinear Eq. (9.20), found numerically by use of a Newton-type procedure at 1000 points for $kd \in [0, 2\pi)$, is plotted in Figs. 9.13–9.20, for both polarizations of the plasmon–polaritons.

From Figs. 9.13 and 9.15 (see also Figs. 9.16–9.19), we notice by comparison with the corresponding plots obtained with the perturbation method that, for longitudinal polarization, the exact solutions for the self-frequencies do not differ significantly from those obtained with the perturbation procedure, but the change suffices to remove the logarithmic divergence from the derivative of the self-frequency. For the transverse polarization, the difference is also not important for the attenuation plot. However, for the transverse polarization self-frequency in the case of the exact solution, we have the quenching of the logarithmic divergence (9.43), in contrast with the approximated version obtained with the perturbation approach. Instead of

9.6 *Exact Solution of Eq. (9.20) for the Plasmon–Polariton Self-Energy* 203

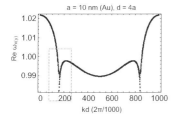
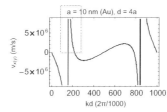
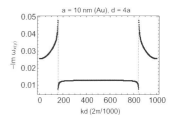

Figure 9.13 The exact solution for the self-frequency and the damping rate of the transverse polarization mode of plasmon–polaritons in a nano-chain (ω is given in ω_1 units); the solution of Eq. (9.20) was found for 1000 points on the segment [0, 2π). In the middle, the plot for the group velocity of the transverse polarization mode is presented; it has hyperbolic-type singularities corresponding to the logarithmic-type singularities of the self-energy (left) mixed with the additional logarithmic-type singularity of the group velocity itself. All singularities are, however, truncated, which is visualized in the broken-line regions in Fig. 9.14 ([45]).

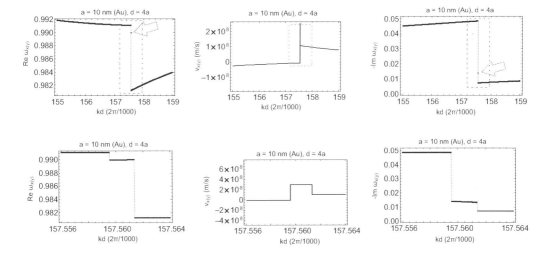

Figure 9.14 Two-level magnified view of the truncated singularity region for the transverse plasmon–polariton mode in a nano-chain in the exact solution of Eq. (9.20); the most enhanced scale is in the lower panels (ω is given in ω_1).

an infinite singularity, we observe in the exact plot for the transverse polarization self-frequencies only a relatively small minimum, resulting then conveniently in a finite group velocity not greater than c. This quenching logarithmic singularity into a small local minimum is presented in Fig. 9.14 (right); on the left, the correction to the discontinuity step in the damping of the transverse mode caused by the logarithmic contribution to Eq. (9.20) is presented.

The exact solution of the dynamic equation (9.20) thus resolves the problem of the risky logarithmic divergent contribution of the transverse far-field part of the dipole interaction and regularizes the final solution for the corresponding

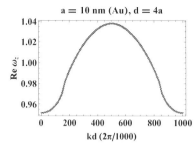
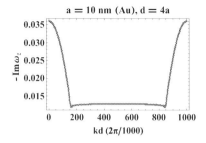

Figure 9.15 The exact solution for the self-frequency and the damping of the longitudinal mode of plasmon–polaritons in a nano-chain (ω is given in ω_1 units) ([45]).

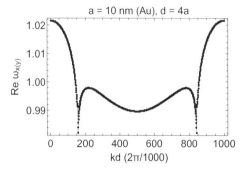
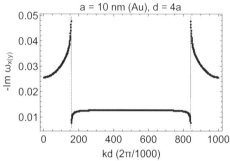

Figure 9.16 The exact solution for the self-frequency and the damping of the transverse mode of plasmon–polaritons in a nano-chain (ω is given in ω_1 units) ([45]).

characteristics of the plasmon–polariton modes. In the vicinity of singularity points in the domain for kd, the group velocity, though still well below the velocity of light, is, however, greater than the group velocities in other regions of the k wave vector domain. This might elucidate former numerical observations [43] of the long-range propagating mode for the transversely polarized plasmon–polariton in a nano-chain. In view of the above analysis, one can argue that the long range of the related mode [43] is connected to a locally higher value of its group velocity, but not to the lowering of its damping.

Even though the real part of the function F_z and thus of $\omega_z(k)$ (Eq. (9.39)) is given by a continuous function, the corresponding group velocity will have a logarithmic singularity, as the derivative of $\omega_z(k)$ with respect to k will contain the sum $[\cos(m(x+\xi_1)) - \cos(m(x-\xi_1))]/m$ (recall that $\xi_1 = \omega_1 d/v$), arising from the last term in Eq. (9.39) after taking the derivative. A similar term is also present in the perturbation formula for $\omega_{x(y)}(k)$ in Eq. (9.40). The origin of these terms for both polarizations is the medium-field contribution to the dipole interaction in the chain. At the points $x \pm \xi_1 = 2\pi p$, p integer, the logarithmic singularity again produces an

9.6 Exact Solution of Eq. (9.20) for the Plasmon–Polariton Self-Energy   205

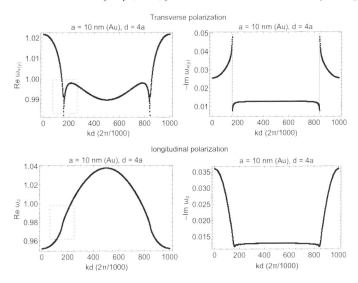

Figure 9.17 Exact solution for the self-frequency and damping rate of transversely (upper) and longitudinally (lower) polarized plasmon–polariton modes in a nanochain (ω is given in ω_1 units). Presented is a solution of Eq. (9.20) at 1000 points in the sector $kd \in [0, 2\pi)$; in the plot for the self-frequency of the transverse mode a truncated logarithmic-type singularity in the self-energy is indicated; for longitudinal polarization the almost vertical local slope of self-frequency is indicated.

artefact whereby the group velocity v_z exceeds c. This precludes the applicability of the perturbation approach, at least close to points of singularity. Therefore, instead of putting $\omega = \omega_1$ in the function F_α on the right-hand side of Eq. (9.16) as in the perturbation method of solution of this equation, one must solve the nonlinear equation directly, in the homogeneous case neglecting $E_{0\alpha}$. As mentioned above, this exact solution can be found numerically by a Newton-type procedure, and both the real and imaginary parts of ω can be determined, point by point, in the whole region $kd \in [0, 2\pi)$. For the longitudinal polarization of the group velocity, the exact solution is presented in Fig. 9.20. The exact solutions for v_z do not exhibit any singularities: the logarithmic singularity of the perturbation term is quenched to only a small local extremum, similarly to what was demonstrated above for the transverse polarization.

The logarithmic-type singularity in the self-energy of the transverse plasmon–polaritons in the chain is the feature that essentially differentiates these modes from the longitudinally polarized modes. This singularity is caused by the sum of the far-field parts of the electric fields of all the nanoparticle dipoles, which influences the charge oscillations in each component of the chain and produces a hyperbolic-type discontinuity in the group velocity only for the transversely polarized modes. In addition to this discontinuity, the medium-field component of the electric dipole

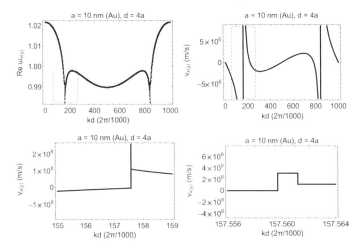

Figure 9.18 Exact solution for the group velocity of the transverse plasmon–polariton mode in a nano-chain for $a = 10$ nm and $d = 4a$; the solution of the nonlinear equation (9.20) removes the logarithmic singularity – the remaining local very narrow extremes are truncated at exactly the value of c. In the near vicinity of the singular point the solution was found for more than 1000 values of kd (lower panels: magnification of the marked areas in the upper panels); the asymmetry at the singular points is caused by the imposition of hyperbolic and logarithmic truncated singularities of the group velocity for the transverse mode.

interaction produces a logarithmic-type singularity in the group velocity for both polarizations (though any singularity in the self-energy). We use here the terms logarithmic-type or hyperbolic-type singularity to distinguish the behaviour of the group velocities obtained by the exact solutions for the self-energies for both polarizations, which are sharpened and truncated at c owing to relativistic constraints imposed on the dynamic equation and manifesting themselves in the form of its solution. The retardation of electric signals prohibits the collective-excitation group velocity from exceeding the velocity of light. This quenching concerns the infinite singularities which occur in the perturbation expressions for the self-energy and then in the perturbation formulae for the group velocities. The relativistic invariance of the dynamic equation for collective dipole plasmon oscillations in a chain prevents, however, the group velocity of particular plasmon–polariton modes from exceeding the velocity of light. Thus, the exact solution of the dynamic equation inherently also possesses this property. Exact self-energies have a suitably regularized dependence with respect to k, such that their derivatives do not exceed c.

It is worth emphasizing, for the sake of completeness, that the inclusion of the magnetic field of the dipoles does not modify this scenario, because the magnetic field contribution to the self-energies is at least two orders lower than the electric field contribution; the reason is that the Fermi velocity of electrons is two orders

9.6 Exact Solution of Eq. (9.20) for the Plasmon–Polariton Self-Energy 207

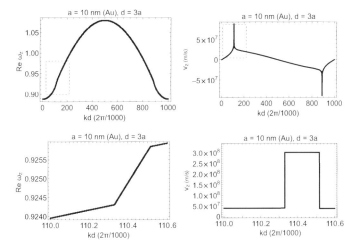

Figure 9.19 Exact solution for the group velocity of the longitudinally polarized mode of a plasmon–polariton in a nano-chain for $a = 10$ nm and $d = 3a$; the almost vertical slope of the self-frequency (right-hand panels) is limited by the velocity of light (lower panels: magnification of the marked areas in the upper panels).

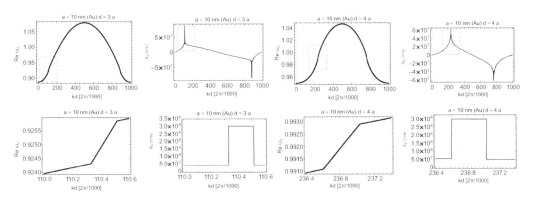

Figure 9.20 The exact solution for the group velocity v_z of the longitudinal mode of plasmon–polaritons in a nano-chain for $a = 10$ nm and $d = 3a, 4a$, respectively, and the corresponding exact solutions for Re ω_z (left). The exact solution of the dynamic equation (9.20) removes the logarithmic singularity in the group velocity. The remaining local very narrow extremes are truncated exactly at c (in close vicinity to the singular points, the solution was found for more than 1000 points kd. In the bottom panels, the plots correspond to the marked regions in the upper panels at an enhanced scale).

lower than the velocity of light, which significantly reduces the Lorentz force. Therefore, the magnetic field of the dipoles [55, 76],

$$B_\omega = ik(\mathbf{D}_\omega \times \mathbf{n}) \left(\frac{ik}{r_0} - \frac{1}{r_0^2} \right) e^{ikr_0}, \qquad (9.44)$$

though contributing to the far-field and medium-field parts of the plasmon–polariton self-energies does not significantly change the similar terms due to the electric field and leads to corrections that are two orders lower and thus practically negligible.

It must be emphasized, however, that the role of the magnetic field may change considerably in the case of magnetic-type metallic chains. If a ferromagnetic material is used to prepare the chain components, one can expect a strengthening of magnetic effects in the surface plasmon collective dynamics for the chain. Thus, the consideration presented, while appropriate for noble metals, need to be lifted to much more complicated magneto-plasmonics for magnetic metals. This is, however, beyond the scope of the present treatise.

The distinction between the truncated singularities in the group velocities for transverse and longitudinal polarizations of plasmon–polaritons as demonstrated above sheds light on the discussion of long-range plasmon–polariton modes in metallic nano-chains studied by numerical simulations with the Green function method for differential equations [173, 176, 174]. The truncated singularities in group velocities indicated above, i.e., the very narrow logarithmic-type truncated singularity for the longitudinal polarization and the wider hyperbolic-type singularity superimposed on a narrow logarithmic-type truncated singularity for the transverse polarization, facilitate an understanding of the peculiarities of numerical studies. The long-range propagation of narrow modes observed in numerical simulations might be linked to the local enhancement of the group velocity of k-modes in singular regions. In simulations [43, 174, 173] it has been assumed that a single selected nanosphere in the chain is initially excited and then the range of propagation for various modes of plasmon–polaritons is observed. A pointlike initial excitation corresponds in the Fourier picture to a uniform excitation of all k-modes, including those in singular regions. The different character of the truncated singularities, being originally infinite divergences in the perturbation series, for the longitudinal and the transverse polarizations results in different features in the corresponding modes (the local narrowing of the group velocity curve is stronger for the longitudinal polarization). Even though the group velocity singularities are truncated at the finite $v = c$ level in different fashions for the transverse and the longitudinal polarizations, the local narrow extrema give the same long-range propagation of closely located modes for both polarizations (about two orders of magnitude longer than for the nonsingular modes).

Summarizing, one can state that in infinite metallic nano-chains, a long-range propagation of plasmon–polariton self-modes can be observed, due to the effectively ideal compensation of the Lorentz friction losses in a particular nanosphere by the energy income from other nanospheres in the chain; this takes place for both polarizations of the collective plasmon modes in a relatively large part of the wave vector k domain (diminishing with increasing d/a). In the case of an inho-

9.6 Exact Solution of Eq. (9.20) for the Plasmon–Polariton Self-Energy

mogeneous equation version of Eq. (9.20) (i.e., when a term of persistent external forcing is added to its right-hand side, i.e., $\varepsilon a^3 \omega_1^2 E_{0\alpha}(k,\omega)$, where ε is the dielectric permittivity of the chain surroundings and $E_{0\alpha}(k,\omega)$ is the external electric field Fourier component), one can expect the completely undamped propagation of the induced plasmon–polaritons corresponding to the forced oscillator properties. Both the low (only ohmic) attenuation of the self-oscillations and the possibility of exciting appropriately formed (by persistent excitation of some part of the chain) wave packets of plasmon–polariton modes in the chain may be responsible for the experimentally observed practically undamped propagation of the collective plasmon signal over relatively large distances in nano-chains of lengths equal to several micrometres.

It should be noted that the almost undamped propagation of surface plasmon–polaritons along metallic nano-chains was first analysed by Markel and then developed by Citrin and, more recently, by other authors. They all tried to solve the problem of plasmon–polariton propagation within the dipole model of particle interactions. This makes the issue relatively simple and clearly formulated, i.e., the corresponding dynamic equation is a linear differential equation and it can be solved using various techniques. Nevertheless, some methods may not have been transparent and additionally may have complicated the solution and interpretation. In the present scheme, we have solved the linear differential equation by a standard technique, that of the Fourier transform (a continuous transform (CFT) in the time domain and a discrete transform (DFT) in the position domain, clearly linking with the problem definition for the periodic chain). All the analysis has been completely and transparently presented, mainly in an analytical manner. This is the basic value of this approach for a wave representation close to the experimental interpretation. Other approaches (e.g., the Z transformation method with the Green function method) give similar results. Unfortunately, none of the methods allowed for a full analytical solution and in each approach some numerical procedure had been applied. As is common in differential equation theory, some small uncertainties in a singular region can cause misleading conclusions. The effective numerical solution for the dispersion may thus have some inherently approximate character connected to the necessary procedure for cutting the infinite series in the numerical treatment. The solution for the Fourier transform of the initial differential equation is transparent in this regard and exhibits some singular points (corresponding to the constructive interference for transversely polarized plasmon–polaritons due to the far-field radiation on the light cone). This singularity is visible in all formulations because of the logarithmic divergence of the infinite sum of $-\cos(nx)/n$. Various conclusions related to this singularity may be formulated, however, but not all are correct, especially if they use perturbative calculus, as directly demonstrated in the present discussion. The only way to resolve this problem is to solve the underlying equation exactly. An exact solution means here a solution of the nonlinear equation with respect to the Fourier-argument algebraic equation point by point. For singularity points all methods agree, even approximate ones. Special attention must be given, however, to the group velocity, which in the framework of the Fourier transform above presented can be explicitly expressed and studied, revealing further sources of singularities: hyperbolic for the transverse polarization (caused by the far-field-zone dipole radiation) and an additional logarithmic singularity (caused by the medium-field-zone radiation for both transverse and longitudinal polarizations). These singularities are hard to notice in numerical experiments, possibly owing to the slight confusion caused by the application of the unavoidable approximation methods of numerical calculus. Resolving the artefact is thus unclear and might cause doubt. The aim of our presentation is to demonstrate directly the character of the above-mentioned singularities in a transparent and highly analytical manner. The singularities in the solution of the differential equation have been explicitly demonstrated as being caused by perturbation methods. Moreover, an exact solution of the problem has been described, free of all nonphysical singularities. It might be emphasized that such a level of transparency is not achieved with other techniques where the problem of singularities is not solved in a final manner. Therefore, a complete and transparent solution of the same initial problem is a novel and important contribution to the debate on plasmon–polaritons in metallic nanostructures.

9.7 Collective Plasmon-Wave-Type Propagation along a Nano-Chain; Near-Field-Zone Approximation

In the case of a metallic nano-chain one has to take into account the mutual effects of the nanospheres in the chain. Assuming that in a sphere located at the point $\mathbf{r}$ we have a dipole $\mathbf{D}(t)$, then at a point $\mathbf{r}_0$ (where the vector $\mathbf{r}_0$ is drawn from the tip of $\mathbf{r}$) this dipole causes electric and magnetic fields in the following form (including electromagnetic retardation) [55, 76]: for the ω-Fourier component of the electric field we have

$$\mathbf{E}_\omega = \mathbf{D}_\omega \left(\frac{(\omega/c)^2}{r_0} + \frac{i\omega/c}{r_0^2} - \frac{1}{r_0^3} \right) e^{i\omega r_0/c}$$
$$+ \mathbf{n}_0 (\mathbf{n}_0 \cdot \mathbf{D}_\omega) \left(-\frac{(\omega/c)^2}{r_0} - \frac{i3\omega/c}{r_0^2} + \frac{3}{r_0^3} \right) e^{i\omega r_0/c}$$

(9.45)

and for the magnetic field Fourier component we have

$$\mathbf{B}_\omega = (i\omega/c)(\mathbf{D}_\omega \times \mathbf{n}_0) \left(\frac{i\omega/c}{r_0} - \frac{1}{r_0^2} \right) e^{i\omega r_0/c}, \qquad (9.46)$$

where $\mathbf{n}_0 = \mathbf{r}_0/r_0$.

9.7.1 Near-Field-Zone Approximation of the Dipole Interaction in a Chain

To examine the role of the terms corresponding to the near-field zone (denominator with r_0^3), the medium-field zone (denominator with r_0^2), and far-field zone (denominator with r_0), let us first confine ourselves to the near-field-zone dipole-type coupling. In this approximation there is only an electric field and its form is given by the static dipole-field formula:

$$\mathbf{E}(\mathbf{r},\mathbf{r}_0,t) = \frac{1}{r_0^3}\left\{3\mathbf{n}_0\left(\mathbf{n}_0\cdot\mathbf{D}\left(\mathbf{r},t-\frac{r_0}{c}\right)\right) - \mathbf{D}\left(\mathbf{r},t-\frac{r_0}{c}\right)\right\}. \quad (9.47)$$

This allows us to write out a dynamical equation for plasmon oscillations for each nanosphere of the chain; the nanospheres will be numbered by integer l (d will denote the separation between nanospheres, $d > 2a$; the vectors $\mathbf{r}$ and $\mathbf{r}_0$ are collinear if the origin is associated with certain nanospheres in the chain). Note additionally that, as follows from numerical studies [41, 182], the dipole approximation for the plasmon interaction in the nanosphere chain is sufficiently accurate for $d > 3a$, when the multipole interaction contribution can be neglected.

This dynamical equation attains the form,

$$\ddot{R}_\alpha(ld,\omega_1 t) + R_\alpha(ld,\omega_1 t) = \sigma_\alpha \frac{a^3}{d^3} \sum_{m=-\infty, m\neq l}^{\infty} \frac{R_\alpha\left(md,\omega_1 t - \frac{\omega_1 d|l-m|}{c}\right)}{|l-m|^3}$$

$$- \frac{2}{\tau_0 \omega_1}\dot{R}_\alpha(ld,\omega_1 t) + \frac{e}{ma\omega_1^2}E_\alpha(ld,\omega_1 t), \quad (9.48)$$

where $R_\alpha(\omega_1 t) = D_\alpha(\omega_1 t/\omega_1)/(eN_e a)$, the dots indicate derivatives with respect to dimensionless time $t' = \omega_1 t$ and

$$\sigma_\alpha = \begin{cases} -1 & \text{for } \alpha = x(y), \\ 2 & \text{for } \alpha = z, \end{cases}$$

is introduced to distinguish the two oscillation polarizations. Thus the index α enumerates the polarizations, longitudinal and transverse, with respect to the chain orientation (the z axis). The first term of the right-hand side in Eq. (9.48) describes the dipole-type coupling in the near-field zone between nanospheres and the other two terms correspond to contributions due to plasmon attenuation (including the Lorentz friction linear term, which can be accounted for as described in the previous section). Note that a similar approach, but including only the near-field dipole coupling in the chain, was utilized by the Atwater group [39, 24].

The summation in the first term on the right-hand side of Eq. (9.48) can be explicitly performed in the manner presented in [185], if one changes to a wave-vector picture, taking advantage of the chain periodicity in analogy to the Bloch states in crystals with the quasimomentum reciprocal lattice, i.e.,

$$R_\alpha(ld,t) = R_\alpha(k,t)\,e^{\mp ikld}, \quad 0 \leq k \leq \frac{2\pi}{d}. \quad (9.49)$$

where $\mp k$ correspond to the two possible orientations of the phase velocity (the time factor is assumed to be $e^{-i\omega t - t/\tau}$). Thus, Eq. (9.48) can be rewritten in the following form (the Lorentz friction term is represented in the same way as in Eq. (6.22)),

$$\ddot{R}_\alpha(k_\alpha, t\omega_1) + \tilde{\omega}_\alpha^2 R_\alpha(k_\alpha, t\omega_1) = -2\dot{R}_\alpha(k_\alpha, t\omega_1)\frac{1}{\tau_\alpha \omega_1}, \quad (9.50)$$

where

$$\tilde{\omega}_\alpha^2 = \left(\frac{\omega_\alpha}{\omega_1}\right)^2 = 1 - 2\sigma_\alpha \frac{a^3}{d^3}\cos(k_\alpha d)\cos\left(\frac{\omega_p d}{c\sqrt{3}}\right), \quad (9.51)$$

$$\frac{1}{\tau_\alpha \omega_1} = \frac{1}{\tau_0 \omega_1} + \left(\frac{1}{3} + \frac{\sigma_\alpha}{12}\right)\left(\frac{\omega_p a}{c\sqrt{3}}\right)^3$$

$$+ \sigma_\alpha \frac{a^3}{d^3}\left(\frac{\omega_p d}{c\sqrt{3}}\right)\left[\frac{\pi^2}{6} - \frac{\pi k_\alpha d}{2} + \frac{(k_\alpha d)^2}{4}\right]. \quad (9.52)$$

Formula (9.52) expresses the attenuation rates for both polarizations. The two components of Eq. (9.52), for $\alpha = x(y)$ and z, give these damping rates explicitly and one notices a remarkable property, that the effective attenuation rates can change their sign depending on the values for d, a, and k. In Fig. 9.21 the regions of negative value for damping rates are marked for both polarizations. These regions shrink with the growth of d/a and with the growth of a itself. For a larger than some critical value, these regions disappear; the longitudinal modes for $a > 35$ nm and the transverse modes for $a > 48$ nm, for Au nanospheres.

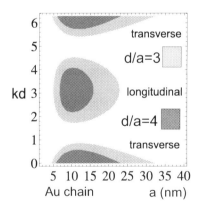

Figure 9.21 The regions with negative values of damping rates for plasmon–polaritons in a chain (for longitudinal polarization and also for transverse polarization) when only near-field dipole coupling between the nanospheres is included. The regions shrink with growth in d/a, for the longitudinal (transverse) modes for $d/a > 6$ (7). The regions completely disappear for the longitudinal modes for $3.5 > a > 35$ nm and for the transverse modes for $3.5 > a > 48$ nm (for Au nanospheres).

9.7 Near-Field-Zone Approximation

Using Eqs. (9.51) one can calculate the group velocity of the plasmon–polariton mode wave packet in the near-field coupling approximation:

$$v_\alpha = \frac{d\omega_\alpha}{dk} = \omega_1 \frac{\sigma_\alpha a^3 \sin(kd) \cos(\omega_1 d/c)}{d^2 \sqrt{1 - 2\sigma_\alpha (a^3/d^3) \cos(kd) \cos(\omega_1 d/c)}}. \quad (9.53)$$

From this formula it follows that the group velocity of the undamped or damped wave-type collective plasmon excitation may attain different values depending on a, d, and k, as is depicted in Fig. 9.22 (upper). With increasing a, this velocity grows

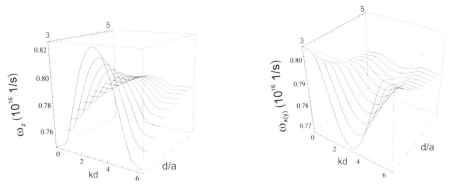

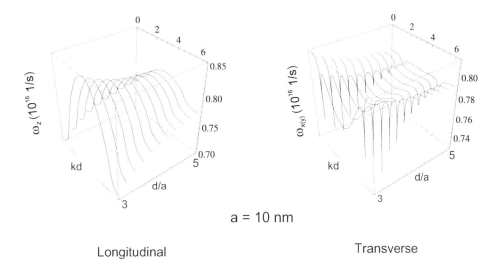

Figure 9.22 The dispersion of plasmon–polaritons in a chain for both polarizations; upper) with the inclusion of only near-field coupling, and (lower) with all near-, medium-, and far-field coupling.

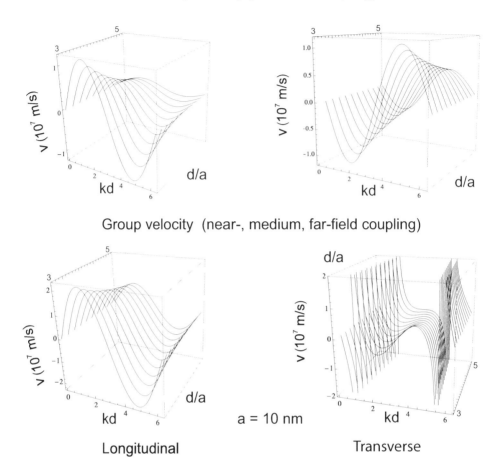

Figure 9.23 The group velocity for both polarizations; with inclusion of only near-field coupling (upper), and with all near-, medium- and far-field coupling (lower).

proportionally and diminishes with the separation of the nanospheres as $(d/a)^{-2}$. In Fig. 9.23 (upper) the dispersion of the collective plasmons in a chain in the same near-field coupling approximation is plotted versus the wave vector and the separation of the nanospheres (for Au nanospheres with radius $a = 10$ nm).

For positive attenuation rates one can expect ordinary damped plasmon–polariton propagation, while for negative damping rates the solution behaves differently, revealing the instability of the linear theory. A negative value of the damping rate, indicating an instability of the system, is, however, an unphysical artefact of the model, in view of energy conservation constraints. In other words, the continuous losses of plasmon oscillation energy due to scattering and radiation is instantly recovered by the electromagnetic influence of the other nanospheres in the chain. When this income overcomes the losses then the wave packet for the corresponding modes would propagate without damping. The near-field coupling model admits such an unphysical scenario for losses lower than a certain threshold. For nanospheres that are too small, with radii lower than about 3.5 nm for Au particles, depending on the constant C in the formula for the scattering damping rate, the scattering attenuation on the particle boundaries, $\sim Cv_F/(2a\omega_1)$, is too high; similarly, the radiation losses, $\sim(1/3 + \sigma_\alpha/12)(\omega_p a/c\sqrt{3})^3$, are above the required threshold, $a > 35$ (48) nm, for the longitudinal (transverse) modes. Thus, outside the regions $3.5 < a < 35$ (48) nm, undamped longitudinal (transverse) modes in the near-field coupling approximation do not occur regardless of the separation in the relative chain. The relative chain separation, d/a, also influences the range of undamped propagation; this is illustrated in Fig. 9.21.

9.7.2 Medium- and Far-Field Corrections to the Near-Field Dipole Interaction

The instability described above in the near-field coupling approach indicates that other effects, essential for plasmon–polariton attenuation, have not been taken into account. The required correction can be linked with the dipole influence of other nanospheres assumed previously in the form (9.47). To rectify this instability of the near-field zone approach, the medium-field-zone and far-field-zone contributions must be taken into account. Thus, using formula (9.45), one arrives at an interdipole interaction contribution to the oscillator equation for the lth nanosphere:

$$e^{\mp ikld}e^{-i\omega t}e^{-\tau t}R_\alpha(k)a^3 \sum_{n=-\infty, n\neq l}^{\infty} e^{\pm iknd}(F_1(|n|) + iF_2(|n|))e^{i|n|\omega d/c} \quad (9.54)$$

or

$$e^{\mp ikld}e^{-i\omega t}e^{-\tau t}R_\alpha(k)2a^3 \sum_{n=1, n\neq l}^{\infty} \cos(knd)(F_1(|n|) + iF_2(|n|))e^{i|n|\omega d/c}, \quad (9.55)$$

where

$$F_1 = \begin{cases} \dfrac{(\omega/c)^2}{|n|d} - \dfrac{1}{|n|^3 d^3} & \text{for } \alpha = x(y), \\ \dfrac{2}{|n|^3 d^3} & \text{for } \alpha = z, \end{cases} \quad (9.56)$$

and

$$F_2 = \begin{cases} \dfrac{\omega/c}{|n|^2 d^2} & \text{for } \alpha = x(y), \\ -\dfrac{2\omega/c}{|n|^2 d^2} & \text{for } \alpha = z. \end{cases} \tag{9.57}$$

Employing the above formulae one can write out the contributions to the real and imaginary parts of the dynamical equation, which give corrections to the frequency ω_α and to the damping rate τ_α, respectively, setting $x = kd$ and $\xi = \omega d/c$,

$$\begin{aligned} 2a^3 \sum_n \cos(nx) F_1 \cos(n\xi) - 2a^3 \sum_n \cos(nx) F_2 \sin(n\xi), \\ -2a^3 \sum_n \cos(nx) F_1 \sin(n\xi) - 2a^3 \sum_n \cos(nx) F_2 \cos(n\xi). \end{aligned} \tag{9.58}$$

These contributions to the frequency of the plasmon–polariton oscillations and to the corresponding damping term have the following explicit forms:

For the case of longitudinal polarization ($\alpha = z$), the corrections to the frequency and to the damping rate are

$$a^3 \left[2 \sum_n \cos(nx) \frac{2}{n^3 d^3} \cos(n\xi) + 2 \sum_n \cos(nx) \frac{2\omega/c}{n^2 d^2} \sin(n\xi) \right] \tag{9.59}$$

and

$$a^3 \left[-2 \sum_n \cos(nx) \frac{2}{n^3 d^3} \sin(n\xi) + 2 \sum_n \cos(nx) \frac{2\omega/c}{n^2 d^2} \cos(n\xi) \right]. \tag{9.60}$$

For the case of transverse polarization ($\alpha = x(y)$), the corrections are

$$a^3 \left[2 \sum_n \cos(nx) \left(\frac{(\omega/c)^2}{nd} - \frac{1}{n^3 d^3} \right) \cos(n\xi) - 2 \sum_n \cos(nx) \frac{\omega/c}{n^2 d^2} \sin(n\xi) \right] \tag{9.61}$$

and

$$a^3 \left[-2 \sum_n \cos(nx) \left(\frac{(\omega/c)^2}{nd} - \frac{1}{n^3 d^3} \right) \sin(n\xi) - 2 \sum_n \cos(nx) \frac{\omega/c}{n^2 d^2} \cos(n\xi) \right]. \tag{9.62}$$

To proceed with estimating these contributions to the oscillator equation, one can apply the iterative perturbation method of solution, which in the first step amounts to substituting ω by the free frequency ω_1 on the right-hand side of the osillatory equation. The sums in the formulae for the damping ratio contributions can be performed accurately [184] (for an arbitrary ω, including $\omega = \omega_1$). As a result one obtains, for $kd - \omega_1 d/c > 0$ and $kd + \omega_1 d/c < 2\pi$, the following.

9.7 Near-Field-Zone Approximation

Table 9.2 *Nanosphere parameters assumed for calculation*

Material is Au		
Bulk plasmon energy	$\hbar\omega_p$	8.57 eV
Bulk plasmon frequency	ω_p	1.302×10^{16}/s
Mie dipole plasmon energy	$\hbar\omega_1$	4.94 eV
Mie frequency	$\omega_1 = \omega_p/\sqrt{3}$	0.752×10^{16}/s
Constant in Eq. (9.4)	C	1
Fermi velocity	v_F	1.396×10^6 m/s
Bulk mean free path	λ_b	53 nm

For the case of transverse polarization,

$$\frac{a}{d}\left(\frac{\omega_1 a}{c}\right)^2\left(\frac{\omega_1 d}{c}\right) + \frac{a^2}{d^2}\left(\frac{\omega_1 a}{c}\right)\left[\frac{\pi^2}{3} - \pi kd + \frac{(kd)^2}{2} + \frac{1}{6}\left(\frac{\omega_1 d}{c}\right)^2\right]$$

$$-\frac{a^2}{d^2}\left(\frac{\omega_1 a}{c}\right)\left[\frac{\pi^2}{3} - \pi kd + \frac{(kd)^2}{2} + \frac{1}{2}\left(\frac{\omega_1 d}{c}\right)^2\right] \quad (9.63)$$

$$= \left(\frac{\omega_1 a}{c}\right)^3 - \frac{1}{3}(\omega_1 a/c)^3 = \frac{2}{3}\left(\frac{\omega_1 a}{c}\right)^3.$$

For the case of longitudinal polarization,

$$-\frac{a^3}{d^3}2\left(\frac{\omega_1 d}{c}\right)\left[\frac{\pi^2}{3} - \pi kd + \frac{(kd)^2}{2} + \frac{1}{6}(\omega_1 d/c)^2\right]$$

$$+\frac{a^2}{d^2}2\left(\frac{\omega_1 a}{c}\right)\left[\frac{\pi^2}{3} - \pi kd + \frac{(kd)^2}{2} + \frac{1}{2}(\omega_1 d/c)^2\right] \quad (9.64)$$

$$= \frac{2}{3}\left(\frac{\omega_1 a}{c}\right)^3.$$

(Note again that in the above formulae substituting ω_1 by ω_a gives the accurate form.)

We see that for both polarizations the instability has disappeared (i.e., the contribution to the damping rate does not change sign, in contrast with the case of the sole near-field-zone contribution [185]). The imaginary term with denominator r^2 (for transverse polarization, a real term with denominator r also contributes) exactly cancels the previous unstable contribution of the real term with denominator r^3.

Simultaneously, the Lorentz friction $2/(\tau\omega_1) = (2/3)(\omega_1 a/c)^3)$ is completely cancelled by the contribution calculated above to the energy income from other nanospheres in the chain; this occurs for both polarizations but only when $kd - \omega_1 d/c > 0$ and $kd + \omega_1 d/c < 2\pi$, as illustrated in Fig. 9.24. In other words, plasmon–polaritons do not radiate energy and any dissipation of energy is due only to electron scattering – this confirms the previous numerical observations

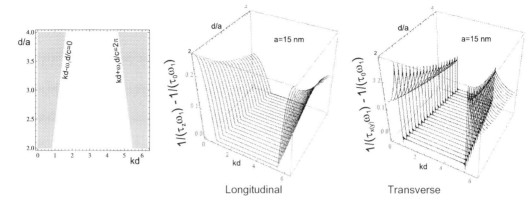

Figure 9.24 The region with completely quenched radiative losses of plasmon–polaritons (left), damping rates beyond the pure scattering contribution (i.e., $\frac{1}{\tau_\alpha \omega_1} - \frac{1}{\tau_0 \omega_1}$) for both polarizations (centre and right); near-, medium-, and far-field dipole coupling included. The small oscillations in the graph for transverse polarization are caused by the slowly convergent sum $\sum_i \sin(ix)/i$; 300 elements of this sum were taken. The sum also gives a discontinuous finite step on the radiatively undamped region border.

[41, 43, 168]. This cancellation of the radiation losses is exact in the first step of the iterative perturbation procedure, when in both terms, corresponding to Lorentz friction and to the energy income from other nanospheres, the frequency is taken as as ω_1. In order to answer the question whether this perfect quenching of radiation holds in general, i.e., for a resonance frequency in the chain ω_α, it can be seen that the accurate energy income is still given by Eqs. (9.63) and (9.64), with ω_1 changed for ω_α. Simultaneously, the linear part of the Lorentz friction attains the exact form, $(2/3)(\omega_\alpha a/c)^3$ (due to the equality $\partial^3 R_\alpha/\partial t^3 = -\omega_\alpha^2 \partial R_\alpha/\partial t$, which is satisfied for $R_\alpha \sim e^{-i\omega_\alpha t}$; the third co-factor with ω_α in the expression for the Lorentz friction results from $1/\tau_\alpha = \omega_\alpha/(\tau_\alpha \omega_\alpha)$). Thus one can argue that the cancellation is perfect in this case also.

In this way the propagation of plasmon–polaritons along a discrete metallic nanostructure (the chain) resembles the well-known phenomenon of plasmon–polariton propagation on the two-dimensional interface between metal and dielectric [1, 5, 6]. The radiatively undamped propagation of plasmon–polaritons along a chain is a similar behaviour, associated with the concentration of electromagnetic energy along the chain and with group velocity about one order lower than c. All these properties are illustrated in Figs. 9.23–9.24, for the parameters listed in Table 9.2. The metallic nano-chain thus behaves like an ideal wave guide for plasmon–polaritons in a way that is suitable for the arrangement of subdiffraction circuits.

One can also calculate the group velocity for both polarization modes. In approximate form, taking only the first term of the quickly convergent sums with

9.7 Near-Field-Zone Approximation

denominators n^3 and n^2 but taking accurately the singular far-field-zone term, one can rewrite the dispersion relations in the perturbation approximation as follows; setting $x = kd$ and $\zeta_1 = \omega_1 d/c$,

$$\omega_z = \omega_1 \left(1 - 4\frac{a^3}{d^3}\cos(x)\cos(\zeta_1) - 4\frac{a^2}{d^2}\left(\frac{\omega_1 a}{c}\right)\cos(x)\sin(\zeta_1)\right)^{1/2}, \quad (9.65)$$

$$\omega_{x(y)} = \omega_1 \left(1 + 2\frac{a^3}{d^3}\cos(x)\cos(\zeta_1) + 2\frac{a^2}{d^2}\left(\frac{\omega_1 a}{c}\right)\cos(x)\sin(\zeta_1)\right.$$
$$\left. + \frac{a}{d}\frac{1}{2}\left(\frac{\omega_1 a}{c}\right)^2 \ln[4(1-\cos(x+\zeta_1))(1-\cos(x-\zeta_1))]\right)^{1/2}. \quad (9.66)$$

The group velocities thus attain the forms

$$v_z = \frac{\partial \omega_z}{\partial k}$$
$$= \omega_1 \frac{2(a^3/d^2)\sin(x)\cos(\zeta_1) + (2a^2/d)(\omega_1 a/c)\sin(x)\sin(\zeta_1)}{\sqrt{1 - (4a^3/d^3)\cos(x)\cos(\zeta_1) - (4a^2/d^2)(\omega_1 a/c)\cos(x)\sin(\zeta_1)}}, \quad (9.67)$$

$$v_{x(y)} = \frac{\partial \omega_{x(y)}}{\partial k} = \omega_1 \frac{\mathcal{E}}{\sqrt{\mathcal{F}}},$$

where

$$\mathcal{E} = -\frac{a^3}{d^2}\sin(x)\cos(\zeta_1) - \frac{a^2}{d}\left(\frac{\omega_1 a}{c}\right)\sin(x)\sin(\zeta_1)$$
$$+ \frac{a}{2}\left(\frac{\omega_1 a}{c}\right)^2 \left[\frac{\sin(x+\zeta_1)}{1-\cos(x+\zeta_1)} + \frac{\sin(x-\zeta_1)}{1-\cos(x-\zeta_1)}\right], \quad (9.68)$$

$$\mathcal{F} = 1 + 2\frac{a^3}{d^3}\cos(x)\cos(\zeta_1) + 2\frac{a^2}{d^2}\left(\frac{\omega_1 a}{c}\right)\cos(x)\sin(\zeta_1)$$
$$+ \frac{a}{d}\frac{1}{2}\left(\frac{\omega_1 a}{c}\right)^2 \ln[4(1-\cos(x+\zeta_1))(1-\cos(x-\zeta_1))].$$

The above expressions are substituted for Eq. (9.53) when the near-field-zone contribution, the medium- and far-field-zone contributions are included. Worth noticing is the hyperbolic singularity in the group velocity formula for transverse polarization, induced by the logarithmic singularity due to the far-field-zone constructive interference of the fields of all particles in the chain. This local increase in the velocity is probably responsible for the long-range fainting mode of plasmon–polariton propagation, discussed previously, in a scenario when a selected single nanosphere is excited [43, 189, 182]. In that case the numerical analysis indicated a long-range signal. Remarkably, the damping rate was not small for this singular mode. According to our approach, at the singular point of dispersion the corresponding damping rate is not singular but grows rapidly (a discontinuous finite jump), see Fig. 9.23. Thus one may suspect that the long-range propagation of local modes noticed numerically corresponds to group velocity enhancement, as presented in Fig. 9.24 (the range of signal propagation is of the order of the damping time multiplied by the group velocity of a particular mode). As the singular point is isolated, the corresponding mode is decaying and is probably impossible to excite in practice, since for each realistic wave packet the summed contributions from both sides of the hyperbolic singularity will cancel, thus reducing the local increase in the wave packet velocity. This seems to be in agreement with experiment, where only wave packets that are not too sharply abrupt in wave-vector space are attainable.

The above analysis of the isolated logarithmic singularity in the transverse modes also explains the observation [43] that a finite length of the chain quenches the long-range propagating mode. It is clear why this is so, as the finite sum is not divergent and reduces the local increase in the group velocity. At the same time, all other details of the dispersion and of the damping rate for both polarizations are robust against any shortening of the chain, which agrees with other numerical studies [41, 168]. Nevertheless, again for the transverse mode, the exact quenching of radiation losses requires the contribution of an infinite number of far-field-zone terms. However, even for a relatively short chain of about 10 nanospheres only small discrepancies occur in the vicinity of boundaries of the nonradiative range defined by $kd - \omega_1 d/c = 0$ and $kd + \omega_1 d/c = 2\pi$. For other details the difference between an infinite number of terms and about 10 terms is negligible.

With regard to the group velocity one can observe its dependence on the chain geometry – the nanosphere radius and chain separation. With increasing radius a the amplitude of the velocity also grows, while it diminishes with enhancement of the separation d. For d/a exceeding about 8 the energy dispersion is almost flat and in this band landscape the only features are singular lines for the transverse modes (in the coordinates kd and d/a) repeating owing to the periodicity, see Fig. 9.25. For such flat bands the group velocity is almost zero except in the vicinity of the singular points. This explains the numerical observation [43] that the long-range modes manifest themselves especially distinctly in the case of large separations in the chain.

Finally let us note that in practice we are usually dealing with finite chains. However, the quick convergence of the sums relating to the interaction contribution to the dipole on a particular nanosphere caused by other nanospheres in the chain, even for a relatively small number of spheres (10) in the chain, gives almost the same result as that for an infinite number of spheres; see Figs. 9.26 and 9.27.

9.7 Near-Field-Zone Approximation

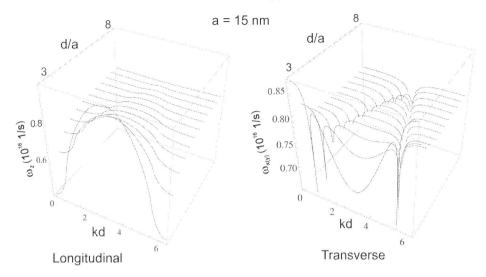

Figure 9.25 Flattening of energy dispersion with growth in chain separation; for d/a exceeding about 6, the only features are the singularity lines for the transverse band, repeating owing to the periodicity; for larger a the flattening begins at larger d/a.

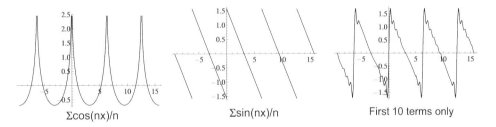

Figure 9.26 The sum $\sum_{n=1}^{\infty} \cos(nx)/n = -\frac{1}{2}\ln[2(1-\cos(x))]$ (left), $\sum_{n=1}^{\infty} \sin(nx)/n = (\pi-x)/2$ for $0 < x < 2\pi$ (centre); the same sum with only the first 10 terms (right).

Furthermore, the singularities described above in the dynamic equation caused by the constructive interference of the dipole radiation in the far- and medium-field zones disappear for a finite chain.

9.7.3 Nonlinear Corrections to the Lorentz Friction

Besides the main contribution to the Lorentz friction field (9.9), there are some small nonlinear corrections to this field, which turn out to be important for collective plasmon propagation in metallic arrays.

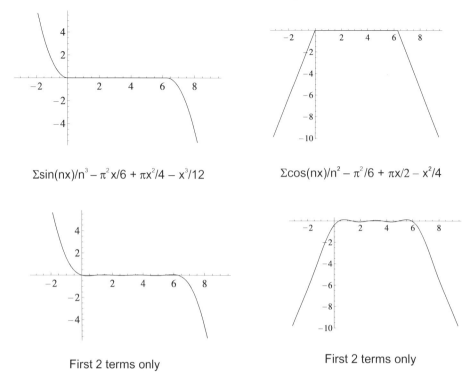

Figure 9.27 The sums $\sum_{n=1}^{\infty} \sin(nx)/n^3 - \pi^2 x/6 + \pi x^2/4 - x^3/12$ (left), $\sum_{n=1}^{\infty} \cos(nx)/n^2 - \pi^2/6 + \pi x/2 + x^2/4$ (right); lower panels the same sums with only the first two terms included.

For a metallic nanosphere with its centre located at $\mathbf{R}_0$, the electric dipole of the electrons (the fluctuation in electron density beyond the uniform distribution compensated by positive jellium) equals

$$\mathbf{D}(\mathbf{R}_0, t) = e \int_V \delta\rho(\mathbf{r}, t) \mathbf{r} d^3 r. \qquad (9.69)$$

This dipole corresponds to surface plasmons and oscillates with the Mie frequency $\omega_1 = \omega_p/\sqrt{3}$, where ω_p is the bulk plasmon frequency. These plasmons are not persisting excitations; they are damped owing to scattering phenomena, with damping rate $1/\tau_0 = v_f/2\lambda_B + C v_F/2a$ and also owing to radiation losses. For large nanospheres the most effective mechanism of plasmon damping is the radiation energy losses, which, as described above, for the case of radiation from the far-field zone can be expressed by the Lorentz friction [77, 55].

Assuming quasiclassically that electrons in the nanosphere have positions $\mathbf{r}_i$ and assuming a static jellium, the dipole of the nanosphere, $\mathbf{D}(\mathbf{R}_0, t) = e \sum_{i=1}^{N_e} \mathbf{r}_i = e N_e \mathbf{r}_e(t)$; here $\mathbf{r}_e = \sum_{i=1}^{N_e} \mathbf{r}_i / N_e$ is the mass centre of the electron system and N_e is

9.7 Near-Field-Zone Approximation

the number of electrons in the nanosphere. In the dynamic case, the velocity of the mass centre $\mathbf{v}_e = \sum_{i=1}^{N_e} \mathbf{v}_i / N_e$.

In order to determine the nonlinear corrections to Eq. (9.9) one can write down the Lorentz friction force acting on a charge eN_e located at the mass centre $\mathbf{r}_e(t)$, expressed in an invariant form [55]:

$$\mathbf{f}_L = \frac{2}{3}(eN_e)^2 \left[\frac{d^2\mathbf{u}}{ds^2} - \mathbf{u} \left(\frac{dU_j}{ds} \right)^2 \right], \qquad j = 1, \ldots, 4, \qquad (9.70)$$

where $ds = cdt\sqrt{1 - v_e^2/c^2}$,

$$U_j = \begin{cases} \mathbf{u} = \dfrac{\mathbf{v}_e}{(c\sqrt{1 - v_e^2/c^2})} \\ u_4 = \dfrac{i}{\sqrt{1 - v_e^2/c^2}} \end{cases}, \qquad U_j^2 = -1.$$

Up to terms of order v_e^2/c^2 with respect to the main term, one can write down an electric field equivalent to the Lorentz friction force:

$$\mathbf{E}_L(t) = \frac{\mathbf{f}_L}{eN_e}$$

$$= \frac{2}{3}(eN_e) \frac{1}{c^3} \left\{ \frac{d^2\mathbf{v}_e}{dt^2} + \frac{1}{c^2} \left[\frac{3}{2} \frac{d^2\mathbf{v}_e}{dt^2} v_e^2 + 3 \frac{d\mathbf{v}_e}{dt} \left(\mathbf{v}_e \cdot \frac{d\mathbf{v}_e}{dt} \right) + \mathbf{v}_e \left(\mathbf{v}_e \cdot \frac{d^2\mathbf{v}_e}{dt^2} \right) \right] \right\}. \qquad (9.71)$$

Next, using the dimensionless variables $t' = t\omega_1$, $\mathbf{R}(t') = \mathbf{r}_e(t)/a$, $\dot{\mathbf{R}}(t') = d\mathbf{r}_e(t)/a\omega_1 dt = \mathbf{v}_e/a\omega_1$, $\ddot{\mathbf{R}}(t') = d^2\mathbf{r}_e(t)/a\omega_1^2 dt^2 = d\mathbf{v}_e/a\omega_1^2 dt$, $\dddot{\mathbf{R}}(t') = d^2\mathbf{v}_e(t), a\omega_1^3 dt^2$ (the dots indicate derivatives with respect to t'), one can write out the dynamical equation in a convenient form. Taking into account that the dipole corresponding to surface plasmons,

$$\mathbf{D} = eN_e a\mathbf{R}, \qquad (9.72)$$

satisfies an equation of oscillatory type, one can write out that equation in the following form (incorporating also the Lorentz friction force),

$$\ddot{\mathbf{R}} + \mathbf{R} + \frac{2}{\tau_0 \omega_1} \dot{\mathbf{R}} = \frac{2}{3} \left(\frac{\omega_p a}{\sqrt{3}c} \right)^3 \left\{ \dddot{\mathbf{R}} + \left(\frac{\omega_p a}{\sqrt{3}c} \right)^2 \right.$$

$$\left. \times \left[\frac{3}{2} \dddot{\mathbf{R}} (\dot{\mathbf{R}} \cdot \dot{\mathbf{R}}) + 3 \ddot{\mathbf{R}} (\dot{\mathbf{R}} \cdot \ddot{\mathbf{R}}) + \dot{\mathbf{R}} (\ddot{\mathbf{R}} \cdot \ddot{\mathbf{R}}) \right] \right\}.$$

$$\qquad (9.73)$$

The terms on the right-hand side of the above equation describe the Lorentz friction including relativistic nonlinear corrections (in the square brackets) beyond the ordinary main linear term $\sim \dddot{\mathbf{R}}$, as previously given by (9.9).

For the case when $1/\tau_0\omega_1$, $\left(\omega_p a/c\sqrt{3}\right)^3 \ll 1$ (fulfilled for nanospheres with radii 5–50 nm for Au or Ag), one can apply the perturbation method of solution, and at zero order one can set $\dddot{\mathbf{R}} + \dot{\mathbf{R}} = 0$. In the next perturbation step one thus substitutes $\dddot{\mathbf{R}} = -\dot{\mathbf{R}}$ and $\ddddot{\mathbf{R}} = -\ddot{\mathbf{R}}$ in the right-hand side of Eq. (9.73).

Nonlinear Correction to Plasmon Radiation Losses of Single Nanosphere

Let us consider first a single metallic nanosphere with dipole-type surface oscillations with dipole **D**. The oscillatory dynamical equation for the dipole, Eq. (9.73), at the first order of perturbation attains the following form (it includes the damping of plasmons due to scattering at a rate $1/\tau_0$ and due to radiation losses coming from the linear term of the Lorentz friction while the right-hand side of Eq. (9.74) below expresses the nonlinear corrections):

$$\ddot{\mathbf{R}} + \mathbf{R} + \left[\frac{2}{\tau_0\omega_1} + \frac{2}{3}\left(\frac{\omega_p a}{\sqrt{3}c}\right)^3\right]\dot{\mathbf{R}} = \frac{2}{3}\left(\frac{\omega_p a}{\sqrt{3}c}\right)^5 \left\{-\frac{5}{2}\dot{\mathbf{R}}\left(\dot{\mathbf{R}}\cdot\dot{\mathbf{R}}\right) + 3\mathbf{R}\left(\dot{\mathbf{R}}\cdot\mathbf{R}\right)\right\}. \quad (9.74)$$

The above nonlinear differential equation can be solved by application of the asymptotic method, as described in [190]. According to this method, one finds the solution of Eq. (9.74) in the following form ($\mathbf{R} = R\mathbf{r}/r$):

$$R(t) = \frac{A_0 e^{-t/\tau}}{\sqrt{1+\frac{9}{8}\gamma A_0^2\left(1-e^{-2t/\tau}\right)}}\cos(\omega_1 t + \theta_0), \quad (9.75)$$

where A_0 and θ_0 are adjusted to the initial conditions,

$$\frac{1}{\tau\omega_1} = \frac{1}{\tau_0\omega_1} + \frac{1}{3}\left(\frac{\omega_p a}{\sqrt{3}c}\right)^3 \approx \frac{1}{3}\left(\frac{\omega_p a}{\sqrt{3}c}\right)^3$$

(which is satisfied for a larger than ~ 15 nm) and $\gamma = \tau\omega_1(1/3)\left(\omega_p a/\sqrt{3}c\right)^5$. From the form of the equation for $1/\tau\omega_1$ it follows that $1/\tau\omega_1$ is always positive. The scattering term, $1/\tau_0 = Cv_F/2a + v_F/2\lambda_B$, is negligible (for nanosphere radii beyond ~ 15 nm) in comparison with the linear contribution of the Lorentz friction, as is demonstrated in Fig. 9.28.

The scale of the nonlinear corrections is given by the coefficient $\gamma \approx 7.3 \times 10^{-4}(a\text{ (nm)})^2$. As this coefficient is small, one can neglect the related contribution in the denominator of the solution (9.75), which results in an ordinary linear solution for damped oscillations. This means that the nonlinear corrections to the

9.7 Near-Field-Zone Approximation

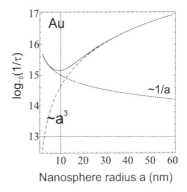

Figure 9.28 Contributions to the damping rate of surface plasmon oscillations in the nanosphere versus nanosphere radius, including the scattering attenuation ($\sim v_F/(2\lambda_B) + C v_F/(2a)$) (the dotted line) and the Lorentz friction damping, $\sim a^3$ (dashed line). For radii greater than ~ 15 nm the second channel dominates the overall damping (upper line). A logarithmic scale for the attenuation ratio has been used.

Lorentz friction have no significance in the case of plasmon oscillations of a single nanosphere. This situation changes, however, in the case of a collective plasmon excitation propagating along a metallic nano-chain, as will be described in the following subsection.

9.7.4 Nonlinear Correction to the Radiation Losses of Plasmon–Polariton in a Nano-Chain

In the case of the dynamics of plasmon–polaritons in a metallic nano-chain, the inclusion of nonlinear correction to the Lorentz friction amounts to accounting for these nonlinear contributions in Eq. (9.48) via the formula for E_L. Instead of Eq. (9.50) we then get the following equation (the Lorentz friction term is represented as in Eq. (9.74)):

$$\ddot{R}_\alpha(k_\alpha, t\omega_1) + \tilde{\omega}_\alpha^2 R_\alpha(k_\alpha, t\omega_1)$$
$$= -2\dot{R}_\alpha(k_\alpha, t\omega_1) \left\{ \frac{1}{\tau_\alpha \omega_1} + \frac{1}{3} \left(\frac{\omega_p a}{c\sqrt{3}} \right)^5 \left[\frac{5}{2} |\dot{R}_\alpha(k_\alpha, t\omega_1)|^2 - 3|R_\alpha(k_\alpha, t\omega_1)|^2 \right] \right\}, \tag{9.76}$$

with the renormalized frequency and damping rate given respectively by Eqs. (9.77) and (9.78) below.

To solve Eq. (9.76) one assumes that in the retardation terms on the right-hand side of this equation, $\omega = \omega_1$, as in the iterative perturbation method of solution. This method was applied earlier to the Lorentz friction, obtaining $\dddot{R} = -\dot{R}$ (note, however, that for $R \sim e^{i\omega t}$ the accurate relation is $\partial^3 R/\partial t^3 = -\omega^2 \partial R/\partial t$). Thus Eq. (9.76) corresponds to the first step of the iterative procedure, where

$$\tilde{\omega}_\alpha^2 = \frac{\omega_\alpha^2}{\omega_1^2} = 1 - a^3 \begin{cases} \mathcal{A}, & \alpha = z, \\ \mathcal{B}, & \alpha = x(y), \end{cases}$$

and

$$\mathcal{A} = 2\sum_n \cos(nkd)\frac{2}{n^3d^3}\cos(n\omega_1 d/c) + 2\sum_n \cos(knd)\frac{2\omega_1/c}{n^2d^2}\sin(n\omega_1 d/c),$$

$$\mathcal{B} = 2\sum_n \cos(nkd)\left(\frac{(\omega_1/c)^2}{nd} - \frac{1}{n^3d^3}\right)\cos(n\omega_1 d/c)$$

$$- 2\sum_n \cos(knd)\frac{\omega_1/c}{n^2d^2}\sin(n\omega_1 d/c). \tag{9.77}$$

Furthermore,

$$\frac{1}{\tau_\alpha \omega_1} = \begin{cases} \frac{1}{\tau_0 \omega_1} & \text{for } kd - \frac{\omega_1}{c}d > 0,\ kd + \frac{\omega_1}{c}d < 2\pi,\ \alpha = z, x(y), \\ \frac{1}{\tau_0 \omega_1} + \frac{1}{3}\left(\frac{\omega_p a}{c\sqrt{3}}\right)^3 + a^3\begin{cases} \mathcal{C}, & \alpha = z, \\ \mathcal{D}, & \alpha = x(y), \end{cases} \end{cases}$$

where

$$\mathcal{C} = -\sum_n \cos(nkd)\frac{2}{n^3d^3}\sin(n\omega_1 d/c) + \sum_n \cos(nkd)\frac{2\omega_1/c}{n^2d^2}\cos(n\omega_1 d/c),$$

$$\mathcal{D} = -\sum_n \cos(nkd)\left(\frac{(\omega_1/c)^2}{nd} - \frac{1}{n^3d^3}\right)\sin(n\omega_1 d/c)$$

$$- \sum_n \cos(nkd)\frac{\omega_1/c}{n^2d^2}\cos(n\omega_1 d/c). \tag{9.78}$$

In Eq. (9.77) one can substitute the point-divergent sum by an exact analytic form:

$$a^3 \sum_{n=1}^\infty \frac{\cos(nkd)(\omega_1/c)^2 \cos(n\omega_1 d/c)}{nd}$$

$$= -\frac{a}{d}\left(\frac{\omega_1 a}{c}\right)^2 \frac{1}{4}\ln[4(1 - \cos(kd + \omega_1 d/c))(1 - \cos(kd - \omega_1 d/c))], \tag{9.79}$$

which displays explicitly the isolated logarithmic singularity in the dispersion formula for the transverse mode. The longitudinal modes are without this singularity; this is linked with the absence of the constructive interference of dipole radiation in the far-field zone, expressed by the term with denominator r_0, which is in a direction collinear with the dipole orientation in the case of the longitudinal polarization.

9.7 Near-Field-Zone Approximation

Applying the same asymptotic methods [190] for the solution of the nonlinear equation (9.76) as in the previous subsection, one can find the corresponding solutions for the regions with positive and negative damping rates, respectively.

For the positive damping rate region, $1/\tau_\alpha \omega_1 > 0$ (see [190]), we have

$$R_\alpha(k,t) = \frac{A_{\alpha 0} e^{-t/\tau_\alpha}}{\sqrt{1 + \gamma_\alpha A_{\alpha 0}^2 \left(1 - e^{-2t/\tau_\alpha}\right)}} \cos(\omega_\alpha t + \theta_0), \qquad (9.80)$$

$$R_\alpha(k,t) \to_{(t \to \infty)} 0,$$

where

$$\gamma_\alpha = |\tau_\alpha \omega_1| \left(\frac{\omega_1 a}{c}\right)^5 \frac{1}{4}\left(\frac{5}{2}\tilde{\omega}_\alpha^2 - 1\right), \qquad \theta_0 = kld + \phi_0.$$

Here ϕ_0 and $A_{\alpha 0}$ are adjusted to the initial conditions. We note from the form of Eq. (9.80) that this is a damped mode vanishing at longer time scales.

For the negative damping rate region, $1/\tau_\alpha \omega_1 < 0$, the solution has a different form (see [190]):

$$R_\alpha(k,t) = \frac{A_{\alpha 0} e^{t/|\tau_\alpha|}}{\sqrt{1 + \gamma_\alpha A_{\alpha 0}^2 \left(e^{2t/|\tau_\alpha|} - 1\right)}} \cos(\omega_\alpha t + \theta_0), \qquad (9.81)$$

$$R_\alpha(k,t) \to_{(t \to \infty)} \frac{1}{\sqrt{\gamma_\alpha}} \cos(\omega_\alpha t + \theta_0),$$

with $\theta_0 = kld + \phi_0$. This solution is stable; it corresponds to an undamped mode which stabilizes at a fixed amplitude $1/\sqrt{\gamma_\alpha}$ at longer time scales, independently of the initial condition, expressed by $A_{\alpha 0}$.

The corresponding dipole oscillations attain in the latter case the form of monochromatic waves propagating along the chain in both directions,

$$D_\alpha = \frac{eN_e a}{\sqrt{\gamma_\alpha}} \frac{1}{2} \cos(\omega_\alpha t \mp kld + \phi_0). \qquad (9.82)$$

From the above discussion it follows that for positive attenuation rates we are dealing with ordinary damped plasmon–polariton propagation, which is not strongly modified from the linear theory (owing to the small value of the factor γ_α). However, in the case of a negative damping rate the solution behaves differently – on longer time scales this solution stabilizes at a constant amplitude independently of the initial conditions. This property characterizes the undamped propagation of plasmon–polaritons along the chain. It should be noted, however, that the existence of undamped modes in a system with scattering losses, and thus with energy dissipation, would contradict energy conservation. Nevertheless, if one assumes that the system is energetically supplied by an external source synchronic to $\dot{\mathbf{R}}$

then, even if its strength were relatively small, it would prevail over the scattering losses and in the whole region of radiative-loss-quenching (the white region in Fig. 9.24 (left)) the total damping rate would be negative.

In such a case we would be dealing with two types of plasmon–polariton: first, ordinary damped modes with a still positive overall attenuation rate (the shaded region in Fig. 9.24, slightly diminished close to its borders with the white region due to the shift caused by the external energy supply); second, undamped modes. The modes from the first region will be extinguished after a distance of the order of the attenuation time multiplied by the group velocity, but the modes from the second region will undergo stable propagation at a fixed amplitude independently of initial conditions. These latter modes express the instability (induced in this case) of the system; thus, the energy for this propagation is not the initial excitation energy but is the energy supplied by the pumping force. In other words, the continuous losses of plasmon oscillation energy due to scattering are instantly recovered by the external pumping. When this income equals the losses, the wave packets of the corresponding modes propagate without damping. This behaviour is typical for other nonlinear oscillation systems and its existence for plasmon–polaritons also would be of practical significance.

Concluding this part of the discussion, the radiatively undamped modes of propagation of collective surface plasmons indicated above seem to match with the experimentally observed long-range propagation of plasmon excitations along finite metallic nano-chains [116, 39, 24, 1]. We have demonstrated the use of the RPA semiclassical model of plasmon oscillations in metallic nanospheres for a description of collective surface plasmon–polariton propagation along a metallic nano-chain. The oscillatory form of the dynamics both for volume and surface plasmons, rigorously described in the RPA semiclassical limit, fits well with the large-nanosphere case, with radii of several nm to several tens of nm; this is confirmed by experimental observations. The most important property of plasmons on large nanospheres is the very strong electromagnetic radiation caused by these excitations, which results in the quick damping of oscillations. Attenuation effects for plasmons are not, however, included in the quantum RPA model. Nevertheless, they can be included in a phenomenological manner, taking advantage of the oscillatory form of the dynamical equations. Some information on plasmon damping can be taken from the microscopic analysis of smaller metallic clusters, with sizes of 1–2 nm (especially those made by the local density approximation (LDA) and time-dependent LDA methods of numerical simulation, employing the Kohn–Sham equation). For larger nanospheres these effects, mainly of a scattering type (but also including Landau damping), are, however, of decreasing significance with growth in radius, since they depend on $1/a$ while the damping of plasmons starts to be dominated by radiation losses growing as a^3.

9.7 Near-Field-Zone Approximation

The radiation effects overwhelming the other energy losses in the case of large nanospheres can be understood in terms of the Lorentz friction, which hampers the movement of charges. Two distinct situations are indicated: first, free radiation to the far-field zone in the dielectric (or vacuum) surroundings of a single nanoparticle; second, the location of an additional charged system in the near-field zone of the plasmons. Such an additional system of charges, acting as an electromagnetic energy receiver in the vicinity of another metallic nanosphere with plasmons, strongly modifies the energy balance of the source and in this way modifies the energy emission in comparison with the free emission in vacuum or in dielectric surroundings. In particular, the Lorentz friction is modified in the presence of an energy receiver in the near-field zone of plasmons, in comparison with simple free emission to the far-field zone. The electromagnetic energy receiver located close to an emitting nanosphere could be a semiconductor (as in the case of metallic modified solar cells) or other metallic nanospheres (as in the case of a metallic nanochain). The latter situation has been analysed above. We showed that, within the near-field coupling approximation [185], collective plasmon–polaritons can propagate along an infinite nano-chain (being collective surface plasmons coupled by the electromagnetic field in the near-field zone); at certain values of nanosphere radius and separation in the chain, these modes are undamped. Moreover, the instability regions in the linear theory of plasmon–polariton dynamics occur, which shows that some corrections must be included to ensure energy conservation. We examined this artefact of near-field-zone coupling and have demonstrated that inclusion of the medium-field and far-field contributions removes the instability and leads to the exact quenching of radiative losses typical for plasmon–polaritons. We analysed the corrections due to the medium- and far-field-zone contributions in full retarded form, including the isolated logarithmic singularity in the dispersion of the transverse modes due to the constructive interference effect in the far-field dipole term. The dispersion, damping rates, and group velocities for longitudinal and transverse polarizations with respect to the chain orientation were analysed with regard to their dependence on nanosphere size and separation in the chain. The results explain some previous, numerical-type observations related to the existence of a decaying mode of plasmon–polaritons due to far-field constructive interference. The wave packet due to a single-point excitation in the chain includes all wave vectors (as in the Fourier picture of a pointlike initial signal), even from the region of the hyperbolic singularity of the group velocity for transverse polarization. As this singularity is isolated and provides opposite-sign contributions on both sides, the resulting far-field propagation (related to the growth in group velocity) may correspond to a decaying mode, though with higher velocity and thus a larger range in comparison with other wave packet components. The manifestation of this behaviour is more visible for larger nanosphere separation, when the group velocity away from the singularity is strongly lowered owing to the flattening of the spectrum, which also fits previous numerical observations.

We examined also nonlinear corrections to the collective plasmon–polariton dynamics along the chain, including the nonlinear contribution to the Lorentz friction force. Even though the related nonlinearity is small, it suffices to regularize the unstable linear approach in the case of an external energy supply. We noted the presence, in this case, of undamped excitations which have a fixed amplitude independently of how small or large the initial conditions were.

10

Plasmon–Polariton Kinetics in a Metallic Nano-chain Located in Absorbing Surroundings

The application of plasmon–polaritons in metallic nanoarrays frequently requires modification of the model concerning the metallic nanosystem surroundings. Plasmons and plasmon–polaritons easily couple with electrically active systems such as metals or semiconductors. This results in properties of plasmon–polaritons that are different from those in dielectric surroundings. In this chapter the theory of plasmon–polariton kinetics along a metallic chain with inclusion of energy-absorbing surroundings is formulated. Taking into account that the coupling channel of plasmon–polaritons with other systems may act in both directions, this opens a wide field of practical applications, e.g., of arbitrarily long-range plasmon–polariton kinetics if energy is supplemented to cover the electron scattering losses in metallic components via the coupling with an external subsystem.

Besides the strong plasmon photovoltaic effect, with growing applications in photovoltaics via the metallization of solar cells at the nanoscale [104, 32, 33, 105, 34, 35], other nano-plasmonic effects are also highly prospective for applications. Plasmon–polaritons propagating in linear arrays of metallic nanoparticles behave as information carriers that are almost lossless [3, 116, 4]. Such perfect wave guides for plasmon–polaritons in the form of metallic nano-chains [166, 40, 38] are considered as promising for subdiffraction light circuits for the new generation of opto-electronics. The transformation of a light signal into a plasmon–polariton with the same frequency but a much shorter wavelength [1, 5, 6] can avoid the diffraction restrictions of conventional opto-electronics. The latter cannot be miniaturized below the light wavelength (of micrometre scale for visible light and higher for infrared) which does not match the nanometre scale of electronics miniaturization. Reducing the wavelength of the information carrier, when changing a photon into a plasmon–polariton, will constitute significant progress toward nanoscale opto-plasmon-electronics.

Plasmon–polaritons also have another exceptional property. They do not irradiate energy when travelling along a wave guide, thus are almost lossless. They do not have any external electromagnetic signature because they do not interact with photons owing to the large incommensurability in momentum for the same energy, and this prevents radiation loss from plasmon–polaritons. Also, they cannot be perturbed by any external electromagnetic field configurations. This property of plasmon–polaritons is of importance for metallic nano-chains located in dielectric surroundings. When a metallic nano-chain is placed on a semiconductor substrate or embedded in semiconductor surroundings, then the situation changes because of the strong coupling of plasmon oscillations in each metallic nanocomponent to the electron band system in the surrounding medium. This near-field-zone coupling causes a plasmon–polariton energy outflow to the semiconductor, resulting in damping of the collective plasmon modes propagating in the chain. Note that the same near-field coupling may be used in the opposite direction to supply energy to plasmon–polaritons in the chain from the semiconductor substrate, with the inverse electron occupation of semiconductor bands; this may be attained, e.g., by application of an appropriate lateral current in the semiconductor. Such an energy supplementation might be convenient to balance the generally small dissipation of plasmon–polariton energy caused by unavoidable ohmic losses. In this chapter we provide a description of the near-field coupling of a plasmon–polariton in a metallic nano-chain to the semiconductor substrate using an application of the Fermi golden rule in order to derive the related damping rate and to analyse the damped propagation of plasmon-polaritons.

The long-range propagation of a plasmon–polariton signal along linear periodic metallic nanostructures has been described theoretically [41, 167, 168] and demonstrated experimentally [24, 170, 39, 171, 166, 169]. The undamped propagation of plasmon–polaritons modes in Au and Ag nano-chains has been observed over distances ~ 0.5 μm. Plasmon–polaritons are the hybridization of plasmons with photons and this results in a lowering of their group velocity below $0.1 \times c$ [170, 171, 166]. The related momentum incommensurability at the same energy of photons and plasmon–polaritons means that the latter do not radiate and so propagate along metallic nano-chains over large distances [40, 169]. However, the residual ohmic losses cause an irreversible dissipation of the energy into Joule heat, limiting the propagation range of plasmon–polaritons.

In the following sections of the present chapter we will describe the propagation of plasmon–polaritons in a metallic nano-chain in close semiconductor surroundings including the coupling of propagating plasmon oscillations to the semiconductor band system in the near-field-zone regime, as schematically presented in Fig. 10.1. This coupling causes an energy transfer between the plasmon–polaritons and the semiconductor electrons. The energy outflow from plasmon–polaritons to the semiconductor band system enhances the damping of the plasmon–polaritons, whereas an opposite direction of energy flow may be utilized to cover the plasmon–polariton Joule heating losses caused by electron scattering in the metallic components. First, we will recall the surface dipole-type

10.1 Plasmon Oscillations in a Single Metallic Nanosphere Including Damping

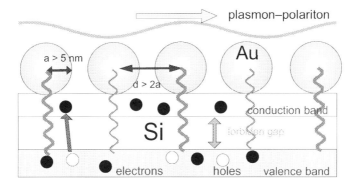

Figure 10.1 When a metallic nano-chain is deposited on or embedded in an absorbing medium (such as a semiconductor), a strong leakage of the plasmon–polariton energy takes place via the coupling in the near-field zone of plasmon dipoles with the band electrons in the surrounding medium.

plasmons in a single metallic nanosphere (within the previously formulated RPA model [23, 77]) including damping effects. The damping of plasmons in a single nanoparticle is caused by the Lorentz friction of the plasmons, which is equivalent to radiation to the far-field zone. Additional damping arises via the channel of energy outflow in the near-field zone due to the coupling of plasmon dipoles to the nearby electrons in the surrounding semiconductor.

Next, we will describe the collective surface plasmon excitations in a nano-chain of metallic particles, i.e., plasmon–polaritons, taking into account both channels for energy losses. The propagation of damped plasmon–polaritons in a metallic nano-chain surrounded by an absorbing medium will be analysed for both longitudinal and transverse polarizations of the plasmon oscillations with respect to the propagation direction and for a variety of geometries, sizes, and material details of the whole system.

10.1 Plasmon Oscillations in a Single Metallic Nanosphere Including Damping

A metallic nanosphere is conventionally modelled as a spherical static positive jellium together with an electrically balanced electron liquid which can locally fluctuate in density, resulting in regular charge oscillations. These are surface plasmons when the oscillations have a translational character, with the electrical local imbalance occurring on the sphere's surface only, and they are volume plasmons when the oscillations have a compressional character along the radius of the nanosphere. It is interesting that the energy of nanosphere volume plasmons for a given mode is always greater than the bulk plasmon energy $\hbar\omega_p = \hbar\sqrt{e^2 n_e/(m\varepsilon_0)}$ (m is the mass of an electron, e is the electron charge, n_e is the density of free electrons in a metal, and ε_0 is the dielectric constant), whereas the energy of the surface plasmon modes is lower than $\hbar\omega_p$ [23]. The latter property makes noble metal (Au, Ag, Cu) nanoparticles attractive for applications as their reduced-energy surface plasmon resonances overlap with the visible light spectrum.

The fluctuations in that local electron density in the nanosphere thus have surface and volume parts as follows:

$$\delta\rho(\mathbf{r},t) = \begin{cases} \delta\rho_1(\mathbf{r},t) & \text{for } r < a, \\ \delta\rho_2(\mathbf{r},t) & \text{for } r \geq a \ (r \to a+), \end{cases} \quad (10.1)$$

These fluctuation modes satisfy equations derived in the framework of the random phase approximation (RPA) [23]:

$$\frac{\partial^2 \delta\rho_1(\mathbf{r},t)}{\partial t^2} = \frac{2}{3}\frac{\epsilon_F}{m}\nabla^2 \delta\rho_1(\mathbf{r},t) - \omega_p^2 \delta\rho_1(\mathbf{r},t), \quad (10.2)$$

and

$$\frac{\partial^2 \delta\rho_2(\mathbf{r},t)}{\partial t^2} = -\frac{2}{3m}\nabla\left\{\left[\frac{3}{5}\epsilon_F n_e + \epsilon_F \delta\rho_2(\mathbf{r},t)\right]\frac{\mathbf{r}}{r}\delta(r-a)\right\}$$

$$-\left[\frac{2}{3}\frac{\epsilon_F}{m}\frac{\mathbf{r}}{r}\nabla\delta\rho_2(\mathbf{r},t) + \frac{\omega_p^2}{4\pi}\frac{\mathbf{r}}{r}\nabla\int d^3 r_1 \frac{1}{|\mathbf{r}-\mathbf{r}_1|}\right.$$

$$\times \left.(\delta\rho_1(\mathbf{r}_1,t)\Theta(a-r_1) + \delta\rho_2(\mathbf{r}_1,t)\Theta(r_1-a))\right]\delta(r-a), \quad (10.3)$$

where Θ is the Heaviside step function defining the static jellium, $n_e(\mathbf{r}) = n_e \Theta(a-r)$, and a is the nanosphere radius. The analysis and solution of the above equations has been performed in detail in previous chapters, resulting in the determination of the plasmon self-mode spectra for both the volume and surface modes.

This RPA treatment does not account, however, for plasmon damping. The damping of plasmons can be included in a phenomenological manner, by adding the term $-(2/\tau_0)\partial \delta\rho(\mathbf{r},t)/\partial t$ to the right-hand side of both Eqs. (10.2) and (10.3), taking advantage of their oscillatory form. The damping ratio $1/\tau_0$ accounts for electron scattering losses (eventually Joule heat losses in the metal of the nanosphere) [24]:

$$\frac{1}{\tau_0} \simeq \frac{v_F}{2\lambda_b} + \frac{C v_F}{2a}. \quad (10.4)$$

In Eq. (10.4), C is a constant of order unity, v_F is the Fermi velocity in the metal, λ_b is the electron mean free path, which is the same as in the bulk metal; included

10.1 Plasmon Oscillations in a Single Metallic Nanosphere Including Damping

are the scattering of electrons on other electrons, on impurities, on phonons in the metallic nanoparticle, and on the nanoparticle boundary [24]. As previously discussed, the latter term in the formula (10.4) accounts for the scattering of electrons on the boundary of the nanoparticle, whereas the former term corresponds to scattering processes similar to those in bulk. Other scattering effects such as Landau damping are important for small nanoparticles [80, 79]) but are negligible for nanoparticle radii larger than 5 nm; here we consider such large nanospheres with radii $\geq$ 5 nm.

The homogeneous equations (10.2) and (10.3) determine the self-frequencies of the plasmon modes. One can write out also dual inhomogeneous equations with an explicit driving factor. This factor would be the time-dependent electric field, e.g., the electric component of the incident electromagnetic wave (of a laser beam or sunlight in photovoltaic applications [30]). The light wavelength in resonance with surface plasmons in the metallic nanosphere (Au, Ag, Cu) is of order 500 nm and greatly exceeds the nanosphere size (with radius 5–50 nm); thus dipole-approximation-regime conditions are satisfied. Hence, the driving field $\mathbf{E}(t)$ of the electromagnetic wave is almost homogeneous over the nanosphere (this corresponds to the dipole approximation) and can only excite the dipole surface mode in the electron liquid of the metallic nanosphere. The dipole-type mode may be described by a function $Q_{1m}(t)$, with $l = 1$ and $m = -1, 0, 1$ the angular momentum quantum numbers, related to the spherical symmetry of the metallic nanoparticle. The function $Q_{1m}(t)$ satisfies the equation

$$\frac{\partial^2 Q_{1m}(t)}{\partial t^2} + \frac{2}{\tau_0}\frac{\partial Q_{1m}(t)}{\partial t} + \omega_1^2 Q_{1m}(t)$$
$$= \sqrt{\frac{4\pi}{3}\frac{en_e}{m}}\left[E_z(t)\delta_{m,0} + \sqrt{2}\left(E_x(t)\delta_{m,1} + E_y(t)\delta_{m,-1}\right)\right], \quad (10.5)$$

where $\omega_1 = \omega_p/\sqrt{3\varepsilon}$ (ω_1 is the dipole surface plasmon Mie-type frequency [18, 17] and ε is the dielectric constant for the nanosphere surroundings). The electron density fluctuations can be written as follows:

$$\delta\rho(\mathbf{r},t) = \begin{cases} 0, & r < a, \\ \sum_{m=-1}^{1} Q_{1m}(t)Y_{1m}(\Omega), & r \geq a \ (r \to a+), \end{cases} \quad (10.6)$$

where $Y_{lm}(\Omega)$ is a spherical harmonic with $l = 1$. The plasmon oscillations given by Eq. (10.6) define a dipole $\mathbf{D}(t)$:

$$\begin{cases} D_x(t) = e\int d^3r\, x\, \delta\rho(\mathbf{r},t) = \sqrt{\frac{2\pi}{3}}eQ_{1,1}(t)a^3, \\ D_y(t) = e\int d^3r\, y\, \delta\rho(\mathbf{r},t) = \sqrt{\frac{2\pi}{3}}eQ_{1,-1}(t)a^3, \\ D_z(t) = e\int d^3r\, z\, \delta\rho(\mathbf{r},t) = \sqrt{\frac{4\pi}{3}}eQ_{1,0}(t)a^3. \end{cases} \quad (10.7)$$

By virtue of Eq. (10.5), the dipole $\mathbf{D}(t)$ thus satisfies the equation

$$\left[\frac{\partial^2}{\partial t^2} + \frac{2}{\tau_0}\frac{\partial}{\partial t} + \omega_1^2\right]\mathbf{D}(t) = \frac{a^3 4\pi e^2 n_e}{3m}\mathbf{E}(t) = \varepsilon a^3 \omega_1^2 \mathbf{E}(t). \qquad (10.8)$$

The damping term in the above equation includes energy dissipation due to electron scattering in the metallic nanosphere, i.e., electron–electron, electron–phonon, and electron–admixture scattering, as well as a contribution from the boundary scattering [24]. There is, however, also an important channel of plasmon damping caused by radiation losses, not included in the formula for τ_0. The radiative losses of plasmon energy in the dielectric surroundings can be expressed by the Lorentz friction [55], i.e., by a fictitious electric field slowing down the motion of charges:

$$\mathbf{E}_L = \frac{2\sqrt{\varepsilon}}{3c^3}\frac{\partial^3 \mathbf{D}(t)}{\partial t^3}. \qquad (10.9)$$

Thus, one can rewrite Eq. (10.8) including the Lorentz friction term as

$$\left[\frac{\partial^2}{\partial t^2} + \frac{2}{\tau_0}\frac{\partial}{\partial t} + \omega_1^2\right]\mathbf{D}(t) = \varepsilon a^3 \omega_1^2 \mathbf{E}(t) + \varepsilon a^3 \omega_1^2 \mathbf{E}_L, \qquad (10.10)$$

or for, the case when $\mathbf{E} = 0$,

$$\left[\frac{\partial^2}{\partial t^2} + \omega_1^2\right]\mathbf{D}(t) = \frac{\partial}{\partial t}\left[-\frac{2}{\tau_0}\mathbf{D}(t) + \frac{2}{3\omega_1}\left(\frac{\omega_p a}{c\sqrt{3}}\right)^3 \frac{\partial^2}{\partial t^2}\mathbf{D}(t)\right]. \qquad (10.11)$$

Now applying a perturbation procedure to solve Eq. (10.11) and treating the right-hand side of this equation as the perturbation, one obtains, in the zeroth step of the perturbation, $[\partial^2/\partial t^2 + \omega_1^2]\mathbf{D}(t) = 0$, from which $(\partial^2/\partial t^2)\mathbf{D}(t) = -\omega_1^2 \mathbf{D}(t)$. Therefore, in the first step of the perturbation, one includes the latter formula on the right-hand side of Eq. (10.11), obtaining

$$\left[\frac{\partial^2}{\partial t^2} + \frac{2}{\tau}\frac{\partial}{\partial t} + \omega_1^2\right]\mathbf{D}(t) = 0, \qquad (10.12)$$

where

$$\frac{1}{\tau} = \frac{1}{\tau_0} + \frac{\omega_1}{3}\left(\frac{\omega_p a}{c\sqrt{3}}\right)^3. \qquad (10.13)$$

In this way we have included the Lorentz friction in the total attenuation rate $1/\tau$. This is justified for nanospheres that are not very large, i.e., when the second term in Eq. (10.13), proportional to a^3, is sufficiently small to fulfil the perturbation procedure constraints. For nanospheres with radius 10–30 nm this approximation is fulfilled and has been verified experimentally for Au and Ag nanospheres [77, 45]. Inclusion of the Lorentz friction in this perturbative manner explains the experimentally observed redshift of the resonance surface

Table 10.1 *Radius a^* of the nanosphere corresponding to the minimal value of surface plasmon damping in dielectric surroundings*

		a^*	
$n = \sqrt{\varepsilon}$	Au	Ag	Cu
$n = 1$ vacuum	8.4	8.44	6.46
$n = 1.4$ water	9.137	9.18	9.202
$n = 2$	9.989	10.037	10.04

plasmon frequency with increasing a. Indeed, the solution of Eq. (10.12) is of the form $\mathbf{D}(t) = \mathbf{A}e^{-t/\tau}\cos(\omega'_1 t + \phi)$, where $\omega'_1 = \omega'_1\sqrt{1 - 1/(\omega_1\tau)^2}$, which gives the experimentally observed redshift of the plasmon resonance due to the $\sim a^3$ growth in the plasmon damping caused by radiation. The Lorentz friction term in Eq. (10.13) dominates the plasmon damping in dielectric surroundings for $a \geq 12$ nm (for Au and Ag) owing to this a^3 dependence. Thus the plasmon damping grows rapidly with a and results in a pronounced redshift of the resonance frequency, in good agreement with the experimental data for $10 < a < 30$ nm (Au and Ag) [77]. Taking into account that $1/\tau_0$ scales as $1/a$, while the Lorentz friction contribution scales as a^3 for $10 < a < 30$ nm, in this size region one encounters a cross-over of the attenuation rate with respect to a. The minimum damping is achieved at

$$a^* = \left(\frac{9}{2}\sqrt{\varepsilon}\frac{v_F c^3}{\omega_p^4}\right)^{1/4}, \tag{10.14}$$

and for $a < a^*$ the damping ratio grows with decreasing a approximately as $\sim 1/a$, while for $a > a^*$ this ratio grows with increasing a in proportion to a^3. The value of a^* can be estimated for Au, Ag, and Cu; see Table 10.1.

10.2 Radiative Properties of Plasmon–Polaritons in a Metallic Nano-Chain Embedded in a Dielectric Medium

When a metallic nanosphere is an element of a chain of similar nanospheres equidistantly distributed along a line, one has to take into account that alongside the radiation losses from that nanosphere, a simultaneous income of energy takes place from the radiation of other nanospheres when a collective plasmon excitation (a plasmon–polariton) propagates along the chain.

The interaction between nanospheres in the chain can be considered as dipole-type coupling. The minimal separation of the nanospheres in the chain is $d = 2a$

(d is the distance between neighbouring sphere centres), and the dipole approximation of the plasmon interaction in the nanosphere chain is sufficiently accurate for $d > 3a$, because then the multipole-interaction contribution can be neglected. Various numerical large-scale calculations of the electromagnetic field distribution in such systems with the inclusion of dipole and also multipole interactions between the plasmonic oscillations in the metallic components have been carried out [41, 167, 168]. A model of interacting dipoles [177, 178] had been developed earlier for the investigation of stellar matter [179, 180] and it was then adopted for metal particle systems [181, 182]. Numerical studies beyond the dipole model [41, 168] indicated that the dipole model is sufficiently accurate when the particle separation is not lower than the particle dimensions, since the multipole contribution to the interaction is then small [183].

An oscillating dipole $\mathbf{D}(t)$ located at a point $\mathbf{r}$ causes an electric field at another point $\mathbf{r}_0$ (we assume that the vector $\mathbf{r}_0$ is drawn from the tip of $\mathbf{r}$) that has the following form (including the relativistic retardation of the electromagnetic signals) [55, 76]:

$$\mathbf{E}(\mathbf{r}, \mathbf{r}_0, t) = \frac{1}{\varepsilon} \left(-\frac{\partial^2}{v^2 \partial t^2} \frac{1}{r_0} - \frac{\partial}{v \partial t} \frac{1}{r_0^2} - \frac{1}{r_0^3} \right) \mathbf{D}\left(\mathbf{r}, t - \frac{r_0}{v}\right) \\ + \frac{1}{\varepsilon} \left(\frac{\partial^2}{v^2 \partial t^2} \frac{1}{r_0} + \frac{\partial}{v \partial t} \frac{3}{r_0^2} + \frac{3}{r_0^3} \right) \mathbf{n}_0 \left(\mathbf{n}_0 \cdot \mathbf{D}\left(\mathbf{r}, t - \frac{r_0}{v}\right) \right), \tag{10.15}$$

where $\mathbf{n}_0 = \mathbf{r}_0/r_0$ and $v = c/\sqrt{\varepsilon}$. The above formula includes terms corresponding to the near-field-zone (denominator r_0^3), medium-field-zone (denominator r_0^2), and far-field-zone (denominator r_0) contributions to the dipole field. Equation (10.15) allows us to write out the dynamical equation for the plasmon oscillations at each nanosphere of the chain, numbered by an integer l. We denote by d the separation between nanospheres in the chain and by a the nanosphere radius. The vectors $\mathbf{r}$ and $\mathbf{r}_0$ are collinear if the origin of the coordinate system is associated with a nanosphere in the chain.

Therefore the equation for the collective surface plasmon oscillation in the lth sphere of the chain is as follows,

$$\left[\frac{\partial^2}{\partial t^2} + \frac{2}{\tau_0} \frac{\partial}{\partial t} + \omega_1^2 \right] D_\alpha(ld, t) = \varepsilon \omega_1^2 a^3 \sum_{m=-\infty, m \neq l}^{m=\infty} E_\alpha\left(md, t - \frac{|l-m|d}{c}\right) \\ + \varepsilon \omega_1^2 a^3 E_{L\alpha}(ld, t) + \varepsilon \omega_1^2 a^3 E_\alpha(ld, t). \tag{10.16}$$

The first term on the right-hand side of Eq. (10.16) describes the dipole coupling between nanospheres in the chain and the next two terms correspond to the contribution to the plasmon attenuation due to Lorentz friction (as described in the

10.2 Radiative Properties of Plasmon–Polaritons

previous section) and to the driving electric field. The index α enumerates the polarizations: longitudinal for $\alpha = z$ and transverse for $\alpha = x(y)$, with respect to the chain orientation (assumed to be in the z direction). According to Eq. (10.15) we have:

$$E_z(md,t) = \frac{2}{\varepsilon d^3}\left(\frac{1}{|m-l|^3} + \frac{d}{v|m-l|^2}\frac{\partial}{\partial t}\right)D_z\left(md, t - |m-l|\frac{d}{v}\right),$$

$$E_{x(y)}(md,t) = -\frac{1}{\varepsilon d^3}\left(\frac{1}{|m-l|^3} + \frac{d}{v|m-l|^2}\frac{\partial}{\partial t} + \frac{d^2}{v^2|l-d|}\frac{\partial^2}{\partial t^2}\right)$$
$$\times D_{x(y)}\left(md, t - |m-l|\frac{d}{v}\right).$$

(10.17)

Taking advantage of the chain periodicity one can assume in analogy to the Bloch states in one-dimensional crystals, with a reciprocal quasimomentum lattice,

$$D_\alpha(ld,t) = D_\alpha(k,t)\,e^{-ikld}, \qquad 0 \le k \le \frac{2\pi}{d}. \qquad (10.18)$$

The above can be asserted in a more formal manner by taking the Fourier picture of Eq. (10.16). As the dipoles are localized at the centres of the nanospheres, the system is discrete, as in the case of phonons in a one-dimensional crystal. One can thus apply a discrete Fourier transform (DFT) with respect to the positions, and a continuous Fourier transform (CFT) with respect to time. A DFT is defined for a finite set of numbers, so one can consider a chain with $2N+1$ nanospheres, i.e., of length $L = 2Nd$. Thus for any discrete characteristics $f(l)$, $l = -N,\ldots,0,\ldots,N$ of the chain, such as the dipole distribution, one is dealing with the DFT picture:

$$f(k) = \sum_{l=-N}^{N} f(l)e^{ikld},$$

where $k = 2\pi n/(2Nd)$, $n = 0,\ldots,2N$. Hence, $kd \in [0, 2\pi]$ owing to the periodicity of the chain. The Born–Karman boundary condition has been imposed, on the whole system, resulting in $k = 2\pi n/(2Nd)$. In order to account for the infinite length of the chain one can take the limit $N \to \infty$, with the result that the variable k becomes quasicontinuous, but $kd \in [0, 2\pi]$ still holds.

The Fourier picture of Eq. (10.16), with a DFT for positions and a CFT for time, is as follows,

$$\left(-\omega^2 - i\frac{2}{\tau_0}\omega + \omega_1^2\right)D_\alpha(k,\omega) = \omega_1^2\frac{a^3}{d^3}F_\alpha(k,\omega)D_\alpha(k,\omega) + \varepsilon a^3\omega_1^2 E_{0\alpha}(k,\omega),$$

(10.19)

with

$$F_z(k,\omega) = 4\sum_{m=1}^{\infty}\left(\frac{\cos(mx)}{m^3}\cos(m\xi) + \xi\frac{\cos(mx)}{m^2}\sin(m\xi)\right)$$
$$+ 2i\left[\frac{1}{3}(\xi)^3 + 2\sum_{m=1}^{\infty}\left(\frac{\cos(mx)}{m^3}\sin(m\xi) - \xi\frac{\cos(mx)}{m^2}\cos(m\xi)\right)\right],$$

$$F_{x(y)}(k,\omega) = -2\sum_{m=1}^{\infty}\left(\frac{\cos(mx)}{m^3}\cos(m\xi) + \xi\frac{\cos(mx)}{m^2}\sin(m\xi)\right.$$
$$\left. - (\xi)^2\frac{\cos(mx)}{m}\cos(m\xi)\right)$$
$$- i\left[-\frac{2}{3}(\xi)^3 + 2\sum_{m=1}^{\infty}\left(\frac{\cos(mx)}{m^3}\sin(m\xi) + \xi\frac{\cos(mx)}{m^2}\cos(m\xi)\right.\right.$$
$$\left.\left. - (\xi)^2\frac{\cos(mx)}{m}\sin(m\xi)\right)\right].$$

(10.20)

The direct calculation of the functions $\text{Im}\,F_z(k,\omega)$ and $\text{Im}\,F_{x(y)}(k,\omega)$ corresponding to radiative damping for the longitudinal and transverse plasmon–polariton polarizations, respectively, was done explicitly in the case of a dielectric medium in Eqs. (9.27) and (9.29), and holds also for an absorbing medium. We showed there that both these functions *vanish* when $0 < kd \pm \omega d/v < 0$ (the corresponding region – the light cone – is indicated in Fig. 9.3). Outside this region the radiative damping expressed by the functions $\text{Im}\,F_\alpha(k,\omega)$ is nonzero, which is illustrated in Figs. 9.4 and 9.5, respectively for the longitudinal and transverse modes.

10.3 Plasmon–Polariton Self-Modes in a Chain in Dielectric Surroundings

The real parts of the functions F_α give the renormalized self-frequency of the plasmon–polaritons in the chain, whereas the imaginary parts give the renormalized damping of these modes. $\text{Re}\,F_\alpha(k,\omega)$ and $\text{Im}\,F_\alpha(k,\omega)$ are functions of k and ω. Applying the perturbation method of solution, within the first-order approximation one can put $\omega = \omega_1$ in $\text{Re}\,F_\alpha$ and also in the residual nonzero $\text{Im}\,F_\alpha$ outside the region $0 < kd \pm \omega_1 d/v < 2\pi$. Let us emphasize, however, that the vanishing of $\text{Im}\,F_\alpha(k,\omega)$ inside the region $0 < kd \pm \omega d/v < 2\pi$ holds for any value of ω [44], thus also for the exact (nonperturbative) solution, as shown numerically in Figs. 10.2–10.5.

10.3 Plasmon–Polariton Self-Modes in a Chain in Dielectric Surroundings   241

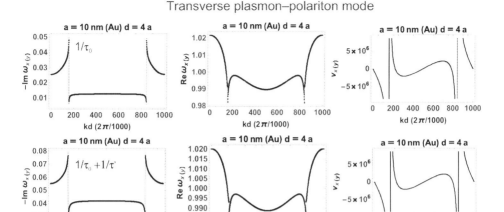

Figure 10.2 The damping rate Im $\omega_{x(y)}(k)$, the self-frequency Re $\omega_{x(y)}(k)$, and the group velocity $v_{x(y)}(k)$ of a plasmon–polariton for transverse polarization in an infinite chain of Au nanospheres with radius $a = 10$ nm and chain separation $d = 4a$ for vacuum surroundings (upper) and semiconductor (Si) surroundings (lower).

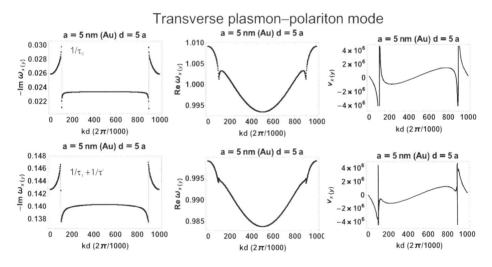

Figure 10.3 The damping rate Im $\omega_{x(y)}(k)$, the self-frequency Re $\omega_{x(y)}(k)$, and the group velocity $v_{x(y)}(k)$ of a plasmon–polariton for transverse polarization in an infinite chain of Au nanospheres with radius $a = 5$ nm and chain separation $d = 5a$ for vacuum surroundings (upper) and semiconductor (Si) surroundings (lower).

242 Plasmon–Polariton Kinetics in Absorbing Surroundings

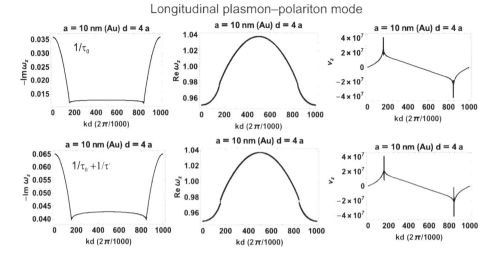

Figure 10.4 The damping rate $\mathrm{Im}\,\omega_{x(y)}(k)$, the self-frequency $\mathrm{Re}\,\omega_{x(y)}(k)$, and the group velocity $v_{x(y)}(k)$ of a plasmon–polariton for longitudinal polarization in an infinite chain of Au nanospheres with radius $a = 10$ nm and chain separation $d = 4a$, for vacuum surroundings (upper) and semiconductor (Si) surroundings (lower).

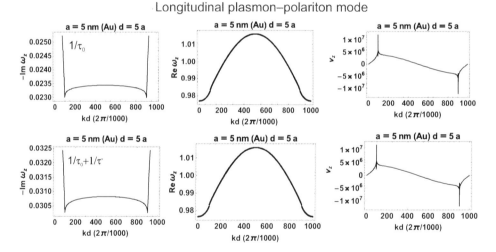

Figure 10.5 The damping rate $\mathrm{Im}\,\omega_{x(y)}(k)$, the self-frequency $\mathrm{Re}\,\omega_{x(y)}(k)$, and the group velocity $v_{x(y)}(k)$ of a plasmon–polariton for longitudinal polarization in an infinite chain of Au nanospheres with radius $a = 5$ nm and chain separation $d = 5a$, for vacuum surroundings (upper) and semiconductor (Si) surroundings (lower).

In the perturbation scheme, one rewrites the dynamic equation (10.19) for plasmon–polariton modes in the chain in the following form:

$$\left(-\omega^2 - i\frac{2}{\tau_\alpha(k)}\omega + \omega_\alpha(k)^2\right) D_\alpha(k,\omega) = \varepsilon a^3 \omega_1^2 E_{0\alpha}(k,\omega), \qquad (10.21)$$

10.3 Plasmon–Polariton Self-Modes in a Chain in Dielectric Surroundings

where the renormalized attenuation rate is given by

$$\frac{1}{\tau_\alpha(k)} = \begin{cases} \dfrac{1}{\tau_0} & \text{for } 0 < kd \pm \omega_1 d/v < 2\pi, \\ \dfrac{1}{\tau_0} + \dfrac{a^3 \omega_1}{2d^3}\,\mathrm{Im}\,F_\alpha(k,\omega_1) & \text{for } kd - \omega_1 d/v < 0 \\ & \text{or } kd + \omega_1 d/a > 2\pi. \end{cases} \quad (10.22)$$

The renormalized bare self-frequency (owing to damping, the true resonance is, however, redshifted, as for a damped oscillator, but this is not given here) is

$$\omega_\alpha^2(k) = \omega_1^2\left(1 - \frac{a^3}{d^3}\mathrm{Re}\,F_\alpha(k,\omega_1)\right). \quad (10.23)$$

Equation (10.21) can be easily solved for both the inhomogeneous case and the homogeneous case (when $E_{0\alpha} = 0$). The general solution of Eq. (10.21) is the sum of the general solution of the homogeneous equation and a single particular solution of the inhomogeneous equation. The first includes the initial conditions and describes damped self-oscillations with frequency

$$\omega'_\alpha = \sqrt{\omega_\alpha^2(k) - \frac{1}{\tau_\alpha^2(k)}}, \quad (10.24)$$

i.e., for each k and α,

$$D'_\alpha(k,t) = A_{\alpha,k} e^{i(\omega'_\alpha t + \phi_{\alpha,k})} e^{-t/\tau_\alpha(k)}, \quad (10.25)$$

with constants $A_{\alpha,k}$ and $\phi_{\alpha,k}$ determined by the initial conditions.

For the inhomogeneous case the particular solution is as follows:

$$D''_\alpha(k,t) = \varepsilon a^3 \omega_1^2 E_{0\alpha}(k) e^{i(\gamma t + \eta_{\alpha,k})} \frac{1}{\sqrt{(\omega_\alpha^2(k) - \gamma^2)^2 + 4\gamma^2/\tau_\alpha^2(k)}}, \quad (10.26)$$

given the assumed single Fourier time component $E_{0\alpha}(k,t) = E_{0\alpha}(k)e^{i\gamma t}$, and

$$tg(\eta_{\alpha,k}) = \frac{2\gamma/\tau(\alpha,k)}{\omega(\alpha,k)^2 - \gamma^2}$$

as for a driven oscillator. Let us emphasize that $E_{0\alpha}(k)$ is a real function: $E_{0\alpha}(ld)^* = E_{0\alpha}(ld) = E_{0\alpha}(-ld)$. An appropriate choice of this function, in practice the choice of the number of externally excited nanospheres in the chain (e.g., by a suitably focused laser beam), allows for the modelling of its Fourier picture $E_{0\alpha}(k)$. This gives the envelope of the wave packet if one inverts the Fourier transform in the solution given by Eq. (10.26) back to the position variable. For the case of the external excitation of only a single nanosphere, the wave packet envelope includes homogeneously all wave vectors $k \in [0, 2\pi]$. The larger the number of nanospheres simultaneously excited, the narrower in k is the wave packet envelope. For $E_{0\alpha}(ld)^* = E_{0\alpha}(ld) = E_{0\alpha}(-ld)$ the Fourier transform has the same properties, i.e., $E_{0\alpha}(k)^* = E_{0\alpha}(k) = E_{0\alpha}(-k)$ and the latter equality can be rewritten, owing to the periodicity $\frac{2\pi}{d}$ of k, as $E_{0\alpha}(-k) = E_{0\alpha}(2\pi/d-k) = E_{0\alpha}(k)$ (this shifts k to the equivalent positively valued domain $k \in [0, 2\pi]$). The inverse Fourier picture of the real part of Eq. (10.26) is

$$D''_\alpha(ld,t) = \int_0^{2\pi/d} dk \cos(kld - \gamma t - \eta_{\alpha,k}) \, \varepsilon a^3 \omega_1^2 E_{0\alpha}(k)$$

$$\times \frac{1}{\sqrt{(\omega_\alpha^2(k) - \gamma^2)^2 + 4\gamma^2/\tau_\alpha^2(k)}}. \tag{10.27}$$

This integral can be rewritten, by virtue of the mean value theorem, in the form

$$D''_\alpha(ld,t) = \frac{2\pi}{d} \cos(k^* ld - \gamma t - \eta_{\alpha,k^*}) \, \varepsilon a^3 \omega_1^2 E_{0\alpha}(k^*)$$

$$\times \frac{1}{\sqrt{(\omega_\alpha(k^*)^2 - \gamma^2)^2 + 4\gamma^2/\tau_\alpha^2(k^*)}}. \tag{10.28}$$

The above expression describes undamped wave motion with frequency γ and the velocity, amplitude, and phase shift determined by k^*. The energy losses are supplemented continuously by the driving force, as for any steady state of a damped and driven oscillator. The amplitude attains its maximal value at resonance, when

$$\gamma = \omega_\alpha(k^*) \sqrt{1 - \frac{2}{(\tau_\alpha(k^*)\omega_\alpha(k^*))^2}}. \tag{10.29}$$

If the chain is subject to a persistent driving force in the form of a time-dependent external electric field applied to some number (even a small number) of the nanospheres, one is dealing with undamped wave packet propagation along the whole chain – the energy supply by the driving factor covers damping losses. These modes, depending on the shaping of the wave packet by the specific choice of chain excitation, may be responsible for the experimentally observed long-range practically undamped plasmon–polariton propagation [169, 171, 39, 40].

The self-modes of Eq. (10.25) are damped and their propagation depends on appropriately prepared initial conditions admitting nonzero values of $A_{\alpha,k}$. The resulting wave packet may embrace the wave-numbers k from some region of $[0, 2\pi]$. If only the wave numbers k for which $0 < kd \pm \omega_1 d/v < 2\pi$ contribute to the wave packet, its damping is only of ohmic type. The value of $1/\tau_0$ decreases with increasing a (see Eq. (10.4)); thus, to achieve a longer range of these damped excitations in the chain, larger spheres are more suitable. The limiting value of $1/\tau_0$ is $v_f/(2\lambda_B) \sim 10^{13}$/s, which gives the maximum range of propagation for these modes of a plasmon–polariton as $\sim 0.1 c\tau_0 \sim 10^{-6}$ m. For the group velocity of the wave packet we assume $\sim 0.1c$ as its maximum value (though this depends on the radius and separation of the nanospheres in the chain). The group velocities calculated for both polarizations are presented in the right-hand panels of Figs. 10.2–10.5 (which give an accurate solution).

Though the analysis presented above is addressed to chains consisting of ideal nanospheres, the conclusions also hold for chains of other shapes of particle and agree with experimental observations at least qualitatively. In [171] the propagation of a plasmon–polariton in a nano-chain of Ag rod-shaped particles (90 : 30 : 30 nm, oriented with the longer axis perpendicular to the chain in order to enhance the near-field coupling [39], with face-to-face separation, 70 nm) is evidenced by observing the luminescence of dye particles located in proximity to the transmitting electromagnetic signal but distant from the pointlike excitation source by about 0.5 μm. The observed behaviour has been supported by FDTD numerical simulations. Several samples of the chain were fabricated by electron beam lithography in the form of a two-dimensional matrix with sufficiently well-separated individual chains. The energy blueshift of the plasmon resonance for the nano-rods in the chain, in comparison with the single particle case is observed to be about 0.1 eV [171]. This agrees with our estimation of the reduction in radiation losses in the chain in comparison with the strong Lorentz friction for a single metallic nanoparticle and the related smaller redshift of damped oscillations. The position of the resonance maximum in the chain is located at a higher energy for the transverse mode than for the longitudinal mode [170], which also agrees with theoretical predictions. In [39] it was indicated that FDTD simulations give lower values of the group velocities for both polarizations and higher attenuation rates in comparison with these quantities as previously estimated [170, 24] using a simplified point-dipole model with near-field–zone interaction only and neglecting retardation effects.

Let us note, however, that the simplified approach including only the near-field contribution to the electric field of the interacting dipoles leads to an artefact, i.e., for some values of the chain parameters d and a an instability of the collective dynamics occurs [185]. This instability is completely removed by the inclusion of the medium- and far-field contributions to the electric field of the dipole, including relativistic retardation [44]. Nevertheless, the dipole interaction model even if it includes, besides the near-field contribution, the medium- and far-field contributions and all retardation effects, still suffers from the absence of the magnetic field component needed for a complete description of the far-field-zone and medium-field-zone wave propagation. As it is demonstrated in [186, 187], for large separations in a chain, the scattering of electromagnetic radiation dominates the signal behaviour in metallic nano-chains, which then act as a Bragg grating for plasmon–polaritons. For ellipsoidal Au nanoparticles (210 : 80 nm) deposited on the top of an Si wave guide, the change of regime from collective plasmon–polariton guiding to the Bragg scattering scenario takes place at distances between nanoparticles exceeding about 1 μm [186]. This proves that the model of dipole coupling in the nano-chain works quite well for a wide region of chain parameters, in practice up to the order of microns for the distance between the metallic elements in the chain. This supports the qualitative argument that the Bragg grating regime is not efficient for subwavelength distances and justifies the applicability of the model considered. The SNOM measurements of near-field coupled plasmon modes in a metallic nano-chain [166], interpreted within the classical field-susceptibility formalism in [188], also support the sufficiency of the dipole approximation for interactions in the chain for the scale considered here, with nanosphere radii a of the order of 10–30 nm and chain separation d not exceeding $\sim 10a$.

10.4 Damping of Surface Plasmons in a Single Metallic Nanoparticle Deposited on a Semiconductor Substrate

The interaction of band electrons with surface plasmons in a metallic nanoparticle deposited on the top surface of a semiconductor (or embedded in a semiconductor medium) causes an energy transfer resulting in an additional damping of the plasmons. The perturbation of the electron band system in the substrate semiconductor due to the presence of dipole surface plasmon oscillations in a metallic nanosphere with radius a deposited on the semiconductor surface (or embedded in it) has the form of the electromagnetic field of an oscillating dipole. The Fourier components of the electric field $\mathbf{E}_\omega$ (the Fourier component of Eq. (9.45)) and the magnetic field $\mathbf{B}_\omega$ produced at a distance $\mathbf{R}$ from the centre of a nanosphere having a surface plasmon dipole with frequency ω have the form [55]

$$\mathbf{E}_\omega = \frac{1}{\varepsilon}\left\{\mathbf{D}_0\left(\frac{k^2}{R}+\frac{ik}{R^2}-\frac{1}{R^3}\right)+\hat{\mathbf{n}}(\hat{\mathbf{n}}\cdot\mathbf{D}_0)\left(-\frac{k^2}{R}-\frac{3ik}{R^2}+\frac{3}{R^3}\right)\right\}e^{ikR} \quad (10.30)$$

and

$$\mathbf{B}_\omega = \frac{ik}{\sqrt{\varepsilon}}[\mathbf{D}_0\times\hat{\mathbf{n}}]\left(\frac{ik}{R}-\frac{1}{R^2}\right)e^{ikR} \quad (10.31)$$

(ε is the dielectric constant). In the case of spherical symmetry, the plasmon dipole is considered as pinned to the centre of the nanosphere, $\mathbf{D} = \mathbf{D}_0 e^{-i\omega t}$. In Eqs. (10.30) and (10.31) we have used the notation for a retarded argument: $i\omega(t - R/c) = i\omega t - ikR$, $\hat{\mathbf{n}} = \mathbf{R}/R$, $\omega = ck$, and momentum $\mathbf{p} = \hbar\mathbf{k}$. Because we are considering the interaction with a closely adjacent layer of the substrate semiconductor, we can neglect the terms with denominators R^2 and R as small in comparison with the term with denominator R^3 – this is the near-field-zone approximation (in which the magnetic field disappears and the electric field is that

10.4 Damping of Surface Plasmons for Single Metallic Nanoparticle

of a static dipole [55]). Therefore the related perturbation potential added to the system Hamiltonian attains the form

$$w = e\psi(\mathbf{R},t) = \frac{e}{\varepsilon R^2}\hat{\mathbf{n}} \cdot \mathbf{D}_0 \sin(\omega t + \alpha) = w^+ e^{i\omega t} + w^- e^{-i\omega t}. \quad (10.32)$$

For the first term we can write

$$w^+ = (w^-)^* = \frac{e}{\varepsilon R^2}\frac{e^{i\alpha}}{2i}\hat{\mathbf{n}} \cdot \mathbf{D}_0.$$

This describes emission, i.e., the case of interest.

According to the Fermi golden rule scheme, the interband transition probability is proportional to

$$w(\mathbf{k}_1, \mathbf{k}_2) = \frac{2\pi}{\hbar}\left|\langle\mathbf{k}_1|w^+|\mathbf{k}_2\rangle\right|^2 \delta(E_p(\mathbf{k}_1) - E_n(\mathbf{k}_2) + \hbar\omega), \quad (10.33)$$

where the Bloch states in the conduction and valence bands are assumed to be plane waves for simplicity,

$$\Psi_\mathbf{k} = \frac{1}{(2\pi)^{3/2}}e^{i\mathbf{k}\cdot\mathbf{R} - iE_{n(p)}(\mathbf{k})t/\hbar},$$

$$E_p(\mathbf{k}) = -\frac{\hbar^2 k^2}{2m_p^*} - E_g, \qquad E_n(\mathbf{k}) = \frac{\hbar^2 k^2}{2m_n^*}$$

(the indices n, p refer to electrons from the conduction and valence bands, respectively; E_g is the forbidden gap).

The matrix element

$$\langle\mathbf{k}_1|w^+|\mathbf{k}_2\rangle = \frac{1}{(2\pi)^3}\int d^3R \frac{e}{\varepsilon 2i}e^{i\alpha}\hat{\mathbf{n}} \cdot \mathbf{D}_0 \frac{1}{R^2}e^{-i(\mathbf{k}_1 - \mathbf{k}_2)\cdot\mathbf{R}} \quad (10.34)$$

can be found analytically by direct integration, which gives the formula (setting $\mathbf{q} = \mathbf{k}_1 - \mathbf{k}_2$),

$$\langle\mathbf{k}_1|w^+|\mathbf{k}_2\rangle = \frac{-1}{(2\pi)^3}\frac{ee^{i\alpha}}{\varepsilon}D_0 \cos\Theta (2\pi) \int_a^\infty dR \frac{1}{q}\frac{d}{dR}\frac{\sin qR}{qR}$$
$$= \frac{1}{(2\pi)^2}\frac{ee^{i\alpha}}{\varepsilon}\frac{\mathbf{D}_0 \cdot \mathbf{q}}{q^2}\frac{\sin qa}{qa}. \quad (10.35)$$

Next we must sum over all initial and final states in both bands. Thus, for the total interband transition probability we have

$$\delta w = \int d^3k_1 \int d^3k_2 \left[f_1(1-f_2)w(\mathbf{k}_1,\mathbf{k}_2) - f_2(1-f_1)w(\mathbf{k}_2,\mathbf{k}_1)\right], \quad (10.36)$$

where f_1, f_2 are the temperature-dependent Fermi–Dirac distribution functions for the initial and final states, respectively. At room temperature, $f_2 \simeq 0$ and $f_1 \simeq 1$, which leads to

$$\delta w = \int d^3k_1 \int d^3k_2 w(\mathbf{k}_1,\mathbf{k}_2). \quad (10.37)$$

After some also analytical integration in the above formula, we arrive at the expression,

$$\delta w = \frac{4}{3} \frac{\mu^2 (m_n^* + m_p^*) 2(\hbar\omega - E_g) e^2 D_0^2}{\sqrt{m_n^* m_p^*} 2\pi \hbar^5 \varepsilon^2} \int_0^1 dx \frac{\sin^2(xa\xi)}{(xa\xi)^2} \sqrt{1-x^2}$$

$$= \frac{4}{3} \frac{\mu^2}{\sqrt{m_n^* m_p^*}} \frac{e^2 D_0^2}{2\pi \hbar^3 \varepsilon^2} \xi^2 \int_0^1 dx \frac{\sin^2(xa\xi)}{(xa\xi)^2} \sqrt{1-x^2}, \quad (10.38)$$

according to assumed band dispersions; m_n^* and m_p^* denote the effective masses of electrons and holes, $\mu = m_n^* m_p^*/(m_n^* + m_p^*)$ is the reduced mass, and the parameter ξ is given by $\sqrt{2(\hbar\omega - E_g)(m_n^* + m_p^*)}/\hbar$. In the limiting cases, for a nanoparticle of radius a we finally obtain

$$\delta w = \begin{cases} \dfrac{4 \mu \sqrt{m_n^* m_p^*} (\hbar\omega - E_g) e^2 D_0^2}{3} & \text{for } a\xi \ll 1, \\[2pt] \dfrac{4 \mu^{3/2} \sqrt{2} \sqrt{\hbar\omega - E_g} e^2 D_0^2}{3 \quad a\hbar^4 \varepsilon^2} & \text{for } a\xi \gg 1. \end{cases} \quad (10.39)$$

In the latter case the following approximation has been applied:

$$\int_0^1 dx \frac{\sin^2(xa\xi)}{(xa\xi)^2} \sqrt{1-x^2} \approx \text{(for } a\xi \gg 1\text{)} \frac{1}{a\xi} \int_0^\infty d(xa\xi) \frac{\sin^2(xa\xi)}{(xa\xi)^2} = \frac{\pi}{2a\xi},$$

whereas in the former case we have $\int_0^1 dx \sqrt{1-x^2} = \pi/4$.

With regard to these two limiting cases, $a\xi \ll 1$ and $a\xi \gg 1$, where $\xi = \sqrt{2(\hbar\omega - E_g)(m_n^* + m_p^*)}/\hbar$, we have already noted in previous chapters that for larger nanospheres, e.g., with $a > 10$ nm, the second regime holds.

Assuming that the energy acquired by the semiconductor band system, $\mathcal{A}$, is equal to the plasmon oscillation energy output (corresponding to plasmon damping), one can estimate the corresponding damping rate of plasmon oscillations. Namely, for a plasmon amplitude decaying in time according to $D_0(t) = D_0 e^{-t/\tau'}$, one finds, for the total transmitted energy,

$$\mathcal{A} = \beta \int_0^\infty \delta w \, \hbar\omega \, dt = \beta \hbar\omega \, \delta w \, \tau'/2$$

$$= \begin{cases} \dfrac{2 \beta \omega \tau' \mu \sqrt{m_n^* m_p^*} (\hbar\omega - E_g) e^2 D_0^2}{3 \quad \hbar^4 \varepsilon^2} & \text{for } a\xi \ll 1, \\[2pt] \dfrac{2 \beta \omega \tau' \mu^{3/2} \sqrt{2} \sqrt{\hbar\omega - E_g} e^2 D_0^2}{3 \quad a\hbar^3 \varepsilon^2} & \text{for } a\xi \gg 1, \end{cases}$$

$$(10.40)$$

where τ' is the damping time rate and β accounts for losses not included in the model, especially those that reduce the energy transfer for the case of a realistic deposition type on the top of the semiconductor layer instead of the fully embedded case. Comparing the value of $\mathcal{A}$ given by Eq. (10.40) with the plasmon energy loss by damping estimated in [23] (the initial energy of the plasmon oscillations which has been transferred step by step to the semiconductor, $\mathcal{A} = D_0^2/(2\varepsilon a^3)$), one finds that

$$\frac{1}{\tau'} = \begin{cases} \dfrac{4\beta\omega\mu\sqrt{m_n^* m_p^*}(\hbar\omega - E_g)e^2 a^3}{3\hbar^4 \varepsilon} & \text{for } a\xi \ll 1, \\ \dfrac{4\beta\omega\mu^{3/2}\sqrt{2}\sqrt{\hbar\omega - E_g}e^2 a^2}{3\hbar^3 \varepsilon} & \text{for } a\xi \gg 1. \end{cases} \quad (10.41)$$

By τ' we denote here the large damping of plasmons due to energy transfer to the semiconductor, which greatly exceeds the internal damping, characterized by τ_0, due to the scattering of electrons inside the metallic nanoparticle [23] ($1/\tau_0 \ll 1/\tau'$).

For example, for nanospheres of Au deposited on an Si layer we obtain for the Mie self-frequency $\omega = \omega_1$,

$$\frac{1}{\tau'(\omega_1)} = \begin{cases} 44.092\beta\, (a^3)^3 \dfrac{\mu}{m} \dfrac{\sqrt{m_n^* m_p^*}}{m}, & \text{for } a\xi \ll 1, \\ 13.648\beta\, (a^2)^2 \left(\dfrac{\mu}{m}\right)^{3/2} & \text{for } a\xi \gg 1, \end{cases} \quad (10.42)$$

where a is in nanometres. For light (heavy) carriers in Si, $m_n^* = 0.19m$ ($0.98m$), $m_p^* = 0.16m$ ($0.52m$), m is the bare electron mass, $\mu = m_n^* m_p^*/(m_n^* + m_p^*)$, $E_g = 1.14$ eV, and $\hbar\omega_1 = 2.72$ eV. For these parameters and for nanospheres with radius a in the range 5–50 nm, the lower case of Eq. (10.42) applies (at $\omega = \omega_1$). The parameter β fitted from the experimental data [33, 23] equals about 0.001.

10.5 Plasmon–Polariton Dynamics in a Metallic Chain Deposited on a Semiconductor

For a metallic nano-chain deposited on (or embedded in) a semiconductor medium, the plasmon oscillations at each nanoparticle of the chain are additionally damped owing to energy transfer to the semiconductor band system as described above. Thus Eq. (10.19) should be generalized to the form:

$$\left(-\omega^2 - i\left(\frac{2}{\tau_0} + \frac{2}{\tau'}\right)\omega + \omega_1^2\right) D_\alpha(k,\omega)$$
$$= \omega_1^2 \frac{a^3}{d^3} F_\alpha(k,\omega) D_\alpha(k,\omega) + \varepsilon a^3 \omega_1^2 E_{0\alpha}(k,\omega), \quad (10.43)$$

where τ' is given by Eq. (10.41). Equations (10.20) still hold however.

The damping rate in the case of a chain completely embedded in the semiconductor medium does not depend on the polarization of the plasmon oscillations; let us consider first this simple case. Using a perturbation scheme for the solution of

Eq. (10.43) we can consider the real part of the right-hand side of this equation as a correction to the frequency of the plasmon–polariton and the imaginary part as a correction to the damping rate. In an explicit form one thus deals with a perturbative solution,

$$\omega(k) = \sqrt{\omega_1^2(1 - \frac{a^3}{d^3}\text{Re } F_\alpha(k,\omega_1)) - \left(\frac{1}{\tau_0} + \frac{1}{\tau'(\omega_1)} - \frac{a^3}{2d^3}\text{Im } F_\alpha(k,\omega_1)\right)^2},$$

$$\frac{1}{\tau} = \frac{1}{\tau_0} + \frac{1}{\tau'(\omega_1)} - \frac{a^3}{2d^3}\text{Im } F_\alpha(k,\omega_1),$$

(10.44)

where $\tau'(\omega_1)$ is given by Eq. (10.42). The exact solution of Eq. (10.43) is presented in Figs. 10.2–10.5 and exhibits rather minor corrections to the perturbative solution, though the singular points need an exact solution to avoid the singularity misbehaviour of the perturbative solution.

Thus in Figs. 10.2–10.5 we present the numerical solution of Eq. (10.43) for the full domain of k and for the real and imaginary parts of $\omega(k)$ which correspond to the perturbation solution given by Eq. (10.44). The numerical solution of the nonlinear equation (10.43) was obtained by a Newton-type high-accuracy method applied at 1000 points of the domain $kd \in [0, 2\pi]$ point by point (the derivative of Re $\omega(k)$ with respect to k, needed in order to find the group velocity, was taken from the interpolated Re $\omega(k)$ function).

Because Im $F_\alpha(k,\omega) = 0$ in the majority of the domain for k, one can expect that the damping of the plasmon–polariton k-mode will go via the channels τ_0 and $\tau'(\omega_1)$ only and not via the Lorentz friction, which is perfectly balanced by the radiation income of energy from other nanoparticles in the chain. The comparison of the contributions of both active channels for plasmon–polariton damping is presented in Figs. 10.2–10.5 for different geometries and size parameters of the chain and for longitudinal and transverse polarizations of the plasmon–polariton oscillations with respect to propagation direction. For the nonhomogeneous equation (10.43) the stationary solution for the k-mode corresponds to the usual damped driven oscillations, when the energy of the driving incident light is transferred to the semiconductor substrate in part whereas the rest is dissipated in Joule heat in the metal material of the chain. The energy transferred to the semiconductor may be transformed eventually into photocurrent, re-emitted as light via exciton recombination, and also partly converted into Joule heat via the scattering of band electrons with phonons, defects, and admixtures (impurities) in the semiconductor medium.

If we are dealing with two channels of plasmon damping, the energy dissipation per time unit for these channels is equal to

$$\int_0^T \frac{dt}{T} \frac{2}{\tau_0} \left(\frac{\partial D_\alpha}{\partial t}\right)^2 \quad \text{and} \quad \int_0^T \frac{dt}{T} \frac{2}{\tau'} \left(\frac{\partial D_\alpha}{\partial t}\right)^2,$$

respectively. After averaging over the period of the stationary solution for the driven and damped oscillator, we arrive at the energy loss per time unit, $(1/\tau_0)\omega^2 A^2$ and $(1/\tau')\omega^2 A^2$, respectively, where ω is the frequency of the driving force and

$$A = \frac{\varepsilon a^3 \omega_1^2 E_0}{\sqrt{(\omega_1^2 - \omega^2)^2 + 4\omega^2(1/\tau_0 + 1/\tau')^2}}.$$

This outflow of energy is compensated during each period T of the driving force by the power of this force averaged over the period. Therefore the ratio of energy losses via the two channels equals $(1/\tau_0)/(1/\tau')$. Because the energy transferred to the semiconductor surroundings (the substrate) may not be converted into Joule heat or radiation, some deficit in energy balance might occur, not exceeding, however, the ratio indicated above. When this ratio is of order $1/3$ (for a realistic value of the parameter β; see Figs. 10.2–10.5) one could observe, e.g., an ostensible shortage in the energy balance at the level of about 75% of the incoming energy.

We have focused attention on the energy losses of plasmon–polaritons caused by the coupling of collective plasmons with another charged system in the near-field zone. The coupling of dipole plasmon modes with closely located band electrons in a semiconductor opens a very quick, and thus effective, channel for energy transfer, which results in the strong damping of plasmon–polaritons in metallic nano-chains deposited on or embedded in semiconductor surroundings. It should be emphasized that this channel of plasmon–polariton energy leakage is not of a radiative type because it happens on a subdiffraction scale with regard to the wavelength of the plasmon resonance. Depending on the nanosphere radii in the chain, different regimes for near-field coupling with the semicoductor substrate occur. For ultrasmall nanoparticles with radii 2–3 nm the dipole coupling with the substrate electrons breaks the translation invariance and lifts the momentum conservation constraints imposed on interband electron transitions in the semiconductor. This greatly enhances the indirect (non-momentum-conserving) interband transition probability, resulting in the strong damping of plasmon–polaritons in the nano-chain. With increasing nanoparticle size this effect gradually weakens but at about 5 nm nanoparticle radius (for Au or Ag) the interband transition probability again grows owing to the increase in the dipole amplitude in proportion to the number of oscillating electrons. Again plasmon–polaritons are strongly damped at this nanoparticle size scale. The damping of plasmon–polaritons due to near-field coupling with the semiconductor surroundings typically exceeds by a factor 3 the Joule heat losses due to electron scattering in metallic nanoparticles, which means that, relatively easily, the latter energy dissipation can be balanced by inverted energy transfer from the surrounding or substrate semiconductor to the metallic nano-chain, resulting in arbitrary long-range plasmon–polariton propagation in such a system.

11

Plasmons in Finite Spherical Ionic Systems

In this chapter the topic of soft plasmonics in finite electrolyte liquid systems bounded by insulating membranes is considered. Such systems can be encountered at the bio-cell organization level, taking into account that the characteristics of ion plasmons fall into the micrometre scale rather than the nanometre scale in metals because the masses of ions are at least three orders of magnitude larger than the mass of an electron. The lower density of ions in electrolytes compared with the density of electrons in metals may reduce the energy of plasmons by several orders of magnitude. Here a fully analytical description of surface and volume plasmons in a finite ionic microsystem is provided, allowing further applications.

In view of the plasmons and plasmon–polaritons found in metallic nanostructures, one can question whether there are similar charge fluctuations in other electrical systems such as electrolytes. A challenging problem that we will discuss here is the possibility of the occurrence of plasmon effects for ionic carriers similar to those for electrons. Many finite ionic systems in the form of an electrolyte closed by membranes are encountered in biological structures and questions arise of to what degree would plasmonic phenomena play a role in them and whether the radiative properties of plasmon fluctuations would also be as significant in ionic systems as they are in metals. It is reasonable that ionic plasmon effects would be found in other regions of energy and wavelength scale compared with those of metallic systems. One could expect that the energy of ionic plasmons would be much lower than that of electrons in metals and that the typical size of plasmonic finite systems would be of micrometre scale. Such 'soft' plasmonics would be linked with the functionality of biological systems, where the electricity is of ionic rather than electronic character. Cell signalling, membrane transfer, and nerve cell conductivity would serve as examples.

For a theoretical model for ions we will adopt, as far as possible, an analogy to metallic nanospheres with plasmon excitations. Ionic systems are much more complicated than a metal crystal structure with free electrons. Therefore, the

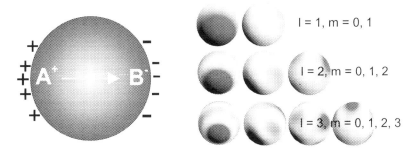

Figure 11.1 Dipole **D**(*t*) creation in a single sphere by the simplest surface plasmon oscillations, due to the symmetrical separation of ions A^+ and B^- (left); more complicated patterns of oscillation of surface plasmon charge distributions can be assigned, in the case of spherical symmetry, with various multipole numbers, *l*, *m*. The different grey-scale intensities indicate distinct values of local charge density from negative to positive (right) (given in colour in [48]).

identification of an appropriate simplification of the approach to ionic systems is of primary significance. The model must be, however, efficient in repeating for ions in electrolytes the plasmon scenario known from metal electrons.

Let us consider a finite spherical ionic system (e.g., a liquid electrolyte artificially enclosed with a membrane) and identify the plasmon excitations of ions in this system. We will determine their energies for various parameters of the ionic system with special attention paid to the radiation properties of the ionic plasmons.

We will analyse ionic plasmon–polariton propagation in electrolyte-sphere chains with a prospective relationship to signalling in biological systems. Bearing in mind that metallic nano-chains serve as very efficient wave guides for electromagnetic signals in the form of collective surface plasmon excitations of wave type called plasmon–polaritons, we will try to model similar phenomena in ionic-sphere chains.

For an initial model we will study a spherical or prolate spheroidal ionic conducting system with balanced charges, in analogy to the jellium model in a metal. In this model the positive-charge background is static and local fluctuations in electron liquid density, negative or positive against the background of the positive jellium, form plasmons in the nanoparticles [23]. In the case of an electrolyte we are dealing, however, with a different situation – in a binary electrolyte both the oppositely charged ions are similarly dynamical as they have similar masses; here there is no rigid jellium.

11.1 Fluctuations of the Charge Density in a Spherical Electrolyte System

The problem of how to establish an adequate model for a multi-ionic system in order to grasp the essential properties of ionic plasmons is challenging. For a simple two-component ionic system (a binary electrolyte) we are dealing with an electrolyte with balanced negative and positive total charges. In equilibrium the charge cancellation is also local, corresponding to a uniform charge distribution of both signs. As there are two kinds of carrier, both carriers form density fluctuations, resulting in the violation of local electric equilibrium. The total compensation of both charge signs requires, however, that any density fluctuation in the negative charge must be accompanied by an in general distant, but ideally equivalent, positive ion fluctuation, and conversely. This means that effectively we have charge density fluctuations of both ion subsystems with respect to an equilibrium homogeneous neutral state. One can thus consider the local effective fluctuations of, let us say, the negative ions as positive or negative in charge value with respect to the uniform charge distribution assumed as ideally cancelled by a positive uniform background – a fictitious jellium in analogy to the case of a metal. In this way we can model a two-component ionic system by two single-component ion systems with double jelliums of opposite sign. These opposite-sign fictitious jelliums cancel each other. However, the oppositely charged ion fluctuations will mutually interact. For simplicity we will assume that the absolute values of the charges and the masses of the oppositely charged ions are the same, but the generalization to different ion characteristics is straightforward.

To simplify the model along the lines described above let us consider a spherical system with radius a with balanced total charge consisting of ions with both sign and with uniform equilibrium density distributions $n^{+(-)}(\mathbf{r}) = n\Theta(a-r)$ ($\Theta(r)$ is the Heaviside step function and a is the sphere radius). We assume, for simplicity, the same absolute value of the charge of each plus- or minus-charged ion. By $m^{+(-)}$ we denote the mass of a positive (negative) ion. Both masses are of the order of $10^{4-5}m_e$, where $m_e = 9.1 \times 10^{-31}$ kg is the mass of the electron. The reduction of the two-component system to two systems each with a fictitious jellium is an approximation, but it may serve for a tentative recognition of the ionic dynamics of the system and an assessment of the scales of its quantitative characteristics, at least. The advantage of such an approach is its close analogy to the description of plasmons in metals where the rigid shape of the system is imposed by the shape of a 'real' jellium – the crystal lattice.

Given the above-mentioned model assumptions we will consider ionic carriers with density oscillating around the zero-valued equilibrium density and balanced by a fictitious charged background of opposite-sign ions (the effect of the simultaneous presence of the opposite-sign ions). Thus the electrolyte is regarded as the sum of two subsystems corresponding to ions of each sign and each with own fictitious jellium. These jelliums mutually cancel. The uniform charge distributions of the ions of each sign also mutually cancel and an approximation is the assumption of the independence of the fluctuations in the two subsystems. In this way, an approximate correspondence with the electron liquid in a metal within the jellium model is established. Hence, the description of fluctuations in the local density of electrons in a metallic nanosphere can be directly used to model fluctuations in the effective ion density, substituting the electron mass by the ion mass and the electronic charge by the ion charge. The dynamical equation for the charged fluid in the spherical ion system can be thus repeated from the case of the metal nanosphere with electrons. The equilibrium density of the effective charged liquid, denoted by n, will be treated as a parameter and will be assumed equal to ηN_0, where η is the molarity of the electrolyte in the sphere and N_0 is the molar electrolyte concentration of ions. The equilibrium density determines the bulk plasmon frequency for the ion system according to the formula, $\omega_p^2 = 4\pi n q^2/m$, where m is the mass of an ion with charge q. As m is much larger than m_e and the concentration of the ions is usually smaller than that for electrons in metals, ω_p can be considerably reduced, even by several orders of magnitude. Note that for electrons in metals $\hbar\omega_p \simeq 10$ eV and ω_p typically falls into the ultraviolet region with corresponding photon energy. In an ionic system the plasmon frequency ω_p can be much lower and is found in the infrared or even lower-energy part of the electromagnetic spectrum.

11.2 Model Definition

The Hamiltonian for a binary ion system has the form

$$\hat{H}_{ion} = -\sum_{i=1}^{N^-}\frac{\hbar^2 \nabla_i^2}{2m^-} - \sum_{j=1}^{N^+}\frac{\hbar^2 \nabla_j^2}{2m^+} - \sum_{i,j}^{N^-,N^+}\frac{q^- q^+}{\varepsilon|\mathbf{r}_i - \mathbf{r}_j|}$$

$$+ \frac{1}{2}\sum_{i,i',i\neq i'}^{N^-}\frac{(q^-)^2}{\varepsilon|\mathbf{r}_i - \mathbf{r}_{i'}|} + \frac{1}{2}\sum_{j,j',j\neq j'}^{N^+}\frac{(q^+)^2}{\varepsilon|\mathbf{r}_j - \mathbf{r}_{j'}|}, \quad (11.1)$$

where $q^{-(+)}$, $M^{-(+)}$, and $N^{-(+)}$ are the charges, masses, and number of $-(+)$ ions, respectively. To analyse this complicated system we propose the following approximation, assuming, for simplicity, $q^- = -q^+ = q$, $N^- = N^+ = N$, $m^- = m^+ = m$. We next add and subtract the same terms, as written below:

$$\hat{H}_{ion} = -\sum_{i=1}^{N}\frac{\hbar^2 \nabla_i^2}{2m} - \sum_{j=1}^{N}\frac{\hbar^2 \nabla_j^2}{2m} - \sum_{i,j}\frac{q^2}{\varepsilon|\mathbf{r}_i - \mathbf{r}_j|}$$

$$+ \frac{1}{2}\sum_{i,i',i\neq i'}^{N}\frac{q^2}{\varepsilon|\mathbf{r}_i - \mathbf{r}_{i'}|} + \frac{1}{2}\sum_{j,j',j\neq j'}^{N}\frac{q^2}{\varepsilon|\mathbf{r}_j - \mathbf{r}_{j'}|}$$

$$-q^2 \sum_j \int \frac{n(\mathbf{r})d^3\mathbf{r}}{\varepsilon|\mathbf{r}_j - \mathbf{r}|} - q^2 \sum_i \int \frac{n(\mathbf{r})d^3\mathbf{r}}{\varepsilon|\mathbf{r}_i - \mathbf{r}|}$$

$$+q^2 \sum_j \int \frac{n(\mathbf{r})d^3\mathbf{r}}{\varepsilon|\mathbf{r}_j - \mathbf{r}|} + q^2 \sum_i \int \frac{n(\mathbf{r})d^3\mathbf{r}}{\varepsilon|\mathbf{r}_i - \mathbf{r}|}, \tag{11.2}$$

where we have introduced formally the fictitious jellium of spherical shape for both types of ions, with density n ideally compensating the opposite charges of the uniformly distributed ions, so that $n(\mathbf{r}) = n\Theta(a - r)$ where a is the sphere radius (the positive and negative jelliums mutually cancel). Assuming that

$$q^2 \sum_j \int \frac{n(\mathbf{r})d^3\mathbf{r}}{\varepsilon|\mathbf{r}_j - \mathbf{r}|} + q^2 \sum_i \int \frac{n(\mathbf{r})d^3\mathbf{r}}{\varepsilon|\mathbf{r}_i - \mathbf{r}|} - \sum_{i,j} \frac{q^2}{\varepsilon|\mathbf{r}_i - \mathbf{r}_j|} \simeq 0, \tag{11.3}$$

which is fulfilled for ion fluctuations that are not too far beyond the uniform distribution (in dilute ion systems), we can separate the original Hamiltonian into a sum, $\hat{H}_{ions} = \hat{H}^- + \hat{H}^+$, where

$$\hat{H}^{-(+)} = \sum_j \left[-\frac{\hbar^2 \nabla_j^2}{2m} - q^2 \int \frac{n(\mathbf{r})d^3\mathbf{r}}{\varepsilon|\mathbf{r}_j - \mathbf{r}|} \right] + \frac{1}{2} \sum_{j \neq j'} \frac{q^2}{\varepsilon|\mathbf{r}_j - \mathbf{r}_{j'}|}. \tag{11.4}$$

The last term on the right-hand side of Eq. (11.4) corresponds to the interaction between ions of the same sign, whereas the second term in the first sum describes the interaction of these ions with the jellium (of opposite sign); ε is the dielectric permittivity of the electrolyte medium. The confined electrolyte system can be shaped by an appropriately formed membrane, as frequently happens in biological systems. Because of the separation of the Hamiltonian (11.1) one can consider the single Hamiltonians (11.4). We will denote by $\Psi_{ion}(t)$ the ion wave function corresponding to the Hamiltonian (11.4).

The form of (11.4) allows for further discussion according to the scheme applied to electrons in metals [23], which we recall below for the sake of completeness. The local density of the chosen type of ion can be written, in analogy to the semiclassical Pines–Bohm RPA approach to electrons in a metal [16, 14], in the following form:

$$\rho(\mathbf{r}, t) = \left\langle \Psi_{ion}(t) \left| \sum_j \delta(\mathbf{r} - \mathbf{r}_j) \right| \Psi_{ion}(t) \right\rangle, \tag{11.5}$$

where $\mathbf{r}_j$ denotes the coordinate of the jth ion. The Fourier picture of the above density has the form,

$$\tilde{\rho}(\mathbf{k}, t) = \int \rho(\mathbf{r}, t) e^{-i\mathbf{k}\cdot\mathbf{r}} d^3 r = \langle \Psi_{ion}(t) | \hat{\rho}(\mathbf{k}) | \Psi_{ion}(t) \rangle, \tag{11.6}$$

where the operator $\hat{\rho}(\mathbf{k}) = \sum_j e^{-i\mathbf{k}\cdot\mathbf{r}_j}$.

11.2 Model Definition

Using the above notation one can rewrite $\hat{H}_{ion}$, (11.1), in the following form, in analogy to the metallic plasmon description presented in previous chapters [16, 23]:

$$\hat{H}_{ion} = \sum_{j=1}^{N}\left[-\frac{\hbar^2 \nabla_j^2}{2m}\right] - \frac{q'^2}{(2\pi)^3}\int d^3k\,\tilde{n}(\mathbf{k})\frac{2\pi}{k^2}\left(\hat{\rho}^+(\mathbf{k}) + \hat{\rho}(\mathbf{k})\right)$$

$$+ \frac{q'^2}{(2\pi)^3}\int d^3k\frac{2\pi}{k^2}\left[\hat{\rho}^+(\mathbf{k})\hat{\rho}(\mathbf{k}) - N\right], \qquad (11.7)$$

where $\tilde{n}(\mathbf{k}) = \int d^3r\, n(\mathbf{r})e^{-i\mathbf{k}\cdot\mathbf{r}}$ (we have taken into account that $4\pi/k^2 = \int d^3r e^{-i\mathbf{k}\cdot\mathbf{r}}/r$), $q'^2 = q^2/\varepsilon$.

Utilizing this form of the effective-ion Hamiltonian one can write down the dynamic equation for the ion density fluctuations in the Heisenberg representation,

$$\frac{d^2\hat{\rho}(\mathbf{k})}{dt^2} = \frac{1}{(i\hbar)^2}\left[\left[\hat{\rho}(\mathbf{k}), \hat{H}_{ion}\right], \hat{H}_{ion}\right], \qquad (11.8)$$

in the following form:

$$\frac{d^2\hat{\rho}(\mathbf{k})}{dt^2} = -\sum_{j} e^{-i\mathbf{k}\cdot\mathbf{r}_j}\left\{-\frac{\hbar^2}{m^2}(\mathbf{k}\cdot\nabla_j)^2 + \frac{\hbar^2 k^2}{m^2}i\mathbf{k}\cdot\nabla_j + \frac{\hbar^2 k^4}{4m^2}\right\}$$

$$-\frac{4\pi q'^2}{m(2\pi)^3}\int d^3p\,\tilde{n}(\mathbf{p})\frac{\mathbf{k}\cdot\mathbf{p}}{p^2}\hat{\rho}(\mathbf{k}-\mathbf{p})$$

$$-\frac{4\pi q'^2}{m(2\pi)^3}\int d^3p\,\hat{\rho}(\mathbf{k}-\mathbf{p})\frac{\mathbf{k}\cdot\mathbf{p}}{p^2}\hat{\rho}(\mathbf{p}). \qquad (11.9)$$

If one takes into account that

$$\hat{\rho}(\mathbf{k}-\mathbf{p})\hat{\rho}(\mathbf{p}) = \delta\hat{\rho}(\mathbf{k}-\mathbf{p})\delta\hat{\rho}(\mathbf{p}) + \tilde{n}(\mathbf{k}-\mathbf{p})\delta\hat{\rho}(\mathbf{p})$$
$$+ \delta\hat{\rho}(\mathbf{k}-\mathbf{p})\tilde{n}(\mathbf{p}) + \tilde{n}(\mathbf{k}-\mathbf{p})\tilde{n}(\mathbf{p})$$

and $\tilde{n}(\mathbf{p})\hat{\rho}(\mathbf{k}-\mathbf{p}) = \tilde{n}(\mathbf{p})\delta\hat{\rho}(\mathbf{k}-\mathbf{p}) + \tilde{n}(\mathbf{p})\tilde{n}(\mathbf{k}-\mathbf{p})$, where $\delta\hat{\rho}(\mathbf{k}) = \hat{\rho}(\mathbf{k}) - \tilde{n}(\mathbf{k})$ describes the operator for local ion density fluctuations above the uniform distribution, one can rewrite Eq. (11.9) as follows,

$$\frac{d^2\delta\hat{\rho}(\mathbf{k})}{dt^2} = -\sum_{j} e^{-i\mathbf{k}\cdot\mathbf{r}_j}\left\{-\frac{\hbar^2}{m^2}(\mathbf{k}\cdot\nabla_j)^2 + \frac{\hbar^2 k^2}{m^2}i\mathbf{k}\cdot\nabla_j + \frac{\hbar^2 k^4}{4m^2}\right\}$$

$$-\frac{4\pi q'^2}{m(2\pi)^3}\int d^3p\,\tilde{n}(\mathbf{k}-\mathbf{p})\frac{\mathbf{k}\cdot\mathbf{p}}{p^2}\delta\hat{\rho}(\mathbf{p})$$

$$-\frac{4\pi q'^2}{m(2\pi)^3}\int d^3p\,\delta\hat{\rho}(\mathbf{k}-\mathbf{p})\frac{\mathbf{k}\cdot\mathbf{p}}{p^2}\delta\hat{\rho}(\mathbf{p}). \qquad (11.10)$$

Hence, for the ion density fluctuation, $\delta\tilde{\rho}(\mathbf{k},t) = \langle\Psi_{ion}|\delta\hat{\rho}(\mathbf{k},t)|\Psi_{ion}\rangle = \tilde{\rho}(\mathbf{k},t) - \tilde{n}(\mathbf{k})$, one finds

$$\frac{\partial^2 \delta\tilde{\rho}(\mathbf{k},t)}{\partial t^2} = -\left\langle \Psi_{ion} \left| \sum_j e^{-i\mathbf{k}\cdot\mathbf{r}_j} \left\{ -\frac{\hbar^2}{m^2}(\mathbf{k}\cdot\nabla_j)^2 + \frac{\hbar^2 k^2}{m^2} i\mathbf{k}\cdot\nabla_j + \frac{\hbar^2 k^4}{4m^2} \right\} \right| \Psi_{ion} \right\rangle$$

$$- \frac{4\pi q'^2}{m(2\pi)^3} \int d^3 p\, \tilde{n}(\mathbf{k}-\mathbf{p}) \frac{\mathbf{k}\cdot\mathbf{p}}{p^2} \delta\tilde{\rho}(\mathbf{p},t)$$

$$- \frac{4\pi q'^2}{m(2\pi)^3} \int d^3 p\, \frac{\mathbf{k}\cdot\mathbf{p}}{p^2} \langle \Psi_{ion}|\delta\hat{\rho}(\mathbf{k}-\mathbf{p})\delta\hat{\rho}(\mathbf{p})|\Psi_{ion}\rangle. \tag{11.11}$$

For small k, in analogy to the semiclassical approximation for electrons [23, 16], the contributions of the second and third components to the first term on the right-hand side of Eq. (11.11) can be neglected as being small in comparison with the first component. The third term on the right-hand side of Eq. (11.11) is also negligible as it involves a product of two $\delta\tilde{\rho}$ (which we assume to be small, $\delta\tilde{\rho}/n \ll 1$). This approach corresponds to the RPA formulated for the bulk metal [16, 14] (note that $\delta\hat{\rho}(0) = 0$ and the coherent RPA contribution of the interaction comprises the last but one term in Eq. (11.11)).

Within the RPA, Eq. (11.11) thus attains the following shape:

$$\frac{\partial^2 \delta\tilde{\rho}(\mathbf{k},t)}{\partial t^2} = \frac{2k^2}{3m} \left\langle \Psi_{ion} \left| \sum_j e^{-i\mathbf{k}\cdot\mathbf{r}_j} \frac{\hbar^2 \nabla_j^2}{2m} \right| \Psi_{vion} \right\rangle \tag{11.12}$$

$$- \frac{4\pi q'^2}{m(2\pi)^3} \int d^3 p\, \tilde{n}(\mathbf{k}-\mathbf{p}) \frac{\mathbf{k}\cdot\mathbf{p}}{p^2} \delta\tilde{\rho}(\mathbf{p},t).$$

Owing to the spherical symmetry,

$$\left\langle \Psi_{ion} \left| \sum_j e^{-i\mathbf{k}\cdot\mathbf{r}_j} \frac{\hbar^2}{m^2}(\mathbf{k}\cdot\nabla_j)^2 \right| \Psi_{ion} \right\rangle \simeq \frac{2k^2}{3m} \left\langle \Psi_{ion} \left| \sum_j e^{-i\mathbf{k}\cdot\mathbf{r}_j} \frac{\hbar^2 \nabla_j^2}{2m} \right| \Psi_{ion} \right\rangle.$$

Thus one can rewrite Eq. (11.12) in the position representation as

$$\frac{\partial^2 \delta\tilde{\rho}(\mathbf{r},t)}{\partial t^2} = -\frac{2}{3m}\nabla^2 \left\langle \Psi_{ion} \left| \sum_j \delta(\mathbf{r}-\mathbf{r}_j) \frac{\hbar^2 \nabla_j^2}{2m} \right| \Psi_{ion} \right\rangle$$

$$+ \frac{\omega_p^2}{4\pi}\nabla\left\{\Theta(a-r)\nabla\int d^3 r_1 \frac{1}{|\mathbf{r}-\mathbf{r}_1|}\delta\tilde{\rho}(\mathbf{r}_1,t)\right\}. \tag{11.13}$$

11.2 Model Definition

In the case of metals the Thomas–Fermi formula is then used to obtain the average kinetic energy [16]:

$$\left\langle \Psi_{ion} \left| -\sum_j \delta(\mathbf{r}-\mathbf{r}_j)\frac{\hbar^2 \nabla_j^2}{2m} \right| \Psi_{ion} \right\rangle \simeq \frac{3}{5}(3\pi^2)^{2/3}\frac{\hbar^2}{2m}(\rho(\mathbf{r},t))^{5/3} \quad (11.14)$$

$$= \frac{3}{5}(3\pi^2)^{2/3}\frac{\hbar^2}{2m}n^{5/3}\Theta(a-r)\left(1+\frac{5}{3}\frac{\delta\tilde{\rho}(\mathbf{r},t)}{n}+\cdots\right).$$

The Thomas–Fermi formula applies, however, to fermionic and degenerate quantum systems such as electrons in metals. For ionic systems, such an estimation of the kinetic energy is inappropriate because the ion concentration is usually much lower than that of electrons in metals and the system is not degenerate even if the ions are fermions; the Maxwell–Boltzmann distribution should be applied instead of the Fermi–Dirac or Bose–Einstein distributions. Independently of whether the statistics of the ions are fermionic or bosonic, the Maxwell–Boltzmann distribution allows the estimation of the average kinetic energy of the ions located inside the spherical electrolyte with the radius a in the following form:

$$\left\langle \Psi_{ion} \left| -\sum_j \delta(\mathbf{r}-\mathbf{r}_j)\frac{\hbar^2 \nabla_j^2}{2m} \right| \Psi_{ion} \right\rangle \simeq (n + \delta\rho(\mathbf{r},t))\Theta(a-r)\frac{3k_B T}{2}, \quad (11.15)$$

where k_B is the Boltzmann constant and T is the temperature. For ions of an extended three-dimensional or linear shape, inclusion of the rotational degrees of freedom results in a factor $6k_B T/2$ or $5k_B T/2$, respectively, instead of $3k_B T/2$ for the pointlike ion model.

Using Eq. (11.15) and taking into account that $\nabla\Theta(a-r) = -(\mathbf{r}/r)\delta(a-r)$, one can rewrite Eq. (11.13) in the following manner:

$$\frac{\partial^2 \delta\tilde{\rho}(\mathbf{r},t)}{\partial t^2} = \left[\frac{k_B T}{m}\nabla^2 \delta\tilde{\rho}(\mathbf{r},t) - \omega_p^2 \delta\tilde{\rho}(\mathbf{r},t)\right]\Theta(a-r)$$

$$- \frac{k_B T}{m}\nabla\left\{[n+\delta\tilde{\rho}(\mathbf{r},t)]\frac{\mathbf{r}}{r}\delta(a-r)\right\}$$

$$- \left[\frac{kT}{m}\frac{\mathbf{r}}{r}\nabla\delta\tilde{\rho}(\mathbf{r},t) + \frac{\omega_p^2}{4\pi}\frac{\mathbf{r}}{r}\nabla\int d^3r_1 \frac{1}{|\mathbf{r}-\mathbf{r}_1|}\delta\tilde{\rho}(\mathbf{r}_1,t)\right]\delta(a-r).$$

(11.16)

In the above formula ω_p is the bulk-ion plasmon frequency, $\omega_p^2 = 4\pi n q'^2/m$. The solution of Eq. (11.16) can be decomposed into two parts relating to distinct domains, inside the sphere and on its surface:

$$\delta\tilde{\rho}(\mathbf{r},t) = \begin{cases} \delta\tilde{\rho}_1(\mathbf{r},t) & \text{for } r < a, \\ \delta\tilde{\rho}_2(\mathbf{r},t) & \text{for } r \geq a \ (r \to a+) \end{cases} \quad (11.17)$$

The two parts of Eq. (11.17) correspond to volume and surface excitations, respectively. They satisfy equations derived from Eq. (11.16)):

$$\frac{\partial^2 \delta\tilde{\rho}_1(\mathbf{r},t)}{\partial t^2} = \frac{k_B T}{m}\nabla^2 \delta\tilde{\rho}_1(\mathbf{r},t) - \omega_p^2 \delta\tilde{\rho}_1(\mathbf{r},t) \qquad (11.18)$$

and (here $\epsilon = 0+$)

$$\frac{\partial^2 \delta\tilde{\rho}_2(\mathbf{r},t)}{\partial t^2} = -\frac{k_B T}{m}\nabla\left\{[n + \delta\tilde{\rho}_2(\mathbf{r},t)]\frac{\mathbf{r}}{r}\delta(a+\epsilon-r)\right\}$$

$$-\left[\frac{k_B T}{m}\frac{\mathbf{r}}{r}\nabla\delta\tilde{\rho}_2(\mathbf{r},t) + \frac{\omega_p^2}{4\pi}\frac{\mathbf{r}}{r}\nabla\int d^3 r_1 \frac{1}{|\mathbf{r}-\mathbf{r}_1|}(\delta\tilde{\rho}_1(\mathbf{r}_1,t)\Theta(a-r_1)\right.$$

$$\left.+\delta\tilde{\rho}_2(\mathbf{r}_1,t)\Theta(r_1-a))\right]\delta(a+\epsilon-r). \qquad (11.19)$$

The Dirac delta in Eq. (11.19) results from the derivative of the Heaviside step function describing the fictitious jellium distribution shape. In Eq. (11.19) an infinitesimal shift, $\epsilon = 0+$, is introduced to fulfil requirements of the Dirac delta definition (its singular point has to be an inner point of an open subset of the domain). This shift is only of a formal character and does not reflect any asymmetry. Some kind of asymmetry is, however, caused by the last term in Eq. (11.16), the gradient-type term, which describes the electric field induced by ion-density fluctuations. The electric field due to the surface charges is zero inside the sphere, and therefore cannot influence the volume excitations. Conversely, the volume-charge-fluctuation-induced electric field can excite surface fluctuations. This asymmetry is visible by comparison of Eqs. (11.18) and (11.19) for the volume and surface plasmons, respectively. The equation for the volume plasmons is completely independent of the surface plasmons, whereas the volume plasmons contribute to the equation for the surface plasmons.

The problem of separating the surface and volume plasmons has been thoroughly analysed for metal. For very small nanoparticles the effect of spill-out of electrons beyond the jellium edge, significant at this size scale, makes the surface fuzzy, resulting in coupling of the volume and surface plasmon oscillations. By direct numerical simulations using the time-dependent local density approximation (TDLDA) [19, 20], it has been proved that the coupling of the volume and surface excitations gradually disappears in larger nanoparticles (with the number of electrons exceeding about 60) [19, 20], which supports the accuracy of the semiclassical RPA description, within which volume plasmons can be separated from the surface plasmons (even though, as mentioned above, the latter can be excited by the former owing to the last term in Eq. (11.19)). This volume–surface coupling, important for ultrasmall-radii metallic nanoparticles, when the shell effects and quantum spill-out are significant, completely disappears, however, with growth in the sphere dimension to several nanometres in the case of metals. We consider here a binary electrolyte system with radius of order a micrometre, when quantum effects specific for ultrasmall clusters are negligible. This gives us the opportunity to formulate an analytical RPA semiclassical description in the form of an oscillator equation, allowing for the phenomenological inclusion of damping effects. The energy dissipation effects, though going beyond the RPA model, turn out to be the overwhelming physical property in the case of larger metallic nanospheres [23, 77] (with $a > 10$ nm for Au or Ag) and also for much larger ionic systems, as we will demonstrate below.

11.3 Solution of RPA Plasmon Equation: Volume and Surface Plasmon Frequencies in a Spherical Finite Ion System

Equations (11.18) and (11.19) have been solved for metallic nanospheres [23] and these solutions can be directly applied to ionic systems. To summarize briefly this analysis, the two parts of the plasma fluctuation can be represented as follows:

$$\delta\tilde{\rho}_1(\mathbf{r},t) = n\left[f_1(r) + F(\mathbf{r},t)\right] \qquad \text{for } r < a,$$
$$\delta\tilde{\rho}_2(\mathbf{r},t) = nf_2(r) + \sigma(\Omega,t)\delta(r+\epsilon-a), \; \epsilon = 0+, \qquad \text{for } r \geq a \; (r \to a+),$$
(11.20)

with initial conditions $F(\mathbf{r},t)|_{t=0} = 0$ and $\sigma(\Omega,t)|_{t=0} = 0$ (Ω represents the spherical coordinate angles) and also the continuity condition $(1+f_1(r))|_{r=a} = f_2(r)|_{r=a}$, the boundary condition $F(\mathbf{r},t)|_{r=a} = 0$, and the neutrality condition $\int \rho(\mathbf{r},t)d^3r = N$.

Neglecting the static contributions $f_{1,2}$ describing the spill-out effect, which decreases sharply with growth in the system size [23], one can write out the time-dependent parts of the ion concentration fluctuations in the form

$$F(\mathbf{r},t) = \sum_{l=1}^{\infty} \sum_{m=-l}^{l} \sum_{i=1}^{\infty} A_{lmn} j_l(k_{nl}r) Y_{lm}(\Omega) \sin(\omega_{li}t), \qquad (11.21)$$

and

$$\sigma(\Omega,t) = \sum_{l=1}^{\infty}\sum_{m=-l}^{l} \frac{B_{lm}}{a^2} Y_{lm}(\Omega)\sin(\omega_{0l}t)$$

$$+ \sum_{l=1}^{\infty}\sum_{m=-l}^{l}\sum_{i=1}^{\infty} A_{lmn}\frac{(l+1)\omega_p^2}{l\omega_p^2-(2l+1)\omega_{li}^2} Y_{lm}(\Omega)n_e \qquad (11.22)$$

$$\times \int_0^a dr_1 \frac{r_1^{l+2}}{a^{l+2}} j_l(k_{li}r_1)\sin(\omega_{li}t),$$

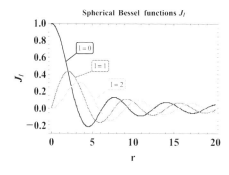 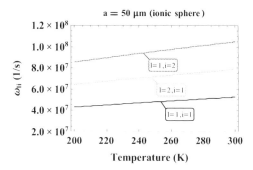

Figure 11.2 The spherical Bessel functions $J_l(r)$ for $l = 0, 1, 2$ displaying charge density fluctuations in a nanosphere along the sphere radius r (arbitrary units) for volume plasmon modes; the angular distribution for these modes is governed by the spherical harmonics $Y_{lm}(\Omega)$ in the same way as for the surface plasmon modes (see Fig. 11.1 (right)). The exemplary temperature dependence of the self-frequencies of the volume plasmon modes ω_{li}, li = 11, 12, 21, for a dilute electrolyte with $n \simeq 10^{14}/m^3$ and ion mass $\sim 10^4 m_e$, $a \sim 50$ μm (right).

where $j_l(\xi) = \sqrt{\pi/2\xi}\, I_{l+1/2}(\xi)$ is a spherical Bessel function, $Y_{lm}(\Omega)$ is a spherical harmonic, $\omega_{li} = \omega_p\sqrt{1 + k_B T x_{li}^2/(\omega_p^2 a^2 m)}$ are the frequencies of the ion volume self-oscillations (the volume plasmon frequencies), x_{li} are the nodes of the Bessel function $j_l(\xi)$ enumerated by $i = 1, 2, 3, \ldots$ (see Fig. 11.2), $k_{li} = x_{li}/a$, and $\omega_{l0} = \omega_p\sqrt{l/(2l+1)}$ are the frequencies of the ion surface self-oscillations (the surface plasmon frequencies). The derivation of the self-frequencies for ionic plasmon oscillations is presented in full detail in the final section of this chapter. The amplitudes A_{lmi} and B_{lm} are arbitrary in the homogeneous problem and are determined by the initial conditions.

The function $F(\mathbf{r}, t)$ describes volume plasmon oscillations, whereas $\sigma(\Omega, t)$ describes the surface plasmon oscillations. Let us emphasize that the first term in Eq. (11.22) corresponds to surface self-oscillations, while the second term describes the surface oscillations induced by the volume plasmons. The frequencies of the surface self-oscillations are equal to

$$\omega_{0l} = \omega_p \sqrt{\frac{l}{2l+1}}, \qquad (11.23)$$

which, for $l = 1$, is the dipole-type surface oscillation frequency, described for a metallic nanosphere by Mie [18], $\omega_{01} = \omega_p/\sqrt{3}$.

11.3.1 Ionic Surface Ion Plasmon Frequencies for an Electrolyte Nanosphere Embedded in a Dielectric Medium with $\varepsilon_1 > 1$

One can now include the influence of dielectric surroundings (in general, distinct from the inner medium of the ionic system under consideration) on the plasmons in

11.4 Damping of Plasmon Oscillations in Ionic Systems

this system. In order to do this let us assume that the ions on the surface ($r = a+$, i.e., $r \geq a$, $r \to a$) interact via Coulomb forces renormalized by the relative dielectric permittivity $\varepsilon_1 > 1$ (which is distinct from the dielectric permittivity ε for the inner medium). Thus a small modification of Eq. (11.19) is in order:

$$\frac{\partial^2 \delta\tilde{\rho}_2(\mathbf{r})}{\partial t^2} = -\frac{2}{3m} \nabla \left\{ \left[\frac{3}{5}\epsilon_F n + \epsilon_F \delta\tilde{\rho}_2(\mathbf{r},t) \right] \frac{\mathbf{r}}{r} \delta(a + \epsilon - r) \right\}$$

$$- \left[\frac{2\,\epsilon_F}{3}\frac{\mathbf{r}}{m\,r}\nabla\delta\tilde{\rho}_2(\mathbf{r},t) + \frac{\omega_p^2}{4\pi}\frac{\mathbf{r}}{r}\nabla \int d^3r_1 \frac{1}{|\mathbf{r}-\mathbf{r}_1|} \Big(\delta\tilde{\rho}_1(\mathbf{r}_1,t)\Theta(a-r_1) \right.$$

$$\left. + \frac{1}{\varepsilon_1}\delta\tilde{\rho}_2(\mathbf{r}_1,t)\Theta(r_1 - a) \Big) \right] \delta(a + \epsilon - r) \tag{11.24}$$

(note that Eq. (11.18) is not affected by the outer medium). The solution of the above equation is of the same form as that for Eq. (11.19) but with renormalized surface plasmon frequencies

$$\omega_{0l} = \omega_p \sqrt{\left(\frac{l}{2l+1}\right)\frac{1}{\varepsilon_1}}. \tag{11.25}$$

11.4 Damping of Plasmon Oscillations in Ionic Systems

The semiclassical RPA treatment of plasmon excitations in finite ion systems as presented above does not account for plasmon damping. The damping of plasmon oscillations can be included, however, in a phenomenological manner, by the addition of an attenuation term to the plasmon dynamic equations; i.e., the term $-(2/\tau_0)\partial\delta\rho(\mathbf{r},t)/\partial t$ is added to the right-hand sides of both Eqs. (11.18) and (11.19), taking advantage of their oscillatory form. The damping rate $1/\tau_0$ accounts for ion scattering losses and can be approximated, in analogy to metallic systems, by the inclusion of energy dissipation caused by its irreversible transformation into heat via various microscopic channels similar to those for ohmic resistivity [24],

$$\frac{1}{\tau_0} \simeq \frac{v}{2\lambda_b} + \frac{Cv}{2a}, \tag{11.26}$$

where a is the sphere radius, v is the mean velocity of the ions, $v = \sqrt{4k_BT/(3m)}$, and λ_b is the ion mean free path in the bulk electrolyte material (it includes the scattering of ions on other ions and on solvent particles and admixtures). The second term in Eq. (11.26) accounts for the scattering of ions at the boundary of the finite ionic system, for a sphere with radius a; the constant C is of the order of unity [24].

In order to express a forcing field which moves ions in the system, an inhomogeneous time-dependent term should be added to the homogeneous equations (11.18) and (11.19). The forcing field may be a time-dependent electric field. If one considers it to be the electric component of the incident electromagnetic wave then a comparison of the resonance wavelength with the system size is in order. As for metallic nanospheres, for finite ionic systems also the surface plasmon resonance wavelength greatly exceeds the system dimension and thus the forcing field is practically uniform across the whole system. Such a perturbation can excite only surface dipole plasmons, i.e., the mode with $l = 1$, which can be described by the function $Q_{1m}(t)$ ($l = 1$ and m are the angular momentum quantum numbers relating to the assumed spherical symmetry of the system). The corresponding dynamical equation for surface plasmons reduced to only the mode $Q_{1m}(t)$ has the following form:

$$\frac{\partial^2 Q_{1m}(t)}{\partial t^2} + \frac{2}{\tau_0}\frac{\partial Q_{1m}(t)}{\partial t} + \omega_1^2 Q_{1m}(t)$$
$$= \sqrt{\frac{4\pi}{3}\frac{qn}{m}}\left[E_z(t)\delta_{m,0} + \sqrt{2}\left(E_x(t)\delta_{m,1} + E_y(t)\delta_{m,-1}\right)\right], \quad (11.27)$$

where $\omega_1 = \omega_p/\sqrt{3\varepsilon_1}$ for a dipole surface plasmon frequency, i.e., the Mie frequency [18] and ε_1 is the relative dielectric permittivity of the system surroundings. Because only the Q_{1m} contribute to the plasmon response to the homogeneous electric field, the effective ion-density fluctuation has the form [23]

$$\delta\rho(\mathbf{r},t) = \begin{cases} 0, & r < a, \\ \sum_{m=-1}^{1} Q_{1m}(t)Y_{1m}(\Omega), & r \geq a, \, r \to a+, \end{cases} \quad (11.28)$$

where $Y_{lm}(\Omega)$ is a spherical harmonic with $l = 1$. One can also explicitly calculate the dipole $\mathbf{D}(t)$ corresponding to the surface plasmon oscillations given by Eq. (11.28):

$$\begin{cases} D_x(t) = q \int d^3 r\, x\, \delta\rho(\mathbf{r},t) = \sqrt{\frac{2\pi}{3}}eQ_{1,1}(t)a^3, \\ D_y(t) = q \int d^3 r\, y\, \delta\rho(\mathbf{r},t) = \sqrt{\frac{2\pi}{3}}eQ_{1,-1}(t)a^3, \\ D_z(t) = q \int d^3 r\, z\, \delta\rho(\mathbf{r},t) = \sqrt{\frac{4\pi}{3}}eQ_{1,0}(t)a^3. \end{cases} \quad (11.29)$$

The dipole $\mathbf{D}(t)$ satisfies the equation (if one rewrites Eq. (11.27)),

$$\left[\frac{\partial^2}{\partial t^2} + \frac{2}{\tau_0}\frac{\partial}{\partial t} + \omega_1^2\right]\mathbf{D}(t) = \frac{a^3 4\pi q^2 n}{3m}\mathbf{E}(t) = \varepsilon a^3 \omega_1^2 \mathbf{E}(t). \quad (11.30)$$

Notice that the dipole (11.29) scales as the system volume, i.e., as a^3, which may be interpreted as indicating that all ions actually contribute to the surface plasmon oscillations. This is connected with the fact that the surface modes correspond

11.4 Damping of Plasmon Oscillations in Ionic Systems

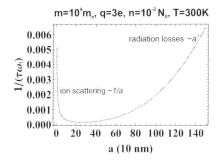
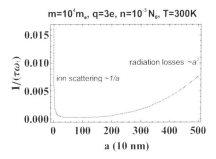

Figure 11.3 The cross-over in the ionic system size dependence of the damping rate for surface plasmons, for $T = 300K$, $m = 10^4 m_e$, $q = 3e$, for $n = 10^{-2} N_0$ (N_0 is the molar concentration of the electrolyte) (left) and for $n = 10^{-3} N_0$ (right); the ion scattering losses, $\sim 1/a$, are of decreasing importance with increasing electrolyte-sphere size (a is the sphere radius), whereas the radiation losses dominate at larger sizes as they initially grow as a^3; in the size region close to the cross-over a perturbative treatment for the Lorentz friction agrees with the exact approach.

to uniform-translation-type oscillations of ions in the system, for which, inside the sphere, the charges of the ions are exactly compensated by oppositely signed ions, whereas an unbalanced charge density occurs only on the surface despite the fact that all the ions oscillate. For the volume plasmons, non-compensated charge-density fluctuation are present also inside the sphere as the volume-plasmon modes have a compressional character with unbalanced charge fluctuations along the sphere radius.

The scattering effects accounted for by the approximate formula (11.26) cause especially strong damping of plasmons for a small system, owing to the nanosphere-edge scattering contribution, proportional to $1/a$. This term is, however, of less significance with increase in radius. We will show that the radiation losses due to the accelerated movement of ions scale as a^3 and, for increasing a, these radiative energy losses quickly dominate the overall plasmon attenuation. The opposite size dependences of the scattering and radiation contributions to the plasmon damping one observes a cross-over in damping with respect to size dependence, as is depicted in Fig. 11.3. One can also determine the radius a^* for which the total attenuation rate for surface plasmons is minimal:

$$a^* = \left(\frac{3^{3/2} C c^3 v}{2 \omega_1 \omega_p^3} \right)^{1/4}.$$

The minimal sizes a^* for two distinct ionic system arrangements are listed in Table 11.1.

Table 11.1 *The ion-system parameters assumed for the calculation of damping rates and self-frequency for dipole surface plasmons*

Material	Ionic system	Sample 1	Sample 2
ion concentration	n (N_0 is one mole of electr.)	$10^{-2} N_0$	$10^{-3} N_0$
effective ion mass	m (m_e electron mass)	$10^4 \, m_e$	$10^4 \, m_e$
charge of effective ion	$q/\sqrt{\varepsilon}$	$3e$	$3e$
temperature	T	300 K	300 K
mean velocity of ions	$v = \sqrt{3kT/m}$	1168 m/s	1168 m/s
bulk plasmon frequency	ω_p	9.3×10^{13}/s	2.93×10^{12}/s
dielectric permittivity of surroundings	ε_1	2	2
Mie frequency	$\omega_1 = \omega_p/\sqrt{3\varepsilon_1}$	3.8×10^{13}/s	1.2×10^{12}/s
constant in Eq. (11.26)	C	2	2
bulk mean free path (room temperature)	λ_b	0.5 μm	0.1 μm
radius for minimal damping	$a^* = \left(\dfrac{3^{3/2} C c^3 v}{2 \omega_1 \omega_p^3}\right)^{1/4}$	2.7×10^{-7} m	8.6×10^{-7} m

The radiation of energy from an oscillating dipole is expressed by the so-called Lorentz friction [55], viewed as an effective electric field slowing down the motion of charges:

$$\mathbf{E}_L = \frac{2\sqrt{\varepsilon}}{3c^3} \frac{\partial^3 \mathbf{D}(t)}{\partial t^3}. \tag{11.31}$$

Hence, we can rewrite Eq. (11.30) including the Lorentz friction term as

$$\left[\frac{\partial^2}{\partial t^2} + \frac{2}{\tau_0} \frac{\partial}{\partial t} + \omega_1^2\right] \mathbf{D}(t) = \varepsilon a^3 \omega_1^2 \mathbf{E}(t) + \varepsilon a^3 \omega_1^2 \mathbf{E}_L, \tag{11.32}$$

or, for $\mathbf{E} = 0$,

$$\left[\frac{\partial^2}{\partial t^2} + \omega_1^2\right] \mathbf{D}(t) = \frac{\partial}{\partial t}\left[-\frac{2}{\tau_0} \mathbf{D}(t) + \frac{2}{3\omega_1}\left(\frac{\omega_p a}{c\sqrt{3}}\right)^3 \frac{\partial^2}{\partial t^2} \mathbf{D}(t)\right]. \tag{11.33}$$

Now one can apply the perturbation method for the solution of Eq. (11.33), treating the right-hand side of this equation as the perturbation. In the zeroth step of the perturbation we have $\left[\partial^2/\partial t^2 + \omega_1^2\right]\mathbf{D}(t) = 0$, from which $(\partial^2/\partial t^2)\mathbf{D}(t) = -\omega_1^2 \mathbf{D}(t)$. Hence, for the first step of the perturbation, we insert the latter formula into the right-hand side of Eq. (11.33), obtaining

$$\left[\frac{\partial^2}{\partial t^2} + \frac{2}{\tau} \frac{\partial}{\partial t} + \omega_1^2\right] \mathbf{D}(t) = 0, \tag{11.34}$$

11.4 Damping of Plasmon Oscillations in Ionic Systems

where

$$\frac{1}{\tau} = \frac{1}{\tau_0} + \frac{\omega_1}{3}\left(\frac{\omega_p a}{c\sqrt{3}}\right)^3. \tag{11.35}$$

Within the first perturbation step, the Lorentz friction can be included in the total attenuation rate $1/\tau$. Nevertheless, this approximation is justified only for sufficiently small perturbations, i.e., when the second term in Eq. (11.35), proportional to a^3, is small enough to fulfil the perturbation restrictions. The related limiting value, $\tilde{a}$, of the ionic system size depends on the ion concentration, charge, mass, and dielectric permittivity, as is exemplified in the next subsection.

The solution of Eq. (11.34) is of the form $\mathbf{D}(t) = Ae^{-t/\tau}\cos(\omega_1' t + \phi)$, where $\omega_1' = \omega_1\sqrt{1 - 1/(\omega_1\tau)^2}$, which gives the redshift of the plasmon resonance due to the strong, $\sim a^3$, growth of the attenuation caused by radiation. The Lorentz friction term in Eq. (11.35) dominates the plasmon damping for $a \leq \tilde{a}$ owing to this a^3 dependence; see Fig. 11.3. The plasmon damping grows rapidly with a and this results in a pronounced redshift of the resonance frequency.

11.4.1 Exact Inclusion of the Lorentz Damping

Now we will consider the dynamic equation for surface plasmons in the ionic spherical system in Eq. (11.33) with the Lorentz friction term, but without using the perturbation method of solution, which results in the substitution of the Lorentz friction term

$$\frac{2}{3\omega_1}\left(\frac{\omega_p a}{v\sqrt{3}}\right)^3 \frac{\partial^3 \mathbf{D}(t)}{\partial t^3}$$

with the approximate formula,

$$-\frac{2\omega_1}{3}\left(\frac{\omega_p a}{v\sqrt{3}}\right)^3 \frac{\partial \mathbf{D}(t)}{\partial t};$$

this was the result of taking on the right-hand side of Eq. (11.33) the zeroth order solution, for which $\frac{\partial^2 \mathbf{D}(t)}{\partial t^2} = -\omega_1^2 \mathbf{D}(t)$. To compare the various contributions to Eq. (11.33) we change to a dimensionless variable: $t \to t' = \omega_1 t$. Then Eq. (11.33) attains the form

$$\frac{\partial^2 \mathbf{D}(t')}{\partial t'^2} + \frac{2}{\tau_0 \omega_1}\frac{\partial \mathbf{D}(t')}{\partial t'} + \mathbf{D}(t') = \frac{2}{3}\left(\frac{\omega_p a}{v\sqrt{3}}\right)^3 \frac{\partial^3 \mathbf{D}(t')}{\partial t'^3}. \tag{11.36}$$

If we solve Eq. (11.36) by the perturbation method we get a renormalized attenuation rate for the effective damping term,

$$\frac{1}{\omega_1 \tau_0} + \frac{1}{3}\left(\frac{\omega_p a}{v\sqrt{3}}\right)^3.$$

This term quickly achieves the value 1, for which the oscillator falls into the overdamped regime. For the system parameters assumed for Fig. 11.3, the attenuation rate reaches 1 at radii 25.5 μm and 8 μm for $n = 10^{-3} N_0$ and for $n = 10^{-2} N_0$, respectively. At these values the frequency $\omega_1' = \omega_1 \sqrt{1 - 1/(\omega_1 \tau)^2}$ goes to zero, which indicates an artefact of the perturbation method. To verify how the exact damped frequency behaves in the system under consideration one needs to solve the dynamical equation without any approximations. As this equation is a third-order linear differential equation, one finds that its solution is in the form $\sim e^{i\Omega t'}$, with analytical expressions for three possible values of the exponent,

$$\Omega_1 = -\frac{i}{3g} - \frac{i 2^{1/3}(1+6gu)}{3g \left(2+27g^2+18gu+\sqrt{4(-1-6gu)^3+(2+27g^2+18gu)^2}\right)^{1/3}}$$

$$- \frac{i\left(2+27g^2+18gu+\sqrt{4(-1-6gu)^3+(2+27g^2+18gu)^2}\right)^{1/3}}{3 \times 2^{1/3} g}$$

$\in \mathrm{Im}(= i\alpha),$

$$\Omega_2 = -\frac{i}{3g}$$

$$+ \frac{i(1+i\sqrt{3})(1+6gu)}{3 \times 2^{2/3} g \left(2+27g^2+18gu+\sqrt{4(-1-6gu)^3+(2+27g^2+18gu)^2}\right)^{1/3}}$$

$$+ \frac{i(1-i\sqrt{3})\left(2+27g^2+18gu+\sqrt{4(-1-6gu)^3+(2+27g^2+18gu)^2}\right)^{1/3}}{6 \times 2^{1/3} g}$$

$$= \omega + i\frac{1}{\tau},$$

$$\Omega_3 = -\frac{i}{3g}$$

$$+ \frac{i(1-i\sqrt{3})(1+6gu)}{3 \times 2^{2/3} g \left(2+27g^2+18gu+\sqrt{4(-1-6gu)^3+(2+27g^2+18gu)^2}\right)^{1/3}}$$

$$+ \frac{i(1+i\sqrt{3})\left(2+27g^2+18gu+\sqrt{4(-1-6gu)^3+(2+27g^2+18gu)^2}\right)^{1/3}}{6 \times 2^{1/3} g}$$

$$= -\omega + i\frac{1}{\tau},$$

(11.37)

11.4 Damping of Plasmon Oscillations in Ionic Systems

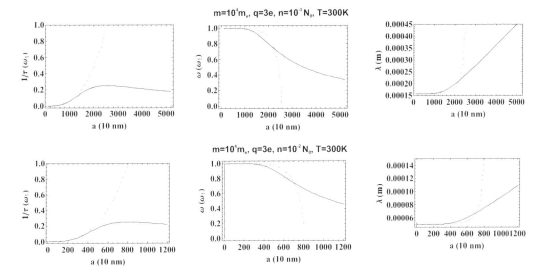

Figure 11.4 For comparison, the damping rate and the resonance frequency (transformed to the resonance wavelength in the right-hand panels). The damping rate and frequency or wavelength of the oscillating solution of Eq. (11.36) are given: exact values (solid line) and approximate values from the perturbation approach (dashed line), both as functions of the ionic finite system radius a.

where

$$u = \frac{1}{\tau_0 \omega_1} \quad \text{and} \quad g = \frac{2}{3}\left(\frac{a\omega_p}{c\sqrt{3\varepsilon_1}}\right)^3.$$

In Fig. 11.4 the damping rate (Im Ω) is plotted and the self-frequency (Re Ω) (in terms of the resonance wavelength with right-hand panels) in terms of the system radius a. For comparison, the approximate perturbative solutions are plotted as dashed lines, whereas the exact solutions of Eq. (11.36) are plotted as solid lines. The dashed lines finish at a_{limit}, when the attenuation rate within the perturbation approach reaches the critical value 1 (then $\lambda \rightarrow \infty$). For the accurate solution of Eq. (11.36) this singular behaviour completely disappears and the oscillating solution, $e^{i\Omega t}$, exists for larger a as well.

We notice that the redshift of the plasmon resonance is strongly overestimated in the framework of the perturbative approach to the Lorentz friction unless $a < \tilde{a}$, where $\tilde{a}$ is sensitive to the ionic system parameters and especially to the ion concentration (as is demonstrated in Fig. 11.4).

Let us emphasize that Eq. (11.36) has in general two types of particular solutions $e^{i\Omega t'}$, with complex self-frequencies Ω. The solutions given by Ω_2 and Ω_3 (see Eq. (11.37)) are of oscillating type with damping ($i\Omega_2$ and $i\Omega_3$ are mutually conjugate: Ω_2 and Ω_3 have real parts of opposite sign but the same imaginary parts;

the latter is positive, giving the damping rate). The second type of solution is given by Ω_1, which turns out to be an unstable exponentially rising solution (a negative imaginary solution). This unstable solution is the well-known artefact in Maxwell electrodynamics (see, e.g., Section 75 in [55]) and corresponds to the infinite self-acceleration of the free charge due to the Lorentz friction force (i.e., to the singular solution of the equation $m\dot{\mathbf{v}} = \text{const.} \times \dddot{\mathbf{v}}$, which is associated with a formal renormalization of the field-mass of the charge – infinite for a pointlike charge and cancelled in an artificial manner by an arbitrary assumed negative infinite non-field mass, resulting in the ordinary mass of, e.g., an electron, which is, however, not defined mathematically in a proper way). This unphysical singular particular solution should be thus discarded. The other, oscillatory-type, solution resembles the solution for an ordinary damped harmonic oscillator, though with a distinct attenuation rate and frequency. The latter are expressed by the analytical formulae for Ω_2 and Ω_3 in Eqs. (11.37) and can be calculated for various a and compared with the corresponding quantities found within the perturbation approach. This comparison is presented in Fig. 11.4. It can be clearly seen that the perturbation approach leads to considerable overestimation of the damping rate for $a > \tilde{a}$. Therefore, we can conclude that the use of the approximate formula for the Lorentz friction damping in the form (11.35) is justified only up to $a \simeq \tilde{a}$, while for $a > \tilde{a}$, these approximate values strongly differ from the exact ones. The value $\tilde{a} < a_{limit}$ depends sharply on the ionic system parameters; approximately, $\tilde{a} \simeq a_{limit}/2$.

11.5 Derivation of Plasmon Frequencies for a Finite Electrolyte System

11.5.1 Volume Ionic Plasmons

In order to determine the self-frequencies of the volume ionic plasmon in the nansphere we must solve Eq. (11.18), the form of the relevant solution being given by Eq. (11.20). For the initial conditions listed below Eq. (11.20) we also assume that $F(\mathbf{r}, t) = F_\omega(\mathbf{r}) \sin(\omega t)$ and, by substitution of this function into Eq. (11.18), we get

$$\Delta F_\omega(\mathbf{r}) + k^2 F_\omega(\mathbf{r}) = 0, \qquad (11.38)$$

where $k^2 = (\omega^2 - \omega_p^2)m/(k_B T)$. This is a well-known Helmholtz differential equation, of which solutions (finite at the origin) can be expressed in terms of the spherical Bessel functions (for the radial dependence of $F_\omega(\mathbf{r})$),

$$F_\omega(\mathbf{r}) = A j_l(kr) Y_{lm}(\Omega), \qquad (11.39)$$

where $j_l(x) = \sqrt{\pi/(2x)} I_{l+1/2}(x)$ is the lth spherical Bessel function, linked to the Bessel function of first kind. The boundary condition, $F(a) = 0$, gives the

11.5 Derivation of Plasmon Frequencies for a Finite Electrolyte System

quantization of k, $k_{li} = x_{li}/a$, where x_{li} is the ith zero of the lth Bessel function (see Fig. 11.2 (left)). Following this quantization one arrives at the corresponding self-frequency quantization,

$$\omega_{li}^2 = \omega_p^2 \left(1 + \frac{k_B T x_{li}}{\omega_p^2 m a^2}\right). \tag{11.40}$$

Thus, the volume ionic plasmons in the sphere are described by the functions

$$\delta\rho_1(\mathbf{r},t) = n \sum_{l=1}^{\infty} \sum_{m=-l}^{m=l} \sum_{i=1}^{\infty} A_{lmi} j_l(k_{li} r) Y_{lm}(\Omega) \sin(\omega_{li} t), \tag{11.41}$$

where the A_{lmi} are arbitrary constants. The component with $l = 0$ vanishes because of the neutrality condition, $\int_0^a r^2 dr d\Omega F(\mathbf{r},t) = 0$ (as $\int d\Omega Y_{lm}(\Omega) = \sqrt{4\pi}\,\delta_{l0}\delta_{m0}$, $d\Omega = \sin\Theta\, d\Theta d\phi$). Note that in ionic systems, the self-frequencies of the volume plasmons in the sphere are temperature dependent – see Eq. (11.40) and Fig. 11.2 (right).

11.5.2 Surface Ionic Plasmons

In order to determine the self-frequencies for surface plasmons, one has to consider Eq. (11.19) and the solution for it given by Eq. (11.20).

The first term on the right-hand side of Eq. (11.19) can be rewritten in the form

$$\frac{kT}{m}\nabla(n+\delta\rho_2)\nabla\Theta(a-r) + \frac{k_B T}{m}(n+\delta\rho_2)\Delta\Theta$$

$$= -\frac{k_B T}{m}\delta(a-r)\frac{\partial}{\partial r}(n+\delta\rho_2) - \frac{k_B T}{m}(n+\delta\rho_2)\frac{1}{r^2}\frac{\partial}{\partial r}[r^2\delta(a-r)]$$

$$= -\frac{k_B T}{m}\frac{1}{r^2}\frac{\partial}{\partial r}\left[(n+\delta\rho_2)r^2\delta(a-r)\right], \tag{11.42}$$

where we have used the formulae $\nabla\Theta(a-r) = (-\mathbf{r}/r)\delta(a-r)$, $(\mathbf{r}/r)\nabla = \partial/\partial r$. The next term on the right-hand side of Eq. (11.19) can be transformed into

$$-\frac{k_B T}{m}\delta(a-r)\frac{\mathbf{r}}{r}\nabla\delta\rho_2 - \frac{\omega_p^2}{4\pi}\delta(a-r)\frac{\mathbf{r}}{r}\nabla\int\frac{d^3 r_1 \delta\rho(\mathbf{r}_1)}{|\mathbf{r}-\mathbf{r}_1|}$$

$$= -\frac{k_B T}{m}\delta(a-r)\frac{\partial}{\partial r}\delta\rho_2 - \frac{\omega_p^2}{4\pi}\delta(a-r)\frac{\partial}{\partial r}\int\frac{d^3 r_1 \delta\rho(\mathbf{r}_1)}{|\mathbf{r}-\mathbf{r}_1|}. \tag{11.43}$$

Equation (11.19) thus attains the form

$$\frac{\partial^2 \rho_2}{\partial t^2} = -\frac{k_B T}{m}\frac{1}{r^2}\frac{\partial}{\partial r}\left[(n+\delta\rho_2)r^2\delta(a-r)\right]$$

$$-\frac{k_B T}{m}\delta(a-r)\frac{\partial}{\partial r}\delta\rho_2 - \frac{\omega_p^2}{4\pi}\delta(a-r)\frac{\partial}{\partial r}\int\frac{d^3 r_1 \delta\rho(\mathbf{r}_1)}{|\mathbf{r}-\mathbf{r}_1|}. \tag{11.44}$$

We suppose the solution of the above equation to be in the form $\delta\rho_2 = \sigma(\Omega, t)\delta(a + 0^+ - r)$, multiply both sides of Eq. (11.44) by r^2, and integrate with respect to r in arbitrary limits, i.e., $\int_l^L r^2 dr \cdots$, such that $a \in (l, L)$ (this integration removes the Dirac deltas), which leads to the equation

$$a^2 \frac{\partial^2 \sigma(\Omega, t)}{\partial t^2} = -\frac{k_B T}{m} \int_l^L dr \frac{\partial}{\partial r}\left[(n + \delta\rho_2)r^2 \delta(a - r)\right]$$

$$-\frac{k_B T}{m}\sigma(\Omega, t) \int_l^L r^2 dr \delta(a - r)\frac{\partial}{\partial r}\delta(a - r)$$

$$-\frac{\omega_p^2}{4\pi} \int_l^L r^2 dr \delta(a - r)\frac{\partial}{\partial r} \int_a^\infty r_1^2 dr_1 \int d\Omega \frac{\delta\rho_2(\mathbf{r}_2)}{|\mathbf{r} - \mathbf{r}_1|}$$

$$-\frac{\omega_p^2}{4\pi} \int_l^L r^2 dr \delta(a - r)\frac{\partial}{\partial r} \int_0^a r_1^2 dr_1 \int d\Omega \frac{\delta\rho_1(\mathbf{r}_1)}{|\mathbf{r} - \mathbf{r}_1|}. \quad (11.45)$$

The two first terms on the right-hand side of the above equation vanish because

$$-\frac{k_B T}{m} \int_l^L dr \frac{\partial}{\partial r}\left[(n + \delta\rho_2)r^2 \delta(a - r)\right] = -\frac{k_B T}{m}\left[(n + \delta\rho_2)r^2 \delta(r - a)\right]\Big|_l^L = 0$$

(11.46)

and

$$\int_l^L r^2 dr \delta(a - r)\frac{\partial}{\partial r}\delta(a - r) = a^2 \int_l^L dr \frac{1}{2}\frac{\partial}{\partial r}\delta^2(a - r)$$

$$= \frac{a^2}{2}\delta^2(a - r)\Big|_l^L$$

$$= \frac{a^2}{2}\lim_{\mu \to 0}\frac{1}{\pi}\frac{\mu}{\mu^2 + (a - r)^2}\delta(a - r)\Big|_l^L = 0. \quad (11.47)$$

The last two terms on the right-hand side of Eq. (11.45) can be transformed using the formula [72]

$$\frac{1}{\sqrt{1 + z^2 - 2z\cos\gamma}} = \sum_{l=0}^\infty P_l(\cos\gamma)z^l \quad \text{for } z < 1,$$

11.5 Derivation of Plasmon Frequencies for a Finite Electrolyte System   273

where $P_l(\cos\gamma) = 4\pi/(2l+1)\sum_{m=-l}^{l} Y_{lm}(\Omega)Y_{lm}^*(\Omega)$ are Legendre polynomials. This formula leads to the following:

$$\frac{\partial}{\partial a}\frac{1}{|\mathbf{a}-\mathbf{r}_1|} = \begin{cases} \sum_{l=0}^{\infty} \frac{la^{l-1}}{r_1^{l+1}} P_l(\cos\gamma) & \text{for } a < r_1, \\ -\sum_{l=0}^{\infty} \frac{(l+1)r_1^l}{a^{l+2}} P_l(\cos\gamma) & \text{for } a > r_1, \end{cases} \quad (11.48)$$

where $\mathbf{a} = a\mathbf{r}/r$, $\cos\gamma = \mathbf{a}\cdot\mathbf{r}_1/(ar_1)$. Employing Eq. (11.48), the last two terms in Eq. (11.45) can be transformed as follows:

$$-\frac{\omega_p^2}{4\pi}\int_l^L r^2 dr\,\delta(a-r)\frac{\partial}{\partial r}\int_a^\infty r_1^2 dr_1\int d\Omega_1 \frac{\delta\rho_2(\mathbf{r}_1)}{|\mathbf{r}-\mathbf{r}_1|}$$

$$= -\frac{\omega_p^2}{4\pi}a^2\int d\Omega_1\int_a^\infty r_1^2 dr_1\,\delta\rho_2(\mathbf{r}_1)\frac{\partial}{\partial a}\frac{1}{\sqrt{a^2+r_1^2-2ar_1\cos\gamma}}$$

$$= -\frac{\omega_p^2}{4\pi}a^2\int d\Omega_1\int_a^\infty r_1^2 dr_1\,\sigma(\Omega_1)\delta(a+0+-r_1)\sum_{l=0}^{\infty}\frac{la^{l-1}}{r_1^{l+1}}P_l(\cos\gamma)$$

$$= -\frac{\omega_p^2}{4\pi}a^2\int d\Omega_1\,\sigma(\Omega_1)\frac{1}{a^2}\sum_{l=0}^{\infty}\frac{4\pi l}{2l+1}\sum_{m=-l}^{l} Y_{lm}(\Omega)Y_{lm}^*(\Omega_1)$$

$$= -\omega_p^2 a^2\sum_{l=0}^{\infty}\sum_{m=-l}^{l}\frac{l}{2l+1}Y_{lm}(\Omega)\int d\Omega_1\,\sigma(\Omega_1)Y_{lm}^*(\Omega_1) \quad (11.49)$$

and

$$-\frac{\omega_p^2}{4\pi}\int_l^L r^2 dr\,\delta(a-r)\frac{\partial}{\partial r}\int_0^a r_1^2 dr_1\int d\Omega_1 \frac{\delta\rho_1(\mathbf{r}_1)}{|\mathbf{r}-\mathbf{r}_1|}$$

$$= -\frac{\omega_p^2}{4\pi}a^2\int d\Omega_1\int_0^a r_1^2 dr_1\,nF(\mathbf{r}_1,t)\frac{\partial}{\partial a}\frac{1}{\sqrt{a^2+r_1^2-2ar_1\cos\gamma}}$$

$$= \frac{\omega_p^2}{4\pi}a^2\int d\Omega_1 nF(\mathbf{r}_1,t)\sum_{l=0}^{\infty}\frac{(l+1)r_1^l}{a^{l+2}}P_l(\cos\gamma)$$

$$= \omega_p^2 n\sum_{l=0}^{\infty}\frac{l+1}{2l+1}Y_{lm}(\Omega)\int_0^a r_1^2 dr_1\frac{r_1^l}{a^l}$$

$$\times \sum_{l_1=1}^{\infty}\sum_{m_1=-l_1}^{l_1}\sum_i A_{lmi} j_{l_1}(k_{l_1i}r_1)\sin(\omega_{l_1i}t)\int d\Omega_1 Y_{lm}^*(\Omega_1)Y_{l_1m_1}(\Omega_1)$$

$$= \omega_p^2 n \sum_{l=0}^{\infty}\sum_{m=-l}^{l}\sum_i \frac{l+1}{2l+1}Y_{lm}(\Omega)A_{lmi}\int_0^a \frac{r_1^{l+2}dr_1}{a^l} j_l(k_{li}r_1)\sin(\omega_{li}t). \quad (11.50)$$

Equation (11.45) thus attains the form

$$\frac{\partial^2 \sigma(\Omega,t)}{\partial t^2} = -\omega_p^2 a^2 \sum_{l=0}^{\infty}\sum_{m=-l}^{l}\frac{l}{2l+1}Y_{lm}(\Omega)\int d\Omega_1 \sigma(\Omega_1)Y_{lm}^*(\Omega_1)$$

$$+ \omega_p^2 n \sum_{l=0}^{\infty}\sum_{m=-l}^{l}\sum_i \frac{l+1}{2l+1}Y_{lm}(\Omega)A_{lmi}\int_0^a \frac{r_1^{l+2}dr_1}{a^l} j_l(k_{li}r_1)\sin(\omega_{li}t).$$

$$(11.51)$$

Assuming now that $\sigma(\Omega,t) = \sum_{l=0}^{\infty}\sum_{m=-l}^{l} q_{lm}(t)Y_{lm}(\Omega)$ and inserting this into the above equation, we obtain

$$\sum_{l=0}^{\infty}\sum_{m=-l}^{l} Y_{lm}(\Omega)\frac{\partial^2 q_{lm}(t)}{\partial t^2} = -\sum_{l=0}^{\infty}\sum_{m=-l}^{l}\frac{\omega_p^2 l}{2l+1}Y_{lm}(\Omega)q_{lm}(t)$$

$$+ \omega_p^2 \sum_{l=1}^{\infty}\sum_{m=-l}^{l}\sum_i \frac{l+1}{2l+1}Y_{lm}(\Omega)A_{lm}$$

$$\times \int_0^a \frac{r_1^{l+2}dr_1}{a^{l+2}} j_l(k_{li}r_1)\sin(\omega_{li}t). \quad (11.52)$$

From the above equation we notice that for $l = 0$ we get $\partial^2 q_{00}/\partial t^2 = 0$ and thus $q_{00}(t) = 0$ (as $q(0) = 0$ and $\lim_{t\to\infty} q(t) < \infty$). For $l \geq 1$ we obtain

$$\frac{\partial^2 q_{lm}(t)}{\partial t^2} = -\frac{\omega_p^2 l}{2l+1}q_{lm}(t) + \sum_i \omega_p^2 \frac{l+1}{2l+1}A_{lm}n \int_0^a \frac{r_1^{l+2}dr_1}{a^{l+2}} j_l(k_{li}r_1)\sin(\omega_{li}t),$$

$$(11.53)$$

which induces the following solution form:

$$q_{lm}(t) = \frac{B_{lm}}{a}^2 \sin\left(\omega_p\sqrt{\frac{l}{2l+1}}t\right)$$

$$+ \sum_i A_{lm}\frac{(l+1)\omega_p^2}{\omega_p^2 - (2l+1)\omega_{li}^2} n \int_0^a \frac{r_1^{l+2}dr_1}{a^{l+2}} j_l(k_{li}r_1)\sin(\omega_{li}t). \quad (11.54)$$

11.5 Derivation of Plasmon Frequencies for a Finite Electrolyte System 275

Finally, $\delta\rho_2(\mathbf{r},t) = \sum_{l=1}^{\infty} \sum_{m=-l}^{l} q_{lm}(t) Y_{lm}(\Omega) \delta(a-r)$. The first term in Eq. (11.54) describes the self-frequencies of surface plasmons, whereas the second corresponds to the surface plasmon oscillations induced by the volume plasmons. This induced part of the surface oscillations is nonzero only when the volume modes are excited and thus their amplitudes A_{lmi} are nonzero. The frequencies of self-oscillation of the surface plasmons equal $\omega_{l0} = \omega_p \sqrt{l/(2l+1)}$, corresponding to various multipole modes (numbered with l). Note that these frequencies are lower than the bulk plasmon frequency ($\omega_p = \sqrt{nq^2 4\pi/m}$ in Gauss units or $\sqrt{nq^2/(\varepsilon_0 m)}$ in SI), whereas the volume plasmon modes oscillate with frequencies higher than ω_p. Worth noting also is the absence of any temperature dependence of the surface ionic plasmon resonances, in contrast to the volume plasmon self-frequencies.

Concluding, we can state that in ionic finite systems we may observe plasmons similar to those in metallic nanoparticles. The structure of the surface and volume plasmons for ions repeats the properties of electronic plasmons in metallic spherical systems with, however, a significant shift of the resonance energy towards lower values, corresponding to the much larger mass of the ions in comparison with the electron mass and the different concentration of ions in the electrolyte. Thus, corresponding to the resonance energy, the electromagnetic wavelength is shifted to the deep infrared or even longer wavelengths, depending on the ion concentration. Typical for metal clusters is the cross-over in the size dependence of plasmon damping between on the one hand the scattering losses, leading to ohmic-type energy dissipation, and on the other hand the radiation losses. This cross-over is also observable in ionic spherical systems with a similar size dependence, though it is shifted toward the micrometre scale for ions whereas for metals it is at the nanometre scale. Of particular interest is the high radiation regime for dipole plasmons in ionic systems with prospective applications in signalling and energy transfer in such systems. An initial strong enhancement in efficiency of the Lorentz friction with the radius growth of an electrolyte sphere is observed on the micrometre scale with typical a^3 radius dependence above some threshold radius whose value depends on the electrolyte parameters. At a certain value of the radius (varying over a wide range and depending on the ion system parameters) this enhancement saturates and then slowly diminishes, which allows definition of the most convenient sizes of finite electrolyte systems for optimizing radiation-mediated transport efficiency.

12

Plasmon–Polaritons in a Chain of Finite Ionic Systems; Model of Saltatory Conduction in Myelinated Axons

In this chapter we apply the theory of ionic plasmons described in the previous chapter to plasmon–polaritons in periodic ionic systems. The complete theory of ionic plasmon–polariton kinetics in a chain of micrometre-size electrolyte spheres confined by a dielectric membrane is formulated and solved. The latter theory is then applied to a problem that has been unclear for several decades, the so-called saltatory conduction of the action potential in the myelinated axons of nerve cells. Contrary to conventional nerve signalling models, the plasmon–polariton model fits well with the unusual properties of saltatory conduction. Moreover, the soft plasmonics application presented regarding signalling in myelinated axons allows for the identification of different information-processing roles of the white and grey matter in the brain and spinal cord. The proposed model of saltatory conduction defines the role of the insulating myelin differently from previous models, which may be helpful in the development of a better understanding of demyelination diseases.

The effective and quick transduction of the action potential through long myelinated axons in the neurons of the peripheral nervous system and in the white matter of the central nervous system undeniably lies behind the motorics and other communication functions of the body. The acceleration of the signal transduction is associated here with jumping of the signal between consecutive so-called Ranvier nodes separated by myelinated sectors approximately 100 µm long. The nature of this signal jumping, called saltatory conduction, has not yet been explained as the timing of the signal jumping exceeds conventional models of electrical signalling in neurons. Despite the in-depth current recognition of nerve function mechanisms, including the action potential spike-formation mechanism on Ranvier nodes [50] and the model of the diffusion-like conduction of ion current in dendrites and nonmyelinated axons according to the cable theory originated by W. Thomson (1854) [52], see Fig. 12.1 (with various modifications [191, 51]), the mechanism of saltatory conduction observed in myelinated axons [192] is still unclear, since it cannot be explained by cable theory, and other possible mechanisms are still being researched [53, 50].

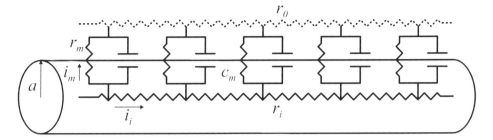

Figure 12.1 Schematic presentation of the cable model of nervous signal transduction in dendrites and in nonmyelinated axons [50, 51]. The capacitor across the neuron membrane, c_m, together with the across-membrane resistivity, r_m, creates a series of coupled RC circuits; the consecutive local charging and discharging of the capacitors c_m results in diffusion-type [52, 192] transduction of the ion current i_i inside the dendrite or nonmyelinated axon. Because of the relatively large longitudinal resistivities $r_{i(o)}$ of the inside and outside cytosol and the values of c_m and r_m across the cell membrane, the velocity of the diffusive current is limited to 1–3 m/s [50–52].

In order to solve the problem of saltatory conduction, a new and original model for this mysterious conduction was developed, on the basis of the kinetic properties of collective plasmon–polariton modes propagating along linear and periodically modified electrolyte systems. For an axon this system is the thin cord of the nerve cell periodically wrapped by Schwann cells creating a periodic and relatively thick myelin sheath (Schwann cells myelinate axons in the peripheral system, whereas in the central neural system axons are myelinated by oligodendrocyte cells [191, 50, 51]). These plasmon–polaritons have been investigated and understood using the well-developed domain of plasmonics [1, 5], especially nano-plasmonics applied to the long-range low-damped propagation of plasmon–polaritons along metallic nano-chains [40], as presented in previous chapters. The main properties of these collective excitations occurring on the conductor–insulator interface [116, 4] owing to the hybridization of the surface plasmons (i.e., the electron density fluctuations on the metal surface) with an electromagnetic wave are as follows: (1) a plasmon–polariton velocity much lower than the velocity of light, corresponding to plasmon–polaritons with wavelengths much shorter than the wavelength of light at the same frequency; (2) the related strong discrepancy between the momenta of plasmon–polaritons and photons with the same energies means that the external electromagnetic waves do not interact with the plasmon–polaritons, i.e., photons cannot be excited or absorbed by plasmon–polaritons owing to momentum conservation constraints; (3) all the electromagnetic field associated with the propagation of plasmon–polaritons is compressed into the tunnel-volume of the chain, especially the spaces between chain elements; (4) all the radiative losses are quenched, and plasmon–polariton attenuation occurs only because of ohmic losses due to oscillating charged carriers (electrons in a metal), which makes metallic nano-chains almost perfect wave guides for plasmon–polaritons; and (5) the long-range and practically undamped propagation of plasmon–polaritons is experimentally observed in metallic nano-chains [40, 39]. The plasmon–polaritons in metallic nanostructures are likely to be exploited for future applications in optoelectronics where the conversion of light signals into plasmon–polariton signals circumvents the diffraction constraints that greatly limit the miniaturization of conventional optoelectronic devices (as the nanoscale of electron confinement inconveniently conflicts with the several-orders-of-magnitude larger scale of the wavelength of light at an energy similar to that of nanoconfined electrons) [4, 41].

Some unique properties of plasmon–polaritons in metallic nano-chains are of great interest, as these properties can be repeated in periodic linear arrangements of electrolyte systems with ions instead of electrons as charge carriers. According to the much larger mass of ions compared with that of electrons and the lower concentration of ions in electrolytes compared with that of electrons in metals, plasmon resonances in finite ionic systems (e.g., in a liquid electrolyte confined to a finite volume by appropriately formed membranes, frequently found in biological cell structures) occur on the scale of micrometres rather than nanometres, as for metals, and at frequencies several orders of magnitude lower (depending on the ion concentration).

For a spherical electrolyte system, the surface and volume plasmons are handled analogously to those of the metallic nanospheres, described in the previous chapter. The ionic surface plasmon frequencies are given for the multipole lth mode by the formula

$$\omega_l = \omega_p \sqrt{\frac{l}{\varepsilon(2l+1)}}$$

and the bulk plasmon frequency is $\omega_p = \sqrt{4\pi q^2 n/m}$ (n is the ion concentration, q and m are the ion charge and mass, respectively, and ε is the relative permittivity of the surroundings). For dipole surface plasmons ($l = 1$), this equation resolves into a Mie-type formula [18, 23], $\omega_1 = \omega_p/\sqrt{3\varepsilon}$. The plasmon oscillations intensively radiate their own energy and are quickly damped due to Lorentz friction losses (i.e., due to the radiation of electromagnetic waves by oscillating charges [55, 76]). For large systems with a large number of ions participating in the plasmon oscillations (thus strengthening the Lorentz friction) these losses are much greater than the ohmic losses due to carrier scattering (the scattering of ions on other ions, solvent and admixture atoms, and the boundary of the system), as illustrated in Fig. 11.3, where a cross-over in the size dependence of the plasmon damping rate is visible (similar to that for metal nanospheres [48, 27]).

Surprisingly, for a linear chain of spherical ionic systems, the radiation losses reduce to zero in exactly the same manner as in metallic chains [193, 43, 44]. The radiation energy losses expressed by the Lorentz friction [55, 76] are ideally compensated by the energy income due to radiation from the other spheres in the chain. As a result, the radiative losses are ideally balanced, and only a relatively small irreversible ohmic energy dissipation remains due to ion scattering. Thus, collective surface dipole plasmon–polaritons can propagate in the chain with strongly reduced damping and, if energy is permanently supplemented to balance the small ohmic losses, this propagation could occur over arbitrarily long distances without any damping.

12.1 Plasmon–Polariton Propagation in Linear Periodic Ionic Systems

We will apply the model of dipole plasmon–polariton excitations in a linear chain of electrolyte spheres to an axon cord periodically wrapped with myelin sheaths, as schematically depicted in Figs. 12.2 and 12.3. The periodicity makes the chain similar to a one-dimensional crystal. Although the cord of an axon is a continuous ion tube, the modelling of an axon by a chain of segments defined by the periodic myelin sheath meets well with plasmon–polariton kinetics, which maintains the same character in discrete chains and in continuous but periodically corrugated wires. The interaction between the chain elements (or segments defined by the myelin sheath) can be regarded as dipole-type coupling. For chains of ionic spheres, the results of the corresponding analysis for metallic chains can be adopted; these results support a dipole model of interaction between the spheres [41, 167, 168]. Note that the model of interacting dipoles [177, 178] was originally developed for the description of ion stellar matter [179, 180] and was then utilized for metal particle systems [181, 182].

The dipole interaction can be understood in terms of the electric and magnetic fields created by an oscillating dipole $\mathbf{D}(\mathbf{r},t)$ at any distant point. If this point is represented by the vector $\mathbf{r}_0$ (with one end fixed at the end of $\mathbf{r}$, where the dipole is placed), then the electric field produced by the dipole $\mathbf{D}(\mathbf{r},t)$ takes the following form, which also includes the relativistic retardation [55, 76]:

$$\mathbf{E}(\mathbf{r},\mathbf{r}_0,t) = \frac{1}{\varepsilon}\left(-\frac{\partial^2}{v^2\partial t^2}\frac{1}{r_0} - \frac{\partial}{v\partial t}\frac{1}{r_0^2} - \frac{1}{r_0^3}\right)\mathbf{D}\left(\mathbf{r},t - \frac{r_0}{v}\right)$$
$$+ \frac{1}{\varepsilon}\left(\frac{\partial^2}{v^2\partial t^2}\frac{1}{r_0} + \frac{\partial}{v\partial t}\frac{3}{r_0^2} + \frac{3}{r_0^3}\right)\mathbf{n}_0\left(\mathbf{n}_0\cdot\mathbf{D}\left(\mathbf{r},t - \frac{r_0}{v}\right)\right), \quad (12.1)$$

with $\mathbf{n}_0 = \mathbf{r}_0/r_0$ and $v = c/\sqrt{\varepsilon}$. The terms with denominators r_0^3, r_0^2, and r_0 are usually referred to as the near-field, medium-field, and far-field components of the

Chain of ionic spheres

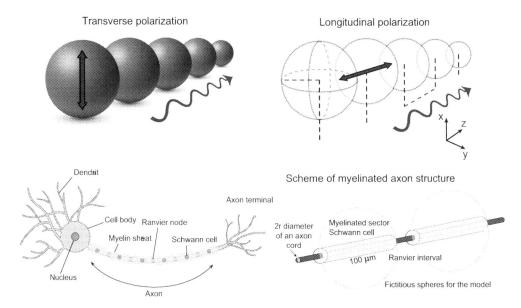

Figure 12.2 In analogy to metallic nano-chains, microchains of finite ionic segments enclosed by dielectric membranes are considered; in these ionic microchains, the plasmon–polarion modes can propagate similarly to those in metallic nano-chains. A myelinated corrugated axon can be modelled as a chain of separated ion segments because the plamon–polaritons can traverse both continuous and discrete linear plasmonic system alignments.

interaction, respectively. Equation (12.1) describes the mutual interactions of the plasmon dipoles at each sphere in the chain. The spheres in the chain are numbered by integers l, and the equation for the surface plasmon oscillation of the lth sphere can be written as follows (where d denotes the separation between the centres of the spheres):

$$\left[\frac{\partial^2}{\partial t^2} + \frac{2}{\tau_0}\frac{\partial}{\partial t} + \omega_1^2\right] D_\alpha(ld,t) = \varepsilon\omega_1^2 a^3 \sum_{m=-\infty,\, m\neq l}^{m=\infty} E_\alpha\left(md, t - \frac{|l-m|d}{v}\right)$$
$$+ \varepsilon\omega_1^2 a^3 E_{L\alpha}(ld,t) + \varepsilon\omega_1^2 a^3 E_\alpha(ld,t). \quad (12.2)$$

Here $\alpha = z$ indicates the longitudinal polarization, whereas $\alpha = x(y)$ indicates a transverse polarization (the chain orientation is assumed to be along the z direction, as illustrated in Fig. 12.2). The first term on the right-hand side of Eq. (12.2) describes the dipole coupling between the spheres, and the other two terms correspond to the plasmon attenuation due to Lorentz friction radiation losses and to the force field arising from an external electric field, respectively;

12.1 Plasmon–Polariton Propagation in Linear Periodic Ionic Systems 281

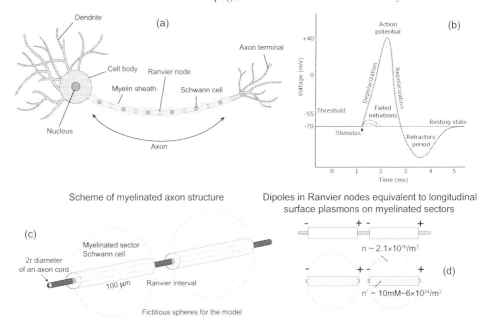

Figure 12.3 (a) Schematic illustration of a long axon, with a chain of periodically repeated myelinated sectors of approximately 100 μm in length separated by unmyelinated Ranvier nodes, corresponding to a number of segments of order 10,000 per metre of axon length; (b) time pattern of the action potential forming on a Ranvier node; (c) periodic fragments of a myelinated axon with a fictitious periodic chain of spherical ionic systems proposed as an effective model; (d) the equivalence of polarized Ranvier nodes and longitudinal surface plasmons on myelinated sectors (the effective concentration of ions n in an auxiliary sphere corresponds to the actual ion concentration n').

$\omega_1 = \omega_p/\sqrt{3\varepsilon}$ is the frequency of the dipole surface plasmons. Ohmic losses are included via the term $2/\tau_0$, as for metals [171] but with the Fermi velocity of electrons in metals substituted by the mean velocity of ions for a nondegenerate classical Boltzmann distribution, regardless of the quantum statistics of the ions, i.e.,

$$\frac{1}{\tau_0} = \frac{v}{2\lambda_B} + \frac{Cv}{2a}, \qquad (12.3)$$

where λ_B is the mean free path of the carriers (ions) in the bulk electrolyte, v is the mean velocity of the carriers at temperature T, $v = \sqrt{3k_BT/m}$, m is the mass of an ion, k_B is the Boltzmann constant, C is a constant of the order of unity (to account for scattering of carriers by the system boundary), and a is the radius of a sphere. The first term in the expression for $1/\tau_0$ approximates ion scattering losses such as those occurring in the bulk electrolyte (collisions with other ions, solvent, and admixture atoms), whereas the second term describes losses due to the scattering of ions on the boundary of a sphere of radius a. According to Eq. (12.1), we can write the quantities that appear in Eq. (12.2) as follows:

$$E_z(md,t) = \frac{2}{\varepsilon d^3}\left(\frac{1}{|m-l|^3} + \frac{d}{v|m-l|^2}\frac{\partial}{\partial t}\right) \times D_z\left(md, t - |m-l|\frac{d}{v}\right),$$

$$E_{x(y)}(md,t) = -\frac{1}{\varepsilon d^3}\left(\frac{1}{|m-l|^3} + \frac{d}{v|m-l|^2}\frac{\partial}{\partial t} + \frac{d^2}{v^2|l-d|}\frac{\partial^2}{\partial t^2}\right) \quad (12.4)$$

$$\times D_{x(y)}\left(md, t - |m-l|\frac{d}{v}\right).$$

Because of the periodicity of the chain, a wave-type collective solution of the dynamical equation in the form of a Fourier component can be assumed for (12.2):

$$D_\alpha(ld,t) = D_\alpha(k,t)\,e^{-ikld}, \qquad 0 \leq k \leq \tfrac{2\pi}{d}. \quad (12.5)$$

In the Fourier version of Eq. (12.2) (i.e., a discrete Fourier transform (DFT) with respect to position and a continuous Fourier transform (CFT) with respect to time) this solution takes a form similar to that of the solution for phonons in one-dimensional crystals. Note that a DFT is defined for a finite set of numbers; therefore, we will consider a chain with $2N+1$ spheres, i.e., a chain of finite length $L = 2Nd$. Then, for any discrete characteristic $f(l)$, $l = -N, \ldots, 0, \ldots, N$, of the chain, such as a selected polarization of the dipole distribution, we must consider the DFT picture,

$$f(k) = \sum_{l=-N}^{N} f(l)e^{ikld}, \qquad k = \frac{2\pi}{2Nd}n, \; n = 0, \ldots, 2N.$$

This means that $kd \in [0, 2\pi)$ because of the periodicity of the equidistant chain. The Born–Karman periodic boundary condition $f(l+L) = f(l)$ is imposed on the entire system, resulting in the form of k given above. For a chain of infinite length, we can take the limit $N \to \infty$, which causes the variable k to become quasicontinuous, although $kd \in [0, 2\pi)$ still holds.

The Fourier representation of Eq. (12.2) takes the following form,

$$\left(-\omega^2 - i\frac{2}{\tau_0}\omega + \omega_1^2\right)D_\alpha(k,\omega) = \omega_1^2\frac{a^3}{d^3}F_\alpha(k,\omega)D_\alpha(k,\omega) + \varepsilon a^3\omega_1^2 E_{0\alpha}(k,\omega),$$

$$(12.6)$$

with, settling as before $x = kd$ and $\xi = \omega d/v$,

12.1 Plasmon–Polariton Propagation in Linear Periodic Ionic Systems

$$F_z(k,\omega) = 4\sum_{m=1}^{\infty}\left(\frac{\cos(mx)}{m^3}\cos(m\xi) + \xi\frac{\cos(mx)}{m^2}\sin(m\xi)\right)$$

$$+ 2i\left[\frac{1}{3}\xi^3 + 2\sum_{m=1}^{\infty}\left(\frac{\cos(mx)}{m^3}\sin(m\xi) - \xi\frac{\cos(mx)}{m^2}\cos(m\xi)\right)\right],$$

$$F_{x(y)}(k,\omega) = -2\sum_{m=1}^{\infty}\left(\frac{\cos(mx)}{m^3}\cos(m\xi) + \xi\frac{\cos(mx)}{m^2}\sin(m\xi)\right.$$

$$\left. - \xi^2\frac{\cos(mx)}{m}\cos(m\xi)\right)$$

$$- i\left[-\frac{2}{3}\xi^3 + 2\sum_{m=1}^{\infty}\left(\frac{\cos(mx)}{m^3}\sin(m\xi)\right.\right.$$

$$\left.\left. + \xi\frac{\cos(mx)}{m^2}\cos(m\xi) - \xi^2\frac{\cos(mx)}{m}\sin(m\xi)\right)\right].$$

(12.7)

As for metallic nano-chains, Im $F_\alpha(k,\omega) \equiv 0$ (for $\alpha = z, x(y)$), which indicates perfect quenching of the radiation losses at any sphere in the chain (meaning that, for each sphere, the amount of energy that comes in from the other spheres is the same as the energy outflow due to Lorentz friction). We can easily verify this property, as the related infinite sums in Eq. (12.7) can be found analytically [72]. Equation (12.6) is highly nonlinear with respect to the complex frequency ω and can be solved both perturbatively in an analytical manner [44] or numerically, which is more accurate. The solutions determined for Re ω and Im ω (i.e., for the self-frequency and damping of the plasmon–polaritons, respectively) can be applied to the axon model as an effective chain of electrolyte spheres with ion concentrations adjusted to the actual neuron parameters.

The problem of the propagation of plasmon–polaritons along metallic nano-chains was addressed by, among others, Citrin [42, 193] in order to obtain an explanation for the radiative losses of plasmon–polaritons in metallic nano-chains in agreement with observations relating to radiation losses in one-dimensional and two-dimensional crystals [194]. Plotted in Fig. 12.4 are the resonance frequency (Re $\omega(k)$), its k-derivative, i.e., the group velocity, and the attenuation rate (Im $\omega(k)$) of the dipole plasmon–polariton modes numbered by the wave vector k, derived by the solution of Eq. (12.6) [44, 45] for ionic chains with typical concentration, chain size, and ion parameters.

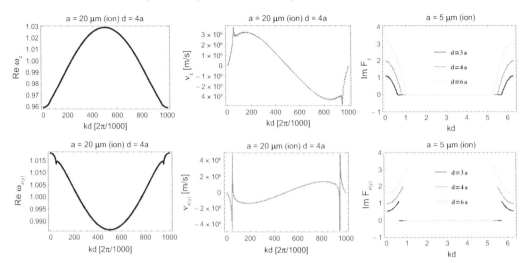

Figure 12.4 Exact solutions for the self-frequencies of the longitudinally and transversely polarized modes of the plasmon–polaritons in an ionic chain (ω in units of ω_1) obtained by solving Eq. (12.6) exactly at 1000 points in the region $kd \in [0, 2\pi)$ (left); the corresponding group velocities for both types of polarization (centre); the functions Im $F_z(k; \omega = \omega_1)$ and Im $F_{x(y)}(k; \omega = \omega_1)$ for infinite chains of electrolyte spheres of radius a at separations $d = 3a$, $4a$, and $6a$; the shift of the singularities towards the band edges with decreasing d/a is noticeable (right) ([49]).

12.2 Plasmon–Polariton Model of Saltatory Conduction: Fitting the Kinetics to the Axon Parameters

The bulk plasmon frequency is $\omega_p = \sqrt{q^2 n 4\pi / m}$ (it is assumed for the model that the ion charge is $q = 1.6 \times 10^{-19}$ C and the ion mass is $m = 10^4 m_e$, where $m_e = 9.1 \times 10^{-31}$ kg is the mass of an electron). For a concentration $n = 2.1 \times 10^{16}/\text{m}^3$, we obtain a Mie-type frequency for ionic dipole oscillations, $\omega_1 \simeq 0.1\omega_p/\sqrt{3\varepsilon_1} \simeq 4 \times 10^6/\text{s}$, where the relative permittivity of water is $\varepsilon_1 \simeq 80$ for frequencies in the MHz range [195] (although for higher frequencies, beginning at approximately 10 GHz, this value decreases to approximately 1.7, corresponding to the optical refractive index of water, $\eta \simeq \sqrt{\varepsilon_1} = 1.33$). For a thin axon cord and thus, in the discrete model, for a strongly prolate inner ionic cord segment, the longitudinal Mie-type frequency is reduced by a factor 0.1 [196, 197] compared with the isotropic spherical case. The axon consists of a cord with a small value of the diameter $2r$, and this thin cord is wrapped with a myelin sheath of length $2a$ per segment; however, for the effective model we consider fictitious electrolyte spheres of radius a. Thus, the auxiliary concentration n of ions in the fictitious spheres corresponds to an ion concentration in the cord of

$$n' = \frac{n 4/3\pi a^3}{2a\pi r^2},$$

12.2 Fitting the Kinetics to the Axon Parameters

which yields a typical concentration of ions in a nerve cell of $n' \sim 10$ mM (i.e., $\sim 6 \times 10^{24}/m^3$). The reason is that in the sphere model all the ions participating in the dipole oscillation correspond to a much smaller volume in the real system, that of the thin cord portion (the insulating myelin sheath consists of a lipid substance without any ions). The insulating, relatively thick, myelin coverage creates the periodically broken channel (a corrugated conductor–insulator structure) required for plasmon–polariton formation and propagation. To reduce the coupling with the surrounding intercellular electrolyte and protect against any leakage of plasmon–polaritons, the myelin sheath must be sufficiently thick, much thicker than what is required merely for electrical insulation. Moreover, to accommodate the conductivity parameters calculated for a spherical geometry to the highly prolate geometry of the real oscillating ionic system, the Mie-type frequency of the longitudinal oscillations must be significantly lower than that for a sphere with a diameter equal to the elongation axis. As a rough estimate, a correction factor of 0.1 was assumed [196, 197]; see Fig. 12.7.

For the resulting Mie-type frequency, $\omega_1 \simeq 4 \times 10^6$/s, one can determine the plasmon–polariton self-frequencies in a chain of spheres of radius $a = 50$ μm (for a Schwann cell length of $2a$) and for small chain separations such that $d/a = 2.01, 2.1$, and 2.2 (corresponding to Ranvier node lengths of 0.5, 5, and 10 μm, respectively) within the approach presented above, via the solution of Eq. (12.6). The derivatives of the obtained self-frequencies with respect to the wave vector k determine the group velocities of the plasmon–polariton modes. The results are presented in Fig. 12.5. We observe that for the ionic system parameters listed above, the group velocity of the plasmon–polaritons reaches 100 m/s for the longitudinal mode with polarization suitable for the prolate geometry, assuming that

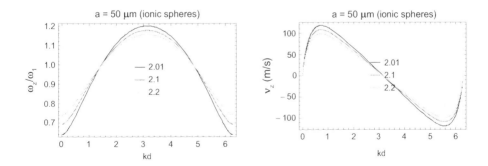

Figure 12.5 Solutions for the self-frequencies and group velocities of the longitudinal mode of a plasmon–polariton in a model ionic chain; ω is presented in units of ω_1; here $\omega_1 = 4 \times 10^6$/s for a chain of spheres with radius $a = 50$ μm and Ranvier separation d, such that $d/a = 2.01, 2.1$, or 2.2, for an equivalent ion concentration in the inner ionic cord of the axon of $n' \sim 10$ mM $= 6 \times 10^{24}/m^3$.

the initial post-synaptic action potential, or that from the cell hillock, predominantly excites longitudinal ion oscillations.

Note that, for $\omega_1 = 4 \times 10^6$/s and $a = 50$ μm, the light cone interference conditions $kd - \omega_1 d/c = 0$ and $kd + \omega_1 d/c = 2\pi$ are fulfilled only for extremely small values of kd and $2\pi - kd$, respectively (of the order of 10^{-6} for $d/a \in [2, 2.5]$). Thus these interference conditions are negligible with regard to the plasmon–polariton ion kinetics (though they are important for metals, as discussed in Chapter 9); the related singularities on the light cone induced by the far- and medium-field contributions to the dipole interaction are pushed to the borders of the k wave vector domain and thus are unimportant for the ionic system considered. Hence, the quenching of the radiative losses (i.e., the perfect balance of the Lorentz friction in each sphere by the radiation income from the other spheres in the chain) for the plasmon–polariton modes in the axon model occurs practically throughout the entire $kd \in [0, 2\pi)$ region. Additionally, the aforementioned singularities [45] are characteristic of infinite chains and therefore cannot fully develop because in the model the nerve electrolyte chains are of a finite length, whereas other effects, such as the quenching of radiation losses, occur for finite chains owing to the very fast convergence of the sums in Eq. (12.7) with denominators m^2 and m^3; in practice, a chain consisting of only 10 elements exhibits almost the same properties as an infinite chain, as was noted in Chapter 9; see Figs. 9.26 and 9.27.

Although the ionic-system chain model for a myelinated axon appears to be a crude approximation to the real axon structure, it can serve for a comparison of the energy and time scales of plasmon–polariton propagation implied by a model with the observed kinetic parameters of nerve signals. In the model, the propagation of a plasmon–polariton through the axon chain, excited by an initial action potential on the first Ranvier node after the synapse or, for the reverse signal direction, in the neuron cell hillock, sequentially ignites the consecutive Ranvier node blocks of Na^+ and K^+ ion gates. The resulting firing of the action potential traverses the axon with a model velocity of approximately 100 m/s, consistent with the velocity actually observed in myelinated axons (and not possible for an ionic diffusive current in the cable model [50, 51]). The plasmon–polariton ignition of consecutive Ranvier nodes enables the creation of the same activation potential pattern, aided by the external energy supply at each Ranvier node block. Because of the nonlinearity of the ion-channel block mechanism, the signal growth saturates at a constant level, and the overall timing of each action potential spike has the stable shape of the local polarization/depolarization scheme in the short fragment of cell membrane that corresponds to the nonmyelinated Ranvier node. The permanent supply of energy associated with the creation of action potential spikes at sequentially firing Ranvier nodes also contributes to the plasmon–

polariton dynamics, ensuring that the amplitude of each Ranvier block dipole excitation is beyond the activation threshold. The external energy supply (through the conventional ATP/ADP cell mechanism) assists the renovation of the action potential pattern at each Ranvier node. This energy supply residually compensates the ohmic thermal losses of the plasmon–polariton mode propagating along the axon and ensures undamped propagation over an unlimited range. Although the entire signal cycle of the action potential on a single Ranvier node block requires several milliseconds (or even longer when one includes the time required to restore steady-state conditions, which, however, on the other hand conveniently blocks the reversing of the signal), subsequent nodes are ignited more rapidly. This occurs because the velocity of the plasmon–polariton wave packet triggers the ignition of consecutive Ranvier nodes, as illustrated in Fig. 12.8. Thus, we are dealing with the firing of the axon, which propagates with the velocity of the ionic plasmon–polariton wave packet. The direction of the velocity of this wave packet corresponds to the semi-infinite geometry of the chain, though in fact the chain is finite and is excited at one of its ends. The firing of the action potential triggered by the plasmon–polariton traverses along the axon in only one direction because the nodes that have already fired have had their Na^+, K^+ gates inactivated (discharged) and require a relatively long time to restore these gates to their original status (the entire Na/K block requires a time of the order of as much as a second and a sufficient energy supply to bring the concentrations to their normal values via cross-membrane active ion pumps against the ion-density gradient). The plasmon–polariton scheme described above for the ignition of the action potential spikes in the ordered chain of Ranvier nodes along an axon is thus consistent with the saltatory conduction observed in myelinated axons. The observation that the firing of the action potential can simultaneously move in two opposite directions if a certain central Ranvier node of a passive axon is ignited, as well as the observation that the maintenance of the firing persists despite small breaks in the axon cord or a few damaged Ranvier nodes, also agrees with the collective wave-type plasmon–polariton model of saltatory conduction, in contrast wth the lack of satisfactory explanations in models based on the cable theory. The maintenance of plasmon–polariton kinetics despite discontinuities in the axon cord agrees well with the discrete chain model of a cord corrugated by the periodic myelin sheath.

In Fig. 12.6 the group velocity of the action potential traversing a firing myelinated axon, in the plasmon–polariton model, is plotted for various diameters of the axon internal cord and for a length of 100 μm for each myelinated sector wrapped by Schwann cells and Ranvier intervals of 0.5 μm, 5 μm, and 10 μm. The dependence of the group velocity on the length of the Ranvier interval is weak (i.e., negligible at the scale considered, which is consistent with the

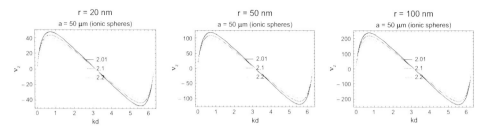

Figure 12.6 For comparison, the group velocities, in m/s, of the longitudinal plasmon–polariton mode with respect to the wave vector $k \in [0, 2\pi/d)$ within the axon model, for Schwann-cell myelinated sectors with length 100 μm. The plots shown are for Ranvier separations 0.5 μm, 5 μm, and 10 μm (represented by $d/a = 2.01$, 2.1, and 2.2 in the figure, respectively) and axon cord radii $r = 20$, 50, and 100 nm.

equivalence of the discrete model for the continuous system if one considers wave-type plasmon–polariton propagation), but the increase in the velocity with increasing internal cord thickness is significant, and is similar to that for real axons with increasing diameters.

To comment on the appropriateness of the chain model for axons, let us note that even though the thin core of the axon is a continuous ion-conducting fibre, the surface electromagnetic field can be closely pinned to the linear conductor in the same way as for the Goubau line (well known in microwave technology) [37, 198]. If periodically wrapped by dielectric sheaths, plasmons resonantly coupled with the electromagnetic field propagate as concentrated plasmon–polaritons along the chain of periodic segments despite the continuity of the fibre. For plasmon–polariton kinetics, the continuity of the conducting fibre is unimportant because here we are dealing with the wave packet of the plasma oscillations and not with the charge carriers themselves. The situation is similar to that for Goubau microwave lines, which also have discontinuous segments; the Goubau lines maintain their transmittance via discrete disconnected elements. The fragments of the thin axon cord wrapped with thick myelin shells with an ion concentration that is typical for neurons of $n' \simeq 10$ mM $\simeq 6 \times 10^{24}/\text{m}^3$ inside the cell can be equivalently modelled by a sphere with diameter equal to the length of the cord fragment and an ion concentration of $n \simeq 2 \times 10^{16}/\text{m}^3$, for an assumed axon cord diameter of 100 nm, conserving the number of ions in the segment participating in the dipole oscillations. Such a model is justified since it has the same dynamical equation structure as that for dipole plasmon fluctuations in a chain of spheres, i.e., Eq. (12.2), and as the modification of this equation for prolate spheroid or elongated cylindrical rod chains. This modification amounts to the substitution of the isotropic ω_1 frequency in Eq. (12.2) by frequencies that are different for each polarization $\omega_{\alpha 1}$, i.e.,

12.2 Fitting the Kinetics to the Axon Parameters

$$\left[\frac{\partial^2}{\partial t^2} + \frac{2}{\tau_{\alpha 0}}\frac{\partial}{\partial t} + \omega_{\alpha 1}^2\right] D_\alpha(ld,t) = A \sum_{m=-\infty,\ m\neq l}^{m=\infty} E_\alpha\left(md, t - \frac{|l-m|d}{v}\right)$$
$$+ AE_{L\alpha}(ld,t) + AE_\alpha(ld,t),$$

where $A = Vnq^2/m$ is a shape-independent factor proportional to the number of ion carriers with concentration n in the volume of the spheroid with semi-axes a, b, c, $V = (4\pi/3)abc = (4\pi/3)a^3$ (the latter for a sphere); see Fig. 12.7. Taking into account that the plasmon frequency in a bulk electrolyte with ion concentration n equals $\omega_p = \sqrt{nq^2 4\pi/m}$, one can rewrite A as follows: $A = abc\omega_p^2/3 = \varepsilon a^3 \omega_1^2$ (the latter for a sphere, for which $\omega_1 = \omega_p/\sqrt{3\varepsilon}$). The ohmic losses can be included via the anisotropic term,

$$\frac{1}{\tau_{\alpha 0}} = \frac{v}{2\lambda_B} + \frac{Cv}{2a^\alpha},$$

where a^α is the dimension (semi-axis) of the spheroid in the direction α ($\alpha = a, b, c$ for a spheroid). The first isotropic term in the expression for $1/\tau_{\alpha 0}$ approximates ion scattering losses such as those occurring in the bulk electrolyte (isotropic losses), whereas the second term describes the losses due to the scattering of ions on the anisotropic boundary of the spheroid. This term can be neglected for longitudinal polarization because the neuron cord is continuous along the z direction. The dipole

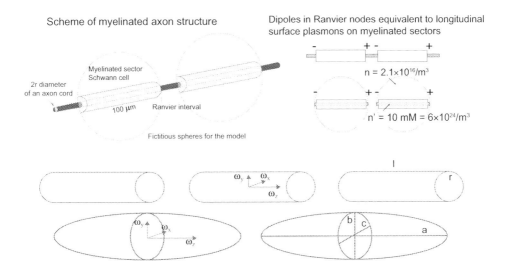

Figure 12.7 The effective-chain model of a continuous axon corrugated by periodic myelin sectors corresponds to a series of rods or prolate spheroids; because of the linearity of the dynamics equation with respect to the dipole field, it is possible to renormalize the equations for both polarizations without changing their structure.

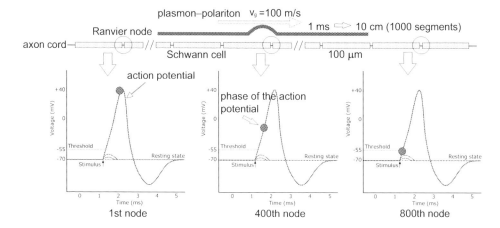

Figure 12.8 Schematic presentation of the firing of a nerve cell triggered by the propagation of a plasmon–polariton wave packet along the periodic structure of a myelinated axon. For a group velocity of the plasmon–polariton wave packet of the order of 100 m/s, within 1 ms the packet traverses 10 cm distance and initiates one by one the formation of the action potential on 1000 Ranvier nodes in sequence. The various time phases are indicated by black dots for exemplary nodes. The left-hand parts of the action potential spikes refer to the depolarization phase, and the right-hand parts to the repolarization phase. The very small peaks are failed initiations and the trough that follows the spike is the refractory period.

coupling is independent of the shape of the chain elements. The mutual independence of dipole oscillations with distinct polarizations described above follows from the linearity of the dynamics equation regardless of the metal or electrolyte conducting elements.

Because the dynamics equation is not affected by the anisotropy of the situation the solutions of the equation for each polarization have the same form as for the spherical case, except that the related frequencies of the dipole oscillations in each direction are modified and that there is a small correction to the orientation-dependent contribution of the scattering ratio (this part is related to the scattering of carriers by the boundaries and is not important for longitudinal polarization when the axon cord is continuous). Thus, we can independently renormalize the equation for the dipole oscillations for each polarization direction, introducing the resonance oscillation frequency for each direction $\omega_{\alpha 1}$ in a phenomenological manner. These frequencies can be estimated numerically, whereas for a sphere, $\omega_1 = \omega_p/\sqrt{3\varepsilon}$; in general, the longer the semi-axis, the lower is the related dipole oscillation frequency. Moreover, the size correction of the boundary scattering term must be taken into account as described above (favourably reducing the longitudinal plasmon–polariton damping for a prolate geometry. This renormalization was done for the model of the axon segment chain considered.

The periodic structure of a myelinated axon does not form a chain of electrolyte spheres but rather is a thin electrolyte cord with periodically distributed

12.2 Fitting the Kinetics to the Axon Parameters

sectors that are wrapped in myelin sheath separated by very short unmyelinated intervals, the Ranvier nodes. The periodic corrugated structure of the dielectric insulation allows, however, for collective plasmon wave-type oscillations, $\sim e^{iqz}$, with q governed by the periodicity, propagating along the cord; these oscillations are wave-type synchronic dipole polarized consecutive Ranvier intervals. Taking into account that the Ranvier nodes are short, these polarized dipoles are equivalent to the synchronically oscillating dipoles of the myelinated sectors. The latter can thus be treated as dipole surface plasmon oscillations propagating along a chain of prolate spheroids with small separations and longitudinal dipole polarization despite the continuous character of the cord, as shown in Fig. 12.3 (right-hand panels). To estimate the characteristics of such collective dipole wave-type excitations along the cord sectors, one can use the model of a chain of spherical electrolyte systems with suitably diluted ion concentration (resulting in the same total ion number as that in the cord fragments) and with decreased resonance frequency similar to that for the longitudinal polarization for highly prolate spheroids (or elongated cylinders). The model allows for the quantitative estimation of the relevant propagation characteristics and verification of whether plasmon–polariton dynamics fit the observed features of saltatory conduction in myelinated axons.

At consecutive Ranvier nodes, the plasmon–polariton wave packet triggers the release of ions through the nonmyelinated membrane and the formation of the action potential (renewed at each Ranvier node). The relatively low polarization induced by the plasmon–polariton initiates the opening of the Na^+ and K^+ across-membrane ion channels at a Ranvier node, which results in a characteristically large action signal due to the transfer of ions through these open channels, caused by the difference in the ion concentrations on opposite sides of the membrane. The entire cycle requires several milliseconds, but the initial increase in polarization caused by the rapid opening of the Na^+ channel occurs on the time scale of a single millisecond. Because the myelin layer wrapped by the Schwann cell prohibits cross-membrane ion transfer, the local polarization/depolarization of the internal cytoplasm of the axon occurs only at the Ranvier nodes, thus strengthening dipole formation in the axon fragment wrapped by the Schwann cell and the amplitude of the wave-type collective dipole oscillations in the whole chain. Each Ranvier node activity consecutively triggered by a plasmon–polariton increases the plasmon–polariton amplitude to a constant value despite the ohmic losses. The plasmon–polariton mode that traverses the axon structure, consisting of periodically polarized electrolyte segments wrapped with myelin sheaths, causes consecutive individual ignitions of the Ranvier nodes in sequence along the axon. This triggering role of the plasmon–polariton thus elucidates how the action signal jumps between neighbouring active Ranvier nodes.

The plasmon–polariton scenario in an ionic chain, even in this simplified version, could share certain features with the ability of nerve systems to achieve efficient and energetically economical electric signalling despite the rather poor ordinary electric conductivity of axons. The plasmon–polariton kinetics in the periodic structure of the axon may provide a convenient explanation for high-performance nerve signalling, providing quick signal propagation despite low ordinary conductivity over a practically unlimited range as energy is continuously supplied to cover the relatively small plasmon–polariton ohmic losses (which are lower than for an ordinary current). Moreover, plasmon–polaritons do not radiate any energy. The energy supply is residually provided by the ATP/ADP mechanism in neuron cells, which energetically contributes to the signal-dependent opening and closing of the Na^+ and K^+ channels in the Ranvier nodes and then to the restoration of steady-state conditions, via the active transport of ions across the membrane against the ionic concentration gradient, which requires an external energy supply. Moreover, the coincidence of the micrometre scale of the periodic structure of the Schwann cells in an axon (approximately 100 μm in length) with the typical requirements for the size of ionic chains supports the plasmon–polariton concept as the explanation for the transduction of the action potential along an axon, in agreement with the tentative quantitative fitting of the model.

It must be emphasized that plasmon–polaritons do not interact with external electromagnetic waves or, equivalently, with photons (even with frequencies adjusted to the plasmon–polariton frequency), as a consequence of the large difference between the group velocity of plasmon–polaritons and the velocity of photons $(c/\sqrt{\varepsilon})$. The resulting large discrepancy between the wavelengths of photons and plasmon–polaritons of the same energy prohibits mutual transformation of these two types of excitations because of momentum and energy conservation constraints. Therefore, signalling by means of collective wave-type dipole plasmon oscillations along a chain, i.e., plasmon–polaritons, can be neither detected nor perturbed by external electromagnetic radiation. This also fits well with neural signalling properties in the peripheral nervous system and in the white myelinated matter in the central nervous system. The ionic surface plasmon frequencies are independent of temperature, although the ionic volume plasmon frequencies are temperature dependent, as has been described in the previous chapter. However, temperature does influence the mean velocity of the ions, $v = \sqrt{3kT/m}$, thereby enhancing the ohmic losses with increasing temperature (see Eq. (12.3)), which in turn strengthens plasmon–polariton damping. Hence, at higher temperatures, higher external energy supplementation is required to maintain the same long-range propagation of plasmon–polaritons with a constant amplitude. This property is also consistent with experimental observations. However, the energy cost is the active ion transport through the membrane to recover the steady state of each

12.3 The Roles of Grey and White Matter in Information Processing

Ranvier node. Only a residual part of this energy is needed to support and maintain undamped plasmon–polariton propagation, which can be included in the scheme for a steady solution of the forced and damped oscillator, applied here to plasmon–polariton dynamics in a way similar to that for a metallic chain activated by a coupled external source (e.g., a coupled quantum dot system [199]).

The energy supply from consecutively excited Ranvier nodes continuously covers plasmon–polariton ohmic losses, which ensures that the amplitude of the action potential spike is renewed at each consctive Ranvier node. This results, however, in a convenient narrowing of the plasmon–polariton k wave packet, strengthened step by step at the Ranvier nodes. This narrowing shifts the packet towards the long-wave limit (smaller k), which follows from the Fourier picture of the Dirac delta (a narrower plasmon–polariton k wave packet is formed as a larger number of chain segments contribute). This favourable property allows for the precise ignition of sequential Ranvier nodes: unlimited propagation of a relatively narrow plasmon–polariton wave packet accurately triggers consecutive Ranvier nodes, which in turn strengthen the packet.

The utilization of radiatively undamped plasmon–polariton propagation in a chain of electrolyte subsystems may explain the efficient and long-range saltatory conduction in myelinated axons in the peripheral neural system and in the white matter of the brain and spinal cord. A plasmon–polariton model of the triggering of action potential firing along an axon myelinated by Schwann cells separated by Ranvier nodes in the peripheral nervous system fits well with the high conduction velocity observed and with the temperature and size dependence of the conduction velocity. This agreement, together with the immunity of plasmon–polaritons to external electromagnetic perturbations and detection, supports the reliability of this new model for saltatory conduction of the action potential, which is very efficient and energy-frugal despite the poor ordinary conductivity of axons.

12.3 Soft Plasmonics Application: The Role of Grey and White Matter in Information Processing in the Brain

The electrical activity of the nervous system refers to the transduction of a series of action potential spikes along the axons and dendrites of neuron cells connected together by electrochemical synapses into a highly entangled web. Not all trajectories for signal travel are accessible (they can be blocked by passive synapses or by their absence). The activation of selected synapses makes paths available for communication for longer or shorter time periods depending on the intensity and repetition of the synapse activation. Via the activation of selected (even in a random manner) synapses, the nervous system creates an entangled pattern or a braided web of trajectories. The role of the braids seems to be more important than the simple connections (the addressing) in the activated neuron network. Classification of the resulting entanglement patterns may be achieved in topological terms using so-called pure braid groups [200], as will be demonstrated in the next paragraph. Elements of these groups display various inequivalent patterns of entanglement of threads (neuron filaments), allowing the distinguishing and precise comparison of complicated filament muddles in a mathematically rigorous topological homotopy way. Two braids are homotopic if one can be transformed into the other by continuous deformation without cutting. For N threads, the number of various topologically inequivalent braid patterns is infinite, though countable [200]; in practice it is limited only by size restrictions such as the finite filament diameter and length. The various braid patterns greatly exceed ordinary addressing limits, for which N filaments give only $N!$ opportunities; this is low in comparison with the infinite number of N-filament braids.

If one electrically activates an entangled network by pushing an alternating charge current along the threads, the unique pattern of braiding with a specific number of loops might serve as a multiloop electromagnetic active circuit such as a distributed coil. The alternating current in this braid coil could generate an individual unique spatially local electromagnetic field configuration that could fall into resonance with another, even a distant, part of the web. Thus, we could associate information messages (elements of memory) imprinted in distinct elements of the pure braid group implemented in a nonlocal manner in fragments of the large neuron web of the brain via previous activation (ceation) and temporal consolidation of a certain subset of synapses. A message could be temporarily stored in the neuron braid corresponding to a certain mathematical braid from the pure braid group, as long as the selected engaged synapses are still active (or are ready to be active). This information can be next invoked by electromagnetic resonance with a similar braid within the brain and temporarily created in another brain structure (e.g., in the hippocampus) as the result of an influx of neuron signals from the surroundings via the senses. The new message tentatively imprinted in a temporary braid in the hippocampus might excite the same braid as was recorded in the past in the cortex via electromagnetic resonance and, in this way, may cause the identification of a new message in the mind. Let us note here that the dendrites in the grey matter prepare thousands of lateral spikes ready to create new synapses between neurons.

To utilize the braids in the proposed electromagnetic recognition process, alternating electric currents are necessary to initiate the resonance response. Such AC electricity may generate the brain waves that are constantly observed via EEG in the grey matter of the cerebral cortex. For the transduction of neuron signals in nonmyelinated axons and in all dendrites in the cortex, the diffusive movement of ions is employed in elongated and branched neuron cells. This electric current-

12.3 The Roles of Grey and White Matter in Information Processing

type communication is described very well by the cable theory originated by W. Thomson (1854) [52] and is widely applied to explain the neural electrophysiology of dendrites and nonmyelinated axons.

Hence, one can expect that the storage and identification of information occurs in the cerebral cortex, taking advantage of the huge information capacity of the entangled neuron filaments, which creates various braid structures enumerated by pure braid-group elements. The topologically precise homotopy resolution and quick and complex access to the stored messages in entangled braids [200–203] via electromagnetic resonance, additionally using a few distinct frequencies of brain waves, might help to understand the functionality of memory and pattern recognition by a mind. All these phenomena occur in the grey matter of the cortex, where signal transduction is accomplished by the diffusion of ions equivalent to an ordinary alternating current, which induces a local electromagnetic field in a coil. Different types of brain wave frequencies might conveniently implement simultaneous independent resonance channels for identification of the multiloop coils created by neuron filament braids [203].

As mentioned above, braid groups were introduced to describe entangled bundles of the classical trajectories of N-particle systems [200]. Two types of braid groups have been investigated and applied; for the first type, full braid groups, the indistinguishability of identical particles is assumed and is applied to characterize the quantum statistics of particles [202]; the second type comprises the so-called pure braid groups for distinguishable particles [200, 202]. The latter are suitable for describing and classifying highly entangled multifilament webs. Both full and pure braid groups are infinite for N particles moving on a plane but are finite when the particles move in three dimensions (3D). In 3D there is room to disentangle the lines representing the particle trajectories, which is impossible in 2D (in 2D, particles must twist around each other and cannot hop beyond or beneath each other). The pure braid group of N distinguishable particles on a plane (in 2D) may be presented as a set of N threads in 3D with their ends fixed to steady points but braided in between in an arbitrary manner. Thus, the 3D web of entangled threads such as neurons in a brain can be mapped onto a 2D pure braid group. The patterns of such braids are unique and cannot be transformed into each other without cutting threads. However, the patterns can be precisely enumerated by pure braid group elements. If we imagine that the N initial and N final points of the threads are fixed in 3D space, then a structure linking them into braids – the 2D N-element pure braid-group elements – reproduces in a one-to-one fashion 3D webs of arbitrary entangled nets of filaments, such as a web of neurons linked via selected synapses in the brain.

12.3.1 Possible Link with the Topological Model of a Neuron Web

The topological characteristics of entangled filament webs, such as neuron webs, may be described in terms of homotopy, reflecting the complexity of the trajectories of many-particle systems [204]. The first homotopy group of the space D, labelled $\pi_1(D)$, is the collection of topologically nonequivalent classes of closed trajectories (i.e., they are nonhomotopic; one class cannot be continuously deformed into another) in the space D. If this space is the configuration space of a system of N particles, each of which has its own trajectory on a manifold M, then the first homotopy group π_1 is called a braid group [204, 201]. The configuration space for N identical particles located on a manifold M (e.g., R^n) is defined as follows: $F_N(M) = M^N \setminus \Delta$, for distinguishable identical particles, where M^N is an N-fold normal product of the manifold M and Δ is a set of diagonal points such that the coordinates of two or more particles coincide, which has to be subtracted in order to conserve the number of particles in the system (particles cannot hide behind one another). The braid groups represent the topologically different available trajectories of a system of N particles (without reference to any specific dynamics). Thus, the pure braid group which we will utilize to describe a neuron web is defined as

$$\pi_1(F_N(M)) = \pi_1(M^N \setminus \Delta). \tag{12.8}$$

When trajectories (representing the neuron filaments) are mutually entangled in multiline braid form, the initial ordering of the particles must be unchanged. The elementary generators of the pure braid group are assigned as l_{ij} (Fig. 12.9) [200]; they correspond to the simplest entanglement of two trajectories of a particle pair, the ith and jth, while the rest of the particle trajectories remain untangled. The generators l_{ij} can be expressed by σ_i (an exchange of neighbouring particles, say the ith and $(i+1)$th, that does not conserve their positions [200]):

$$l_{ij} = \sigma_{j-1}\sigma_{j-2} \cdots \sigma_{i+1}\sigma_i^2 \sigma_{i+1}^{-1} \cdots \sigma_{j-2}^{-1}\sigma_{j-1}^{-1}, \qquad 1 \geq i \geq j \geq N - 1. \tag{12.9}$$

The pure braid group can be considered as an abstract mathematical group generated by l_{ij} defined by the following relations [200, 202] (see Fig. 12.10):

$$l_{rs}^{-1} l_{ij} l_{rs} = \begin{cases} l_{ij}, & i < r < s < j, \\ l_{ij}, & r < s < i < j, \\ l_{rj} l_{ij} l_{rj}^{-1}, & r < i = s < j, \\ l_{rj} l_{sj} l_{ij} l_{sj}^{-1} l_{rj}^{-1}, & i = r < s < j, \\ l_{rj} l_{sj} l_{rj}^{-1} l_{sj}^{-1} l_{ij} l_{sj} l_{rj} l_{sj}^{-1} l_{rj}^{-1}, & r < i < s < j. \end{cases} \tag{12.10}$$

We propose to identify a fragment of the neuron web, connected via active synapses predefined by a prior record of certain information in the filament web, as a certain pattern from the pure braid group. In particular, one can use binary

12.3 The Roles of Grey and White Matter in Information Processing

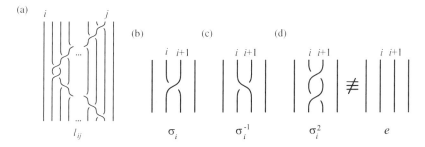

Figure 12.9 (a) Generator l_{ij} of the pure braid group; (b) the exchange of neighbouring particles, σ_i; (c) its inverse, σ_i^{-1}; (d) and the reason why braids on a plane are complicated.

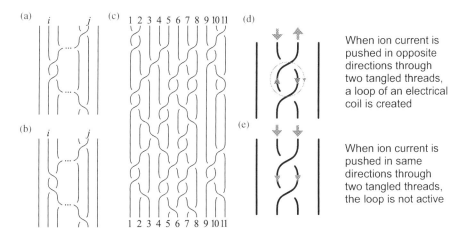

Figure 12.10 (a) Generator l_{ij} and (b) its inverse l_{ij}^{-1} of the pure braid group; (c) an example of the $N = 11$ pure braid; (d) the creation of an electromagnetic active loop of coil by entanglement of threads in a pure braid; (e) the creation of a nonactive loop.

code to record any message in a three-filament bundle (two of the generators of the corresponding $N = 3$ pure braid group serve to code the 0 and 1 bits, and the third generator encodes the end of a word). It is possible to code an N-letter alphabet message in an Nth-order pure braid group [203]. It has been demonstrated that the ratio of the information efficiency of an Nth-order pure braid group to the resources needed to organize a physical web (such as a neuron web with N entangled filaments) attains a maximum at $N = 20$–30, which agrees with the number of phonics in the majority of phonic languages [202]. This result may suggest that ordinary words in a language are coded in a relatively small number of filaments in entangled bunches.

In the braid scenario it was proposed that information is coded in the brain in entanglement patterns in pieces of the neuron web selected by the activation of certain synapses. This coding is a topologically nonlocal type of coding and is more efficient than the conventional addressing of synapse registers (N-synapse addressing offers only $N!$ various connections with another N-synapse register, whereas the number of distinct braids of threads linking these registers is infinite, limited only by the physical constraints of finite size, diameter, and length of the threads).

To identify neuron braids we propose electromagnetic resonance utilizing the ion currents that flow in the neuron filaments of the grey matter. A different role is played by the white matter, with axon cords corrugated by the periodic myelin sheaths. The effective and fast saltatory transduction of action potentials through the long myelinated axons of the peripheral nerves and in the white matter of the central nervous system undeniably lies behind the motorics and other communication functions of the body. This saltatory transduction requires at least 100 times faster signalling than that offered by diffusion-type ion flows, according to the cable theory. This high-speed signalling is observed in myelinated axons and is described as saltatory conduction of the action potential [192]. As mentioned above in myelinated axons, the action potential spikes occur at the Ranvier nodes, small unmyelinated breaks periodically distributed along a myelinated axon [50]. The spikes are renewed on the subsequent series of these nodes according to the well-known Hodgkin–Huxley membrane model [50]. However, the signal rapidly hops in between the Ranvier nodes, along the myelinated sectors of an axon; this strongly accelerates the transduction of the action signal along the periodically myelinated axon. The velocity of this saltatory conduction is at least two orders faster than the diffusion ion current. Such an acceleration in transduction is certainly required for long-distance signalling in the peripheral nervous system but is also required for quick communication in the white matter of the brain and spinal cord. In the brain, the white matter is located in inner structures beneath the cortex, but in the spinal cord the reverse is the case. This difference might be connected to the different roles of the grey matter in the cerebral cortex and in the spinal cord. The cortex is the place where memory is stored and where new messages from the surroundings accessed via the senses are identified or compared with the memory, whereas the grey matter in the spinal cord serves to identify the internal body information necessary to control body functions. The latter must be precise and errorless; thus, the grey matter in the spinal cord should be isolated from the surroundings in the best possible way, here by an outer white matter layer. It has been demonstrated that an external electromagnetic field, especially the magnetic field component, can influence and transcranially perturb the identification process in the brain cortex; this would be, however, very dangerous in the spinal cord, thus in the latter the grey structure must be better isolated than the cortex in the brain.

12.3 The Roles of Grey and White Matter in Information Processing

The white matter participates in communication but is not used in the recognition of braids with stored information, which is the role of the grey matter. Signal transduction in myelinated axons is not achieved by any electrical current and thus cannot participate in the electromagnetic identification of braids.

This scenario is consistent with the plasmon–polariton wave-type character of saltatory conduction in myelinated axons presented above. Saltatory conduction is apparently of a wave type because conduction is maintained even if a myelinated axon is divided into two disjoint pieces with the ends separated slightly by an insulating barrier that would be impossible for a diffusive ion current to cross.

We have discussed a model for saltatory conduction based on the kinetic properties of collective plasmon–polariton modes propagating along linear and periodically myelinated electrolyte systems, which, for axons, is the thin cord of the nerve cell periodically wrapped by Schwann cells creating thick myelin sheaths (Schwann cells myelinate axons in the peripheral nervous system, whereas in the central nervous system the white matter is built from axons myelinated by oligodendrocyte cells). Plasmon–polaritons were previously investigated and understood based on the well-developed nano-plasmonics of metal chains [1, 5]. The long-range low-damped propagation of plasmon–polaritons has been demonstrated along a metallic nano-chain [40]. Let us recall the main properties of these collective excitations occurring in metallic nano-chains on a conductor–insulator interface [116, 4] by the hybridization of surface plasmons (i.e., fluctuations of the charge density of the electron fluid on the metal surface) with the electromagnetic wave.

1. The much lower velocity of plasmon–polaritons in comparison to the velocity of light gives a much shorter plasmon–polariton wavelength than the wavelength of light for a given frequency; this causes a strong discrepancy in the momenta of plasmon–polaritons and photons with the same energy. Thus, plasmon–polaritons do not interact with electromagnetic waves, i.e., photons cannot be excited or absorbed by plasmon–polaritons.
2. There are no radiative losses of plasmon–polaritons; plasmon–polariton attenuation is due only to ohmic losses, which makes metallic nano-chains almost perfect wave guides for plasmon–polaritons,
3. The long-range and practically undamped propagation of plasmon–polaritons is observed experimentally [40, 39].

Certain unique properties of plasmon–polaritons in metallic nano-chains are of special interest, as these properties can be repeated in periodic linear alignments of electrolyte systems with ions instead of electrons as the charged carriers. Owing to the larger mass of ions compared with that of electrons and the lower concentration of ions in electrolytes compared with that of electrons in metals, plasmon resonances in finite ionic systems (e.g., in liquid electrolytes confined to a finite volume by appropriately formed membranes, which are frequently found in biological cell organization) occur on the scale of micrometres rather than nanometres and at frequencies or energies several orders of magnitude lower (depending on the ion concentration).

For spherical electrolyte microsystems, the surface and volume plasmons are handled analogously to those in metallic nanospheres, as demonstrated in the previous chapter. The ionic surface plasmon frequencies are given for the lth multipole mode by the formula $\omega_l = \omega_p \sqrt{l/(\varepsilon(2l+1))}$, where the bulk plasmon frequency $\omega_p = \sqrt{4\pi q^2 n/m}$ (n is the ion concentration, q and m are the ion charge and mass, respectively, and ε is the relative dielectric permittivity of the surroundings). For dipole-type surface plasmons (for which $l = 1$), the frequency is given by a Mie-type formula, [18, 23] $\omega_1 = \omega_p/\sqrt{3\varepsilon}$.

In a linear chain of ionic systems, the radiation losses are reduced to zero in exactly the same manner as in metallic nano-chains [193, 43, 44]. The high radiation energy losses expressed by the Lorentz friction at each chain element [55, 76] are ideally compensated by the radiation from the rest of the chain. Since the radiative losses are ideally balanced by mutual radiation exchange, only the relatively low irreversible ohmic energy dissipation remains. The collective plasmon–polaritons can thus propagate in the ionic chain with strongly reduced damping as in an almost perfect wave guide. If energy is slightly supplemented to compensate the ohmic losses, the propagation can reach arbitrarily long distances without any damping, which is observed in saltatory conduction in myelinated axons; moreover, the plasmon–polaritons in these axons can be neither detected nor perturbed by an external electromagnetic field.

The advantages of plasmon–polariton kinetics that benefit the myelinated neuron signalling properties observed are listed as follows.

- Plasmon–polaritons cannot be perturbed nor detected by electromagnetic waves (EEG signals are related to grey matter activity with ordinary conduction, according to the cable theory, which produces weak electromagnetic fields, whereas myelinated white matter does not generate EEG signals).
- Immunity to electromagnetic perturbation is convenient in the long links of myelinated neurons in the peripheral nervous system (motorics and sensing cannot be disturbed by an external electromagnetic field).
- The electromagnetic field of plasmon–polaritons is compressed along the tight tunnel around the axon cord.
- The myelin sheath must be sufficiently thick, not for insulation but for the creation of a dielectric tunnel in the surrounding intercellular electrolyte that allows for plasmon–polariton formation (reducing the myelin sheath thickness perturbs the formation of plasmon–polaritons and lowers the plasmon–polariton group velocity, resulting e.g., in multiple sclerosis).

12.3 The Roles of Grey and White Matter in Information Processing

- Increasing temperature causes increasing ohmic losses but does not reduce the plasmon–polariton group velocity (however, more energy ADP/ATP, residual to the action potential spike formation, is needed to compensate the increased ohmic losses at higher temperatures).
- No radiative losses occur for plasmon–polaritons; if plasmon–polaritons are residually aided by spike formation (renovation) at consecutive Ranvier nodes, an arbitrary long-range plasmon–polariton kinetics is possible with constant dipole-signal amplitude.
- The plasmon–polariton group velocity increases with the cross section of the axon cord.
- The plasmon–polariton group velocity for real neuron electrolytes and size parameters fits the observed velocity of saltatory conduction.
- Wave-type plasmon–polariton kinetics explains the maintenance of signal transduction despite small breaks and gaps in the neuron cord or damage of some Ranvier nodes (these observations cannot be explained within the cable theory or by the compression-soliton model).
- One-direction kinetics occurs if propagation starts from one end of an axon, but two-direction kinetics occurs if a plasmon–polariton is ignited from a certain central point of a passive axon.

Taking into account the properties listed above of plasmon–polariton conduction, one might thus propose a nonlocal topological braid-group-type approach to information processing in the brain, considering the specific and different electricity of the grey and white matter of neurons. The web of neuron filaments in grey matter would serve to encode the information in the entanglement of filaments selected by the activation of certain synapses, because the electricity of dendrites and of nonmyelinated axons (in the grey matter) is of an electrical current type, providing an electromagnetic signature when an AC signal is transduced through the neuron network. The method of electromagnetic resonance identification of the information stored in braided segments of the neuron web can be thus suggested. This method favours the grey matter with ordinary ion currents of the diffusive type (as in the cable model) but excludes the myelinated white matter, which has wave-type plasmon–polariton signalling without any electromagnetic signature. The saltatory conduction in myelinated axons in the white matter is not active with respect to electromagnetic resonance, as is typical for plasmon–polaritons. A model of ionic plasmon–polariton propagation to explain saltatory conduction in myelinated axons has thus been proposed. This wave-type propagation of the action potential is actually passive with respect to the electromagnetic response and cannot contribute to the braid information processing in the grey matter, though it conveniently fulfils a communication role in the peripheral and central nervous systems. The plasmon–polariton model explains the efficient, very fast, and long-range saltatory conduction in myelinated axons in the peripheral neural system and in the white matter of the brain and spinal cord, where the velocity and energy-efficiency of communication are of the most significance. The coincidence with observations supports the appropriateness of this new model proposed for saltatory conduction of the action potential, which, on the other hand, fits in a complementary way the different role and functioning of the grey matter. Both myelinated and nonmyelinated structures share between them different but complementary roles in the whole nervous system.

References

[1] W. L. Barnes, A. Dereux, and T. W. Ebbesen, 'Surface plasmon subwavelength optics', *Nature* **424**, 824, 2003.
[2] H. A. Atwater and A. Polman, 'Plasmonics for improved photovoltaic devices', *Nat. Materi.* **9**, 205, 2010.
[3] S. A. Maier, *Plasmonics: Fundamentals and Applications*, Springer, Berlin, 2007.
[4] F. J. G. de Abajo, 'Optical excitations in electron microscopy', *Rev. Mod. Phys.* **82**, 209, 2010.
[5] J. M. Pitarke, V. M. Silkin, E. V. Chulkov, and P. M. Echenique, 'Theory of surface plasmons and surface-plasmon polaritons', *Rep. Prog. Phys.* **70**, 1, 2007.
[6] P. Berini, 'Long-range surface plasmon polaritons', *Adv. Opt. Photonics* **1**, 484, 2009.
[7] L. Jacak, A. Wojs, and P. Hawrylak, *Quantum Dots*, Springer, Berlin, 1998.
[8] W. Jacak, J. Krasnyj, L. Jacak, and R. Gonczarek, *Decoherence of Orbital and Spin Degrees of Freedom in QDs*, WUT University Press, Wrocław, 2009.
[9] A. A. Abrikosov, L. P. Gorkov, and I. E. Dzialoshinskii, *Methods of Quantum Field Theory in Statistical Physics*, Dover Publications, New York, 1975.
[10] E. M. Lifshitz and L. P. Pitaevskii, *Statisticeskaja fizika, czast 2*, Nauka, Mascow, 1978.
[11] A. L. Fetter and J. D. Walecka, *Quantum Theory of Multi-Particle Systems*, PWN, Warsaw, 1988.
[12] A. A. Abrikosov, *Wvedenie w Teoriu Normalnych Metalov*, Nauka, Moscow, 1972.
[13] D. Pines and P. Nozières, *The Theory of Quantum Liquids*, W. A. Benjamin, New York, 1966.
[14] D. Pines and D. Bohm, 'A collective description of electron interactions: II. Collective vs. individual particle aspects of the interactions', *Phys. Rev.* **85**, 338, 1952.
[15] D. Pines and D. Bohm, 'A collective description of electron interactions: III. Coulomb interactions in a degenerate electron gas', *Phys. Rev.* **92**, 609, 1953.
[16] D. Pines, *Elementary Excitations in Solids*, ABP Perseus Books, Massachusetts, 1999.
[17] C. F. Bohren and D. R. Huffman, *Absorption and Scattering of Light by Small Particles*, Wiley, New York, 1983.
[18] G. Mie, 'Beiträge zur Optik trüber medien, speziell kolloidaler Metallösungen', *Ann. Phys.* **330**(3), 377–445, 1908.
[19] M. Brack, 'The physics of simple metal clusters: self-consistent jellium model and semiclassical approaches', *Rev. Mod. Phys.* **65**, 667, 1993.

[20] W. Ekardt, 'Size-dependent photoabsorption and photoemission of small metal particles', *Phys. Rev. B* **31**, 6360, 1985.

[21] V. V. Kresin, 'Collective resonances and response properties of electrons in metal clusters', *Phys. Rep.* **220**, 1, 1992.

[22] S. Link and M. A. El-Sayed, 'Size and temperature dependence of the plasmon absorption of colloidal gold nanoparticles', *J. Phys. Chem. B* **103**, 4212, 1999.

[23] J. Jacak, J. Krasnyj, W. Jacak, R. Gonczarek, A. Chepok, and L. Jacak, 'Surface and volume plasmons in metallic nanospheres in semiclassical RPA-type approach; near-field coupling of surface plasmons with semiconductor substrate', *Phys. Rev. B* **82**, 035418, 2010.

[24] M. L. Brongersma, J. W. Hartman, and H. A. Atwater, 'Electromagnetic energy transfer and switching in nanoparticle chain arrays below the diffraction limit', *Phys. Rev. B* **62**, R16356, 2000.

[25] M. Vollmer and U. Kreibig, 'Optical properties of metal clusters', *Springer Ser. Mat. Sci.* **25**, 1995.

[26] W. A. Jacak, 'Lorentz friction for surface plasmons in metallic nanospheres', *J. Phys. Chem. C* **119**(12), 6749–6759, 2015.

[27] W. A. Jacak, 'Size-dependence of the Lorentz friction for surface plasmons in metallic nanospheres', *Opt. Express* **23**, 4472–4481, 2015.

[28] K. Kluczyk and W. Jacak, 'Damping-induced size effect in surface plasmon resonance in metallic nano-particles: comparison of RPA microscopic model with numerical finite element simulation (COMSOL) and Mie approach', *J. Quant. Spectrosc. Radiat. Transfer* **78**, 168, 2016.

[29] K. Kolwas, A. Derkachova, and M. Shopa, 'Size characteristics of surface plasmons and their manifestation in scattering properties of metal particles', *J. Quant. Spectrosc. Radiat. Transfer* **110**(14), 1490–1501, 2009.

[30] W. Jacak, E. Popko, A. Henrykowski, E. Zielony, G. Luka, R. Pietruszka *et al.*, 'On the size dependence and the spatial range for the plasmon effect in photovoltaic efficiency enhancement', *Sol. Energy Mater. Sol. Cells* **147**, 1, 2016.

[31] M. Jeng, Z. Chen, Y. Xiao, L. Chang, J. Ao, Y. Sun *et al.*, 'Improving efficiency of multicrystalline silicon and CIGS solar cells by incorporating metal nanoparticles', *Materials* **8**(10), 6761–6771, 2015.

[32] S. Pillai, K. R. Catchpole, T. Trupke, G. Zhang, J. Zhao, and M. A. Green, 'Enhanced emission from Si-based light-emitting diodes using surface plasmons', *Appl. Phys. Lett.* **88**, 161102, 2006.

[33] D. M. Schaadt, B. Feng, and E. T. Yu, 'Enhanced semiconductor optical absorption via surface plasmon excitation in metal nanoparticles', *Appl. Phys. Lett.* **86**, 063106, 2005.

[34] M. Westphalen, U. Kreibig, J. Rostalski, H. Lüth, and D. Meissner, 'Metal cluster enhanced organic solar cells', *Sol. Energy Mater. Sol. Cells* **61**, 97, 2000.

[35] A. J. Morfa, K. L. Rowlen, T. H. Reilly, M. J. Romero, and J. Lagemaat, 'Plasmon-enhanced solar energy conversion in organic bulk heterojunction photovoltaics', *Appl. Phys. Lett.* **92**, 013504, 2008.

[36] M. A. Green and S. Pillai, 'Harnessing plasmonics for solar cells', *Nat. Photon* **6**, 130–132, 2012.

[37] G. Goubau, 'Surface waves and their application to transmission lines', *J. Appl. Phys.* **21**, 119, 1950.

[38] P. A. Huidobro, M. L. Nesterov, L. Martin-Moreno, and F. J. Garcia-Vidal, 'Transformation optics for plasmonics', *Nano Lett.* **10**, 1985, 2010.

[39] S. A. Maier, P. G. Kik, and H. A. Atwater, 'Optical pulse propagation in metal nanoparticle chain waveguides', *Phys. Rev. B* **67**, 205402, 2003.
[40] S. A. Maier and H. A. Atwater, 'Plasmonics: localization and guiding of electromagnetic energy in metal/dielectric structures', *J. Appl. Phys.* **98**, 011101, 2005.
[41] D. S. Citrin, 'Plasmon polaritons in finite-length metal-nanoparticle chains: the role of chain length unravelled', *Nano Lett.* **5**, 985, 2005.
[42] D. Citrin, 'Coherent excitation transport in metal-nanoparticle chains', *Nano. Lett.* **4**, 1561, 2004.
[43] V. A. Markel and A. K. Sarychev, 'Propagation of surface plasmons in ordered and disordered chains of metal nanospheres', *Phys. Rev. B* **75**, 085426, 2007.
[44] W. Jacak, 'On plasmon polariton propagation along metallic nano-chain', *Plasmonics* **8**, 1317, 2013. DOI: 10.1007/s11468-013-9528-8.
[45] W. Jacak, 'Exact solution for velocity of plasmon–polariton in metallic nano-chain', *Optics Express* **22**, 18958, 2014.
[46] W. Jacak, J. Krasnyj, and A. Chepok, 'Plasmon–polariton properties in metallic nanosphere chains', *Materials* **8**, 3910, 2015.
[47] W. Jacak, 'On plasmon polariton propagation along metallic nano-chain', *Plasmonics* **8**, 1317, 2013.
[48] W. Jacak, 'Plasmons in finite spherical electrolyte systems: RPA effective jellium model for ionic plasma excitations', *Plasmonics*, **11**, 637–651, 2016. Epub 2015, DOI: 10.1007/s11468-015-0064-6.
[49] W. Jacak, 'Propagation of collective surface plasmons in linear periodic ionic structures: plasmon polariton mechanism of saltatory conduction in axons', *J. Phys. Chem. C* **119**(18), 10015, 2015.
[50] D. Debanne, E. Campanac, A. Bia?owąs, E. Carlier, and G. Alcaraz, 'Axon physiology', *Physiol. Rev.* **91**, 555, 2011.
[51] S. Brzychczy and R. Poznaski, *Mathematical Neuroscience*, Academic Press, San Diego, 2011.
[52] W. Thomson, 'On the theory of the electric telegraph', *Proc. R. Soc. London* **7**, 382, 1854.
[53] T. Heimburg and A. D. Jackson, 'On soliton propagation in biomembranes and nerves', *Proc. Natl Acad. Sci. USA* **102**, 9790, 2005.
[54] N. W. Ashcroft and N. D. Mermin, *Solid State Theory*, Holt, Rinehart, and Winston, New York, 1976.
[55] L. D. Landau and E. M. Lifshitz, *Field Theory*, Nauka, Moscow, 1973.
[56] L. D. Landau, *Zh. Eksp. Teor. Fiz.* **30**, 1058, 1956.
[57] W. P. Silin, *Zh. Eksp. Teor. Fiz.* **33**, 495, 1957.
[58] P. Nozières, *Theory of Interacting Fermi Systems*, W. A. Benjamin, New York, 1964.
[59] P. S. Kiriejew, *Physics of Semiconductors*, PWN, Warsaw, 1969.
[60] L. D. Landau, *Zh. Eksp. Teor. Fiz.* **32**, 59, 1957.
[61] L. D. Landau, *Zh. Eksp. Teor. Fiz.* **84**, 262, 1958.
[62] I. Y. Pomeranchuck, *Zh. Eksp. Teor. Fiz.* **35**, 524, 1958.
[63] P. Nozières and J. M. Luttinger, *Phys. Rev.* **127**, 1423, 1962.
[64] P. Nozières and J. M. Luttinger, *Phys. Rev.* **127**, 1431, 1962.
[65] L. Jacak, *Nonlinear Topics in the Theory of Fermi Liquids*, vol. 11 of *Monografie*, Oficyna Wydawnicza PWr, Wrocław, 1987.
[66] G. N. Eliashberg, *Zh. Eksp. Teor. Fiz.* **42**, 1658, 1962.
[67] A. I. Larkin and A. N. Migdal, *Zh. Eksp. Teor. Fiz.* **44**, 1703, 1963.
[68] J. Czerwonko, *Acta Phys. Polon.* **37**, 335, 1967.
[69] J. Czerwonko, *Zh. Eksp. Teor. Fiz.* **71**, 1099, 1976.

[70] N. F. Mott and H. Jones, *The Theory of Metals and Alloys*, Oxford University Press, New York, 1936.
[71] M. Brack, 'Multipole vibration of small alkali-metal spheres in a semiclassical description', *Phys. Rev. B* **39**, 3533, 1989.
[72] I. S. Gradshteyn and I. M. Ryzhik, *Table of Integrals, Series and Products*, Academic Press, Boston MA, 1994.
[73] J. Wu, S. C. Mangham, V. Reddy, M. Manasreh, and B. Weaver, 'Surface plasmon enhanced intermediate band based quantum dots solar cell', *Sol. Energy Mater. Sol. Cells* **102**, 44–49, 2012.
[74] N. Kalfagiannis, P. Karagiannidis, C. Pitsalidis, N. Panagiotopoulos, C. Gravalidis, S. Kassavetis et al., 'Plasmonic silver nanoparticles for improved organic solar cells', *Sol. Energy Mater. Sol. Cells* **104**, 165–174, 2012.
[75] L. D. Landau and L. M. Lifshitz, *Quantum Mechanics. Nonrelativistic Theory*, Pergamon Press, New York, 1965.
[76] J. D. Jackson, *Classical Electrodynamics*, John Wiley and Sons, New York, 1998.
[77] W. Jacak, J. Krasnyj, J. Jacak, R. Gonczarek, A. Chepok, L. Jacak et al., 'Radius dependent shift of surface plasmon frequency in large metallic nanospheres: theory and experiment', *J. Appl. Phys.* **107**, 124317, 2010.
[78] P. J. Compaijen, V. A. Malyshev, and J. Knoester, 'Surface-mediated light transmission in metal nanoparticle chains', *Phys. Rev. B* **87**, 205437, 2013.
[79] W. Ekardt, 'Anomalous inelastic electron scattering from small metal particles', *Phys. Rev. B* **33**, 8803, 1986.
[80] C. Yannouleas, R. A. Broglia, M. Brack, and P. F. Bortignon, 'Fragmentation of the photoabsorption strength in neutral and charged metal microclusters', *Phys. Rev. Lett.* **63**, 255, 1989.
[81] A. Rubio and L. Serra, 'Dielectric screening effects on the photoabsorption cross section of embedded metallic clusters', *Phys. Rev. B* **48**, 18222, 1993.
[82] C. Sönnichsen, T. Franzl, T. Wilk, G. von Plessen, and J. Feldmann, 'Plasmon resonances in large noble-metal clusters', *New J. Phys.* **4**, 93, 2002.
[83] S. Link and M. A. El-Sayed, 'Shape and size dependence of radiative, non-radiative and photothermal properties of gold nanocrystals', *Int. Rev. Phys. Chem.* **19**, 409, 2000.
[84] K. Kolwas and A. Derkachova, 'Damping rates of surface plasmons for particles of size from nano- to micrometers; reduction of the nonradiative decay', *J. Quant. Spectrosc. Radiat. Transfer* **114**, 45, 2013.
[85] J. Mock, M. Barbic, D. Smith, D. Schultz, and S. Schultz, 'Shape effects in plasmon resonance of individual colloidal silver nanoparticles', *J. Chem. Phys.* **116**(15), 6755–6759, 2002.
[86] C. Noguez, 'Surface plasmons on metal nanoparticles: the influence of shape and physical environment', *J. Phys. Chem. C* **111**(10), 3806–3819, 2007.
[87] S. Dickreuter, J. Gleixner, A. Kolloch, J. Boneberg, E. Scheer, and P. Leiderer, 'Mapping of plasmonic resonances in nanotriangles', *Beilstein J. Nanotechnol.* **4**(1), 588–602, 2013.
[88] V. E. Ferry, J. N. Munday, and H. A. Atwater, 'Design considerations for plasmonic photovoltaics', *Advanced Mat.* **22**(43), 4794–4808, 2010.
[89] J. A. Schuller, E. S. Barnard, W. Cai, Y. C. Jun, J. S. White, and M. L. Brongersma, 'Plasmonics for extreme light concentration and manipulation', *Nature Mat.* **9**(3), 193–204, 2010.
[90] R. Gans, 'Über die Form ultramikroskopischer Goldteilchen', *Ann. Phys.* **342**(5), 881–900, 1912.

[91] A. Taflove and S. C. Hagness, 'Computational electrodynamics', Artech House, 2000.

[92] J. M. Montgomery, T.-W. Lee, and S. K. Gray, 'Theory and modeling of light interactions with metallic nanostructures', *J. Phys.: Condensed Matter* **20**(32), 323201, 2008.

[93] F. G. de Abajo and A. Howie, 'Retarded field calculation of electron energy loss in inhomogeneous dielectrics', *Phys. Rev. B* **65**(11), 115418, 2002.

[94] C. G. Khoury, S. J. Norton, and T. Vo-Dinh, 'Investigating the plasmonics of a dipole-excited silver nanoshell: Mie theory versus finite element method', *Nanotechnology* **21**(31), 315203, 2010.

[95] V. Myroshnychenko, J. Rodríguez-Fernández, I. Pastoriza-Santos, A. M. Funston, C. Novo, P. Mulvaney *et al.*, 'Modelling the optical response of gold nanoparticles', *Chem. Soc. Rev.* **37**(9), 1792–1805, 2008.

[96] P. K. Jain, K. S. Lee, I. H. El-Sayed, and M. A. El-Sayed, 'Calculated absorption and scattering properties of gold nanoparticles of different size, shape, and composition: applications in biological imaging and biomedicine', *J. Phys. Chem. B* **110**(14), 7238–7248, 2006.

[97] J. Jin, *The Finite Element Method in Electromagnetics*, John Wiley & Sons, New York, 2014.

[98] C. G. Khoury, S. J. Norton, and T. Vo-Dinh, 'Plasmonics of 3-D nanoshell dimers using multipole expansion and finite element method', *ACS Nano.* **3**(9), 2776–2788, 2009.

[99] X. Cui and D. Erni, 'Enhanced propagation in a plasmonic chain waveguide with nanoshell structures based on low- and high-order mode coupling', *JOSA A* **25**(7), 1783–1789, 2008.

[100] P. B. Johnson and R.-W. Christy, 'Optical constants of the noble metals', *Phy. Rev. B* **6**(12), 4370, 1972.

[101] W. Haiss, N. T. Thanh, J. Aveyard, and D. G. Fernig, 'Determination of size and concentration of gold nanoparticles from uv-vis spectra', *Anal. Chem.* **79**(11), 4215–4221, 2007.

[102] P. N. Njoki, I.-I. S. Lim, D. Mott, H.-Y. Park, B. Khan, S. Mishra *et al.*, 'Size correlation of optical and spectroscopic properties for gold nanoparticles', *J. Phys. Chem. C* **111**(40), 14664–14669, 2007.

[103] A. L. Koh, K. Bao, I. Khan, W. E. Smith, G. Kothleitner, P. Nordlander *et al.*, 'Electron energy-loss spectroscopy (EELS) of surface plasmons in single silver nanoparticles and dimers: influence of beam damage and mapping of dark modes', *ACS Nano.* **3**(10), 3015–3022, 2009.

[104] K. Okamoto, I. Niki, A. Scherer, Y. Narukawa, and Y. Kawakami, 'Surface plasmon enhanced spontaneous emission rate of InGaN/ GaN quantum wells probed by time-resolved photoluminescence spectroscopy', *Appl. Phys. Lett.* **87**, 071102, 2005.

[105] S. P. Sundararajan, N. K. Grandy, N. Mirin, and N. J. Halas, 'Nanoparticle-induced enhancement and suppression of photocurrent in a silicon photodiode', *Nano. Lett.* **8**, 624, 2008.

[106] L. Hong, X. Rusli, H. Wang, L. Zheng, H. He, H. Xu *et al.*, 'Design principles for plasmonic thin film GaAs solar cells with high absorption enhancement', *J. Appl. Phys.* **112**, 054326, 2012.

[107] D. Derkacs, S. H. Lim, P. Matheu, W. Mar, and E. T. Yu, 'Improved performance of amorphous silicon solar cells via scattering from surface plasmon polaritons in nearby metallic nanoparticles', *Appl. Phys. Lett.* **89**, 093103, 2006.

[108] S. Kim, S. Na, J. Jo, D. Kim, and Y. Nah, 'Plasmon enhanced performance of organic solar cells using electrodeposited Ag nanoparticles', *Appl. Phys. Lett.* **93**, 073307, 2008.

[109] M. Kirkengen, J. Bergli, and Y. M. Galperin, 'Direct generation of charge carriers in c-Si solar cells due to embedded nanoparticles', *J. Appl. Phys.* **102**(9), 093713, 2007.

[110] J. Lee and P. Peumans, 'The origin of enhanced optical absorption in solar cells with metal nanoparticles embedded in the active layer', *Opt. Express* **18**, 10078, 2010.

[111] M. Losurdo, M. M. Giangregorio, G. V. Bianco, A. Sacchetti, P. Capezzuto, and G. Bruno, 'Enhanced absorption in Au nanoparticles/a-Si:H/c-Si heterojunction solar cells exploiting Au surface plasmon resonance', *Sol. Energy Mater. Sol. Cells* **93**, 1749, 2009.

[112] K. R. Catchpole, S. Mokkapati, F. Beck, E. Wang, A. McKinley, A. Basch et al., 'Plasmonics and nanophotonics for photovoltaics', *MRS Bulletin* **36**, 461–467, 2011.

[113] N. Guillot and M. L. de la Chapelle, 'The electromagnetic effect in surface enhanced Raman scattering: enhancement optimization using precisely controlled nanostructures', *J. Quant. Spectrosc. Radiat. Transfer* **113**(18), 2321, 2012.

[114] E. L. Ru, E. Blackie, M. Meyer, and P. Etchegoin, 'Surface enhanced Raman scattering enhancement factors: a comprehensive study', *J. Phys. Chem. C* **111**, 13794, 2007.

[115] I. Smolyaninov, A. A. Zayats, and O. Keller, 'The effect of the surface enhanced polariton field on the tunneling current of an STM', *Phys Lett. A* **200**, 438, 1995.

[116] A. V. Zayats, I. I. Smolyaninov, and A. A. Maradudin, 'Nano-optics of surface plasmon polaritons', *Phys. Rep.* **408**, 131, 2005.

[117] H. R. Stuart and D. G. Hall, 'Enhanced dipole–dipole interaction between elementary radiators near a surface', *Phys. Rev. Lett.* **80**, 5663, 1998.

[118] K. Okamoto, I. Niki, A. Shvartser, Y. Narukawa, T. Mukai, and A. Scherer, 'Surface plasmon enhanced spontaneous emission rate of InGaN/GaN QW probed by time-resolved photolumimnescence spectroscopy', *Nat. Mater.* **3**, 601, 2004.

[119] C. Wen, K. Ishikawa, M. Kishima, and K. Yamada, 'Effects of silver particles on the photovoltaic properties of dye-sensitized TiO_2 thin films', *Sol. Cells* **61**, 339, 2000.

[120] F. Masia, W. Langbein, and P. Borri, 'Measurement of the dynamics of plasmons inside individual gold nanoparticles using a femtosecond phase-resolved microscope', *Phys. Rev. B* **85**, 235403, 2012.

[121] T. L. Temple and D. M. Bagnall, 'Optical properties of gold and aluminium nanoparticles for silicon solar cell applications', *J. Appl. Phys.* **109**, 084343, 2011.

[122] M. L. Trolle and T. G. Pedersen, 'Indirect optical absorption in silicon via thin-film surface plasmon', *J. Appl. Phys.* **112**, 043103, 2012.

[123] W. Ekardt, 'Dynamical polarizability of small metal particles: self-consistent spherical jellium model', *Phys. Rev. Lett.* **52**, 1925, 1984.

[124] W. Jacak, J. Krasnyj, J. Jacak, W. Donderowicz, and L. Jacak, 'Mechanism of plasmon mediated enhancement of PV efficiency', *J. Phys. D: Appl. Phys.* **44**, 055301, 2011.

[125] M. Kauranen and A. Zayats, 'Nonlinear plasmonics', *Nat. Photonics* **6**, 737, 2012.

[126] K. Tsia, S. Fathpour, and B. Jalali, 'Energy harvesting in silicon wavelength converters', *Opt. Express* **14**, 12327, 2006.

[127] I. Volovichev, 'New non-linear photovoltaic effect in uniform bipolar semiconductor', *J. Appl. Phys.* **116**, 193701, 2014.

[128] Y. Chen, D. M. Bagnall, H. Koh, K. Park, K. Hiraga, Z. Zhu et al., 'Plasma assisted molecular beam epitaxy of ZnO on c-plane sapphire: growth and characterization', *J. Appl. Phys.* **84**, 39128, 1998.

[129] R. Yousefi, F. Jamali-Sheini, and A. K. Zak, 'A comparative study of the properties of ZnO nano/microstructures grown using two types of thermal evaporation set-up conditions', *Chem. Vapor Depos.* **18**, 215–220, 2012.

[130] E. Guziewicz, I. A. Kowalik, M. Godlewski, K. Kopalko, V. Osinniy, A. Wójcik et al., 'Extremely low temperature growth of ZnO by atomic layer deposition', *J. Appl. Phys.* **103**, 033515, 2008.

[131] L. Chen, 'Photodiodes – from fundamentals to applications', in: *Si-Based ZnO Ultraviolet Photodiodes*, ISBN 978-953-51-0895-5.

[132] H. S. Bae and S. Im, 'Ultraviolet detecting properties of ZnO-based thin film transistors', *Thin Solid Films* **75**, 469, 2004.

[133] M. Dutta and D. Basak, 'p-ZnO/n-Si heterojunction: sol-gel fabrication, photoresponse properties, and transport mechanism', *Appl. Phys. Lett.* **92**, 21212, 2008.

[134] D. B. S. Mridha and M. Dutta, 'Photoresponse of n-ZnO/p-Si heterojunction towards ultraviolet/visible lights: thickness dependent behavior', *J. Mater. Sci. Mater. Electron.* **20**, 376, 2009.

[135] A. Djurišić and Y. H. Leung, 'Optical properties of ZnO nanostructures', *Small* **2**, 944, 2006.

[136] K. Bhandari, J. Collier, R. Ellingson, and D. Apul, 'Energy payback time (EPBT) and energy return on energy invested (EROI) of solar photovoltaic systems: a systematic review and meta-analysis', *Ren. Sust. Energy Rev.* **47**, 133, 2015.

[137] H. R. Stuart and D. G. Hall, 'Island size effect in nanoparticles photodetectors', *Appl. Phys. Lett.* **73**, 3815, 1998.

[138] P. Matheu, S. Lim, D. Derkacs, C. McPheeters, and E. Yu, 'Metal and dielectric nanoparticle scattering for improved optical absorption in photovoltaic devices', *Appl. Phys. Lett.* **93**(11), 113108, 2008.

[139] S. Lim, W. Mar, P. Matheu, D. Derkacs, and E. Yu, 'Photocurrent spectroscopy of optical absorption enhancement in silicon photodiodes via scattering from surface plasmon polaritons in gold nanoparticles', *J. Appl. Phys.* **101**(10), 104309, 2007.

[140] L. Luo, C. Xie, X. Wang, Y. Yu, C. Wu, H. Hu et al., 'Surface plasmon resonance enhanced highly efficient planar silicon solar cell', *Nano Energy* **9**, 112–120, 2014.

[141] S. Pillai, K. Catchpole, T. Trupke, and M. Green, 'Surface plasmon enhanced silicon solar cells', *J. Appl. Phys.* **101**(9), 093105, 2007.

[142] T. Temple, G. Mahanama, H. Reehal, and D. Bagnall, 'Influence of localized surface plasmon excitation in silver nanoparticles on the performance of silicon solar cells', *Solar Energy Mater. Solar Cells* **93**(11), 1978–1985, 2009.

[143] C. Uhrenfeldt, T. Villesen, A. Têtu, B. Johansen, and A. Nylandsted Larsen, 'Broadband photocurrent enhancement and light-trapping in thin film Si solar cells with periodic Al nanoparticle arrays on the front', *Opt. Express* **23**(11), A525, 2015.

[144] W. Ho, C. Hu, C. Yeh, and Y. Lee, 'External quantum efficiency and photovoltaic performance of silicon cells deposited with aluminum, indium, and silver nanoparticles', *Japanese J. Appl. Phys.* **55**(8S3), 08RG03, 2016.

[145] K. Kluczyk, C. David, J. Jacak, and W. Jacak, 'On modeling of plasmon-induced enhancement of the efficiency of solar cells modified by metallic nano-particles', *Nanomaterials* **9**, 3, 2019. DOI:10.3390/nano9010003.

[146] D. Aspnes and A. Studna, 'Dielectric functions and optical parameters of Si, Ge, GaP, GaAs, GaSb, InP, InAs, and InSb from 1.5 to 6.0 eV', *Phys. Rev. B* **27**(2), 985–1009, 1982.

[147] M. Ghidelli, L. Mascaretti, B. Bricchi, A. Zapelli, V. Russo, C. Casari *et al.*, 'Engineering plasmonic nanostructured surfaces by pulsed laser deposition', *Appl. Surface Sci.* **434**, 1064–1073, 2018.

[148] B. Bricchi, M. Ghidelli, L. Mascaretti, A. Zapelli, V. Russo, C. Casari *et al.*, 'Integration of plasmonic Au nanoparticles in TiO_2 hierarchical structures in a single-step pulsed laser co-deposition', *Mater. Design* **156**, 311–319, 2018.

[149] J. Borges, T. Kubart, S. Kumar, K. Leifer, M. Rodrigues, N. Duarte *et al.*, 'Microstructural evolution of Au/TiO_2 nanocomposite films: the influence of Au concentration and thermal annealing', *Thin Solid Films* **580**, 77–88, 2015.

[150] P. Tonui, E. Arbab, and G. Mola, 'Metal nano-composite as charge transport co-buffer layer in perovskite based solar cell', *J. Phys. Chem. Solids* **126**, 124–130, 2019.

[151] E. Arbab and G. Mola, 'Metals decorated nanocomposite assisted charge transport in polymer solar cell', *Mater. Sci. Semiconductor Proc.* **91**, 1–8, 2019.

[152] F. Bella, P. Renzi, C. Cavallo, and C. Gerbaldi, 'Caesium for perovskite solar cells: an overview', *Chemistry – A European J.* **24**, 12183–12205, 2018.

[153] J. Hao, H. Hao, J. Li, L. Shi, T. Zhong, C. Zhang *et al.*, 'Light trapping effect in perovskite solar cells by the addition of Ag nanoparticles, using textured substrates', *Nanomaterials* **8**(10), 815, 2019.

[154] S. Galliano, F. Bella, G. Piana, G. Giacona, G. Viscardi, C. Gerbaldi *et al.*, 'Finely tuning electrolytes and photoanodes in aqueous solar cells by experimental design', *Sol. Energy* **163**, 251–255, 2018.

[155] D. Pintossi, G. Iannaccone, A. Colombo, F. Bella, M. Välimäki, K. Väisänen *et al.*, 'Luminescent downshifting by photon-induced solar–gel hybrid coatings: accessing multifunctionality on flexible organic photovoltaics via ambient temperature material processing', *Adv. Electron. Mater.* **2**, 1600288, 2016.

[156] P. Pearce, A. Mellor, and N. Ekins-Daukes, 'The importance of accurate determination of optical constants for the design of nanometallic light-trapping structures', *Sol. Energy Mater. Sol. Cells* **191**, 133–140, 2019.

[157] Y. Zhang, B. Cai, and B. Jia, 'Ultraviolet plasmonic aluminium nanoparticles for highly efficient light incoupling on silicon solar cells', *Nanomaterials* **6**, 95, 2016, DOI:10.3390/nano6060095.

[158] D. Song, H. Kim, J. Suh, B. Jun, and W. Rho, 'Multi-shaped Ag nanoparticles in the plasmonic layer of dye-sensitized solar cells for increased power conversion efficiency', *Nanomaterials* **7**(6), 136, 2017.

[159] K. Kluczyk, C. David, and W. Jacak, 'On quantum approach to modeling of plasmon photovoltaic effect', *J. Optical Soc. America B* **34**, 2115, 2017.

[160] E. Runge and E. K. U. Gross, 'Density-functional theory for time-dependent systems', *Phys. Rev. Lett.* **52**, 997, 1984.

[161] P. Hohenberg and W. Kohn, 'Inhomogeneous electron gas', *Phys. Rev. Lett.* **136**, 864, 1964.

[162] R. Esteban, A. Zugarramurdi, P. Zhang, P. Nordlander, F. J. G. Vidal, A. G. Borisov *et al.*, 'A classical treatment of optical tunneling in plasmonic gaps: extending the quantum corrected model to practical situations', *Faraday Discuss.* **178**, 151, 2015.

[163] T. V. Teperik, P. Nordlander, J. Aizpurua, and A. G. Borisov, 'Quantum effects and nonlocality in strongly coupled plasmonic nanowire dimers', *Opt. Express* **21**, 27306, 2013.

[164] F. J. G. de Abajo, 'Nonlocal effects in the plasmons of strongly interacting nanoparticles, dimers, and waveguides', *J. Phys. Chem. C* **112**, 17983, 2008.

[165] K. Kluczyk, L. Jacak, C. David, and W. Jacak, 'Microscopic electron dynamics in metal nanoparticles for photovoltaic systems', *Materials* **11**, 1077, 2018.

[166] S. A. Maier, P. G. Kik, H. A. Atwater, S. Meltzer, E. Harel, B. E. Koel *et al.*, 'Local detection of electromagnetic energy transport below the diffraction limit in metal nanoparticle plasmon waveguides', *Nat. Mater.* **2**, 229, 2003.

[167] L. L. Zhao, K. L. Kelly, and G. C. Schatz, 'The extinction spectra of silver nanoparticle arrays: influence of array structure on plasmon resonance wavelength and width', *J. Phys. Chem. B* **107**, 7343, 2003.

[168] S. Zou, N. Janel, and G. C. Schatz, 'Silver nanoparticle array structures that produce remarkably narrow plasmon lineshapes', *J. Chem. Phys.* **120**, 10871, 2004.

[169] J. R. Krenn, A. Dereux, J. C. Weeber, E. Bourillot, Y. Lacroute, J. P. Goudonnet *et al.*, 'Squeezing the optical near-field zone by plasmon coupling of metallic nanoparticles', *Phys. Rev. Lett.* **82**, 2590, 1999.

[170] S. A. Maier, M. L. Brongersma, P. G. Kik, and H. A. Atwater, 'Observation of near-field coupling in metal nanoparticle chains using far-field polarization spectroscopy', *Phys. Rev. B* **65**, 193408, 2002.

[171] S. A. Maier, P. G. Kik, L. A. Sweatlock, H. A. Atwater, J. J. Penninkhof, A. Polman *et al.*, 'Energy transport in metal nanoparticle plasmon waveguides', *Mat. Res. Soc. Symp. Proc.* **777**, T7.1.1, 2003.

[172] H. Ditlbacher, A. Hohenau, D. Wagner, U. Kreibig, M. Rogers, F. Hofer *et al.*, 'Silver nanowires as surface plasmon resonators', *Phys. Rev. Lett.* **95**, 257403, 2005.

[173] I. L. Rasskazov, S. V. Karpov, and V. A. Markel, 'Nondecaying surface plasmon polaritons in linear chains of silver nanospheroids', *Opt. Lett.* **38**, 4743, 2013.

[174] A. A. Govyadinov and V. A. Markel, 'From slow to superluminal propagation: dispersive properties of surface plasmon polaritons in linear chains of metallic nanospheroids', *Phys. Rev. B* **78**, 035403, 2008.

[175] Y. Hadad and B. Z. Steinberg, 'Green's function theory for infinite and semi-infinite particle chains', *Phys. Rev. B* **84**, 125402, 2011.

[176] V. A. Markel and A. K. Sarychev, 'Comment on Green's function theory for infinite and semi-infinite particle chains', *Phys. Rev. B* **86**, 037401, 2012.

[177] S. B. Singham and C. F. Bohren, 'Light scattering by an arbitrary particle: a physical reformulation of the coupled dipole method', *Opt. Lett.* **12**, 10, 1987.

[178] T. Jensen, L. Kelly, A. Lazarides, and G. C. Schatz, 'Electrodynamics of noble metal nanoparticles and nanoparticle clusters', *J. Cluster Sci.* **10**, 295, 1999.

[179] B. T. Draine, 'The discrete-dipole approximation and its application to interstellar graphite grains', *Astrophys. J.* **333**, 848, 1988.

[180] E. M. Purcell and C. R. Pennypacker, 'Scattering and absorption of light by nanospherical dielectric grains', *Astrophys. J.* **186**, 705, 1973.

[181] B. T. Draine and P. J. Flatau, 'Discrete-dipole approximation for scattering calculations', *J. Opt. Soc. America A* **11**, 1491, 1994.

[182] V. A. Markel, 'Coupled-dipole approach to scattering of light from a one dimensional periodic dipole structure', *J. Mod. Opt.* **40**, 2281, 1993.

[183] P. Anger, P. Bharadwaj, and L. Novotny, 'Enhancement and quenching of single-molecule fluorescence', *Phys. Rev. Lett.* **96**, 113002, 2006.

[184] I. S. Gradstein and I. M. Rizik, *Tables of Integrals*, Fizmatizdat, Moscow, 1962.

[185] W. Jacak, J. Krasnyj, J. Jacak, A. Chepok, L. Jacak, W. Donderowicz *et al.*, 'Undamped energy transport by collective surface plasmon oscillations along metallic nanosphere chain', *J. Appl. Phys.* **108**, 084304, 2010.

[186] M. Fevrier, P. Gogol, J.-M. Lourtioz, and B. Dagens, 'Metallic nanoparticle chains on dielectric waveguides: coupled and uncoupled situations compared', *Opt. Express* **21**, 24505, 2013. DOI:10.1364/OE.21.024504.

[187] A. Apuzzo, M. Fevrier, R. Salas-Montiel, A. Bruyant, A. Chelnokov, G. Lerondel *et al.*, 'Observation of near-field dipolar interactions involved in a metal nanoparticle chain waveguide', *Nano Lett.* **13**, 1000, 2013. doi: 10.1021/nl304164y.

[188] C. Girard and R. Quidant, 'Near-field optical transmittance of metal particle chain waveguides', *Opt. Express* **12**, 6141, 2004.

[189] V. A. Markel, 'Divergence of dipole sums and the nature of non-Lorentzian exponentially narrow resonances in one-dimensional periodic arrays of nanospheres', *J. Phys. B: Atom., Mol. Opt. Phys.* **38**, L115, 2005.

[190] N. N. Bogolubov and J. A.Mitropolkyj, *Asymptotical Methods in the Theory of Nonlinear Oscillations*, Nauka, Moscow, 2005.

[191] I. A. Lazarevich and V. B. Kazantsev, 'Dendritic signal transmission induced by intracellular charge inhomogeneities', *Phys. Rev. E.* **88**, 062718, 2013.

[192] R. S. Lillie, 'Factors affecting transmission and recovery in passive iron nerve model', *J. Gen. Physiol.* **7**, 473, 1925.

[193] D. Citrin, 'Plasmon–polariton transport in metal-nanoparticle chains embedded in a gain medium', *Opt. Letters* **31**, 98, 2006.

[194] V. M. Agranovich and O. A. Dubovskii, 'Effect of retarded interaction on the exciton spectrum in one-dimensional and two-dimensional crystals', *JETP Lett.* **3**, 223, 1966.

[195] T. Meissner and F. Wentz, 'The complex dielectric constant of pure and sea water from microwave satellite observations', *IEEE THRS* **42**, 1836, 2004.

[196] T. A. El-Brolossy, T. Abdallah, M. B. Mohamed, S. Abdallah, K. Easawi, S. Negm et al., 'Shape and size dependence of the surface plasmon resonance of gold nanoparticles studied by photoacoustic technique', *Eur. Phys. J. Special Topics* **153**, 361, 2008.

[197] L. M. Liz-Marzan, 'Tuning nanorod surface plasmon resonances', *SPIE Newsroom* (10.1117/2.1200707.0798), 2007.

[198] A. Sommerfeld, 'Über die fortpflanzung elektrodynamischer Wellen langs eines Drahts', *Ann. Phys. Chem.* **67**, 233, 1899.

[199] E. Andrianov, A. A. Pukhov, A. V. Dorofeenko, A. P. Vinogradov, and A. A. Lisyansky, 'Stationary behavior of a chain of interacting spasers', *Phys. Rev. B* **85**, 165419, 2012.

[200] J. S. Birman, *Braids, Links and Mapping Class Groups*, Princeton University Press, 1974.

[201] D. Mermin, 'The topological theory of defects in ordered media', *Rev. Mod. Phys.* **51**, 591, 1979.

[202] J. Jacak, R. Gonczarek, L. Jacak, and I. Jóźwiak, *Application of Braid Groups in 2D Hall System Physics*, World Scientific, Singapore, 2012.

[203] H. Stokowski and J. Jacak, 'Multi-character alphabet coding using braid group formalism', *Comput. Math. Appl.*, submitted.

[204] E. Spanier, *Algebraic Topology*, Springer, Berlin, 1966.

Index

acoustic sound, 26
Avogadro number, 21, 39

Bethe–Salpeter equation, 29, 31, 32
binary electrolyte, 12, 253, 254, 261
Bloch theory, 17, 19, 111, 187, 239
Bohm, 2, 3, 35, 64, 100, 101, 256
Boltzmann constant, 20, 21, 281
Born–Karman conditions, 18, 239, 282
Bravais webs, 17–20
bulk metal, 3, 4, 35, 43, 49, 58, 65, 98, 155

cable model, 13, 286, 301
CIGS cells, 9, 133, 143, 145, 148, 162
collective excitations, 25, 32
commutator for Heisenberg equation, 37, 45
COMSOL, ix, 3, 85, 87, 90, 96, 151, 154, 157, 163, 172
Cooper pair, 30, 34
crystal lattice cell, 1, 2, 9, 12, 17

dephasing, 1
dielectric function, 3, 67, 79, 82, 86, 93, 154
Drude–Lorentz model, 3, 67, 86, 150, 166

effective mass, 26, 28, 32, 110
extinction cross section, 3, 82, 87, 96

Fermi energy, 2, 21, 33, 39, 98
Fermi liquid, 22, 25, 26, 30, 32
Fermi sea, 2, 21
Fermi surface, 2, 21, 23, 24, 26, 27, 29, 33, 98
Fermi velocity, 12, 13, 67, 104, 206
four-momentum, 29, 31
Fourier transform, 36, 40, 43, 47, 49, 188, 190, 196, 209, 239
Fresnel–Maxwell electrodynamics equations, 3, 155

Goubau transmission lines, 9, 288
Green functions for metal, 2, 80
group velocity of plasmon–polariton, 10, 24, 179, 181, 197, 200, 204, 213, 220, 244

Hamiltonian for plasmons, 34, 35, 37, 43, 44, 46, 135, 255, 256
harmonic damped oscillator (plasmons), 6, 71, 83
Heisenberg equation, 4, 46
hyperbolic singularity (plasmon–polariton), 10, 201, 202, 219, 229

jellium, 4, 36, 38, 40, 42, 43, 45, 48, 99, 102, 222, 234, 254, 256
Joule heat, 10, 14, 61, 232, 234, 250, 278

kinetical energy, 4, 5, 12, 45
Kohn–Sham approach, 43, 48, 63, 98, 155

Landau damping, 24, 25, 30, 67
Landau interaction, 26, 29, 32
light cone, 10, 11, 201
linear response theory of Kubo, 32
logarithmic singularity, plasmon–polariton, 10, 11, 181, 200–202, 204
longitudinal polarization, 192, 202, 205, 208, 216, 226, 280
Lorentz friction, 6, 60–62, 70, 71, 75, 77, 81, 84, 95, 105, 151, 155, 192, 211, 223, 229, 237, 270, 300
Lorentz invariance, 11
Luttinger theorem, 21, 28

Matsubara Green functions, 27, 29
metallic clusters, 3, 4, 42, 43, 62, 69, 80, 178, 180, 253
metallic nanoparticles, 1, 62–64, 69, 71, 76, 79, 83, 85, 87, 95, 96, 99, 101, 109, 119, 137, 198, 235, 275
metamaterials, 3, 98
Mie energy, 4, 134
Mie-type calculation, 6, 80, 85, 300

313

momentum conservation, 39, 100, 131, 134, 151
multipole modes of plasmons, 4, 84, 91, 275, 276
myelinated axons, 13, 301

nano-chains, 9–11, 61
nanosphere, 5, 62, 69, 77, 81, 101, 109, 187, 192
near-field zone, 7, 8, 198, 211, 215, 229, 245, 251
neurons, 13, 276, 288, 295, 301
Newton-type procedure, 10, 202, 205, 250

ordinary photoeffect, 7, 8, 100, 112, 113, 118, 131, 137, 154
overdamped regime, 6, 73, 108, 114, 115, 143

peripheral nervous system, 13
perturbative solution, 10, 61, 71, 76, 84, 107, 185, 250
phonon, 1
photodiode, 8, 118, 123, 149
Pines, 2–4, 35, 256
plasmon–polariton, 9, 283, 285, 286, 288, 290, 292, 300, 301
plasmon-optoelectronics, 9, 10
plasmonic effect, 9, 119, 122, 163
plasmons, 2, 6, 34, 96, 109, 137, 229, 252, 260
Pomeranchuck instability, 25
pseudomomentum, 19

quasiparticles, 2, 5, 23

random phase approximation (RPA) approach to plasmons, 3, 5, 38, 51, 64, 66, 67, 82, 99, 181
Ranvier nodes, 15, 276, 286, 291, 293
retarded or advanced Green function, 27, 115
Ritchie frequency, 4

saltatory conduction, 4, 14, 276, 287, 298, 302
scattering, 3, 29, 39, 61, 69, 79, 83, 88, 151, 178, 224, 245, 278

Schrödinger equation, 17
semi-metal, 20
semiconductor quantum dots, 1
semiconductor substrate, 6, 97, 100, 110, 113, 120, 131, 140, 155, 173, 232
singlet spin state, 34
solar cell, 7, 9, 59, 97, 120, 133, 147, 154
spill-out, 4, 62, 80, 94, 99, 178
spin waves, 27, 32
stellar kernels, 2
subdiffraction, 1, 12, 60
superfluid He^3, 34
superluminal propagation, 10, 179
surface plasmons, 4, 5, 8, 11, 53, 59, 66, 68, 71, 77, 79, 87, 95, 99, 106, 109, 115, 119, 124, 148, 153, 160, 187, 229, 237, 246, 264, 291

Thomas–Fermi 5/3 formula, 5, 12
Thomas–Fermi wave vector, 4, 40, 66, 99, 101
Thomson, 13, 276, 295
time-dependent local density approximation (TDLDA) method, 48, 51, 63, 101, 180, 260
transverse polarization, 10, 91, 93, 181, 189, 194, 198, 200, 203, 205, 216, 240, 280
triplet superfluid, 34

ultraviolet radiation, 2, 34, 139

vertical interband transitions, 8, 100, 118, 131, 133, 151, 158
vibrational and rotational excitations, 13, 259
volume plasmons, 41, 51, 52, 66, 68, 103, 234, 260, 262, 265, 275

wave function, 18, 64, 102, 111, 124, 256

zero sound, 25, 27, 32

Printed in the United States
By Bookmasters